CAMBRIDGE MONOGRAPHS ON APPLIED AND COMPUTATIONAL MATHEMATICS

25 **Algebraic Geometry and Statistical Learning Theory**

The *Cambridge Monographs on Applied and Computational Mathematics* reflect the crucial role of mathematical and computational techniques in contemporary science. The series publishes expositions on all aspects of applicable and numerical mathematics, with an emphasis on new developments in this fast-moving area of research.

State-of-the-art methods and algorithms as well as modern mathematical descriptions of physical and mechanical ideas are presented in a manner suited to graduate research students and professionals alike. Sound pedagogical presentation is a prerequisite. It is intended that books in the series will serve to inform a new generation of researchers.

The series includes titles in the *Library of Computational Mathematics*, published under the auspices of the Foundations of Computational Mathematics organisation.

Also in this series:

Simulating Hamiltonian Dynamics, *B. Leimkuhler and Sebastian Reich*

Collocation Methods for Volterra Integral and Related Functional Differential, *Hermann Brunner*

Topology for Computing, *Afra J. Zomorodian*

Scattered Data Approximation, *Holger Wendland*

Matrix Preconditioning Techniques and Applications, *Ke Chen*

Spectral Methods for Time-Dependent Problems, *Jan Hesthaven, Sigal Gottlieb and David Gottlieb*

The Mathematical Foundations of Mixing, *Rob Sturman, Julio M. Ottino and Stephen Wiggins*

Curve and Surface Reconstruction, *Tamal K. Dey*

Learning Theory, *Felipe Cucker and Ding Xuan Zhou*

Algebraic Geometry and Statistical Learning Theory

SUMIO WATANABE
Tokyo Institute of Technology

CAMBRIDGE
UNIVERSITY PRESS

University Printing House, Cambridge CB2 8BS, United Kingdom

One Liberty Plaza, 20th Floor, New York, NY 10006, USA

477 Williamstown Road, Port Melbourne, VIC 3207, Australia

4843/24, 2nd Floor, Ansari Road, Daryaganj, Delhi - 110002, India

79 Anson Road, #06-04/06, Singapore 079906

Cambridge University Press is part of the University of Cambridge.

It furthers the University's mission by disseminating knowledge in the pursuit of education, learning and research at the highest international levels of excellence.

www.cambridge.org
Information on this title: www.cambridge.org/9780521864671

First published 2009
Reprinted 2010

A catalogue record for this publication is available from the British Library

Library of Congress Cataloging in Publication data
Watanabe, Sumio, 1959–
Algebraic geometry and statistical learning theory / Sumio Watanabe.
p. cm.
Includes bibliographical references and index.
ISBN 978-0-521-86467-1 (hardback)
1. Computational learning theory – Statistical methods. 2. Geometry, Algebraic.
I. Title.
Q325.7.W38 2009
006.3′1 – dc22 2009011366

ISBN 978-0-521-86467-1 Hardback

Contents

Preface

In this book, we introduce a fundamental relation between algebraic geometry and statistical learning theory.

A lot of statistical models and learning machines used in information science, for example, mixtures of probability distributions, neural networks, hidden Markov models, Bayesian networks, stochastic context-free grammars, and topological data analysis, are not regular but singular, because they are non-identifiable and their Fisher information matrices are singular. In such models, knowledge to be discovered from examples corresponds to a singularity, hence it has been difficult to develop a mathematical method which enables us to understand statistical estimation and learning processes.

Recently, we established singular learning theory, in which four general formulas are proved for singular statistical models. Firstly, the log likelihood ratio function of any singular model can be represented by the common standard form even if it contains singularities. Secondly, the asymptotic behavior of the evidence or stochastic complexity is clarified, giving the result that the learning coefficient is equal to the maximum pole of the zeta function of a statistical model. Thirdly, there exist equations of states which express the universal relation of the Bayes quartet. We can predict Bayes and Gibbs generalization errors using Bayes and Gibbs training errors without any knowledge of the true distribution. And lastly, the symmetry of the generalization and training errors holds in the maximum likelihood and *a posteriori* estimators. If one-point estimation is applied to statistical learning, the generalization error is equal to the maximum value of a Gaussian process on a real analytic set.

This book consists of eight chapters. In Chapter 1, an outline of singular learning theory is summarized. The main formulas proved in this book are overviewed without mathematical preparation in advance. In Chapter 2, the definition of a singularity is introduced. Resolution of singularities is the essential theorem on which singular learning theory is constructed. In Chapter 3, several basic concepts in algebraic geometry are briefly explained: ring and ideal, correspondence between algebra and geometry, and projective spaces. The algorithm by which a resolution map is found using recursive blow-ups is also described. In Chapter 4, the relation between the singular integral and the zeta function of a singular statistical model is clarified, enabling some inequalities used in Chapter 6 to be proved. In Chapter 5, function-valued random variables are studied and convergence in law of empirical processes is proved. In Chapter 6, the four main formulas are proved: the standard form of the likelihood ratio function, the asymptotic expansion of the stochastic complexity, the equations of states in a Bayes quartet, and the symmetry of generalization and training errors in one-point estimation. In Chapters 7 and 8, applications of singular learning theory to information science are summarized and discussed.

This book involves several mathematical fields, for example, singularity theory, algebraic geometry, Schwartz distribution, and empirical processes. However, these mathematical concepts are introduced in each chapter for those who are unfamiliar with them. No specialized mathematical knowledge is necessary to read this book. The only thing the reader needs is a mathematical mind seeking to understand the real world.

The author would like to thank Professor Shun-ichi Amari for his encouragement of this research. Also the author would like to thank Professor Bernd Sturmfels for his many helpful comments on the study and this book.

In this book, the author tries to build a bridge between pure mathematics and real-world information science. It is expected that a new research field will be opened between algebraic geometry and statistical learning theory.

Sumio Watanabe

1
Introduction

In this book, we study a system which perceives the real world. Such a system has to estimate an information source by observation. If the information source is a probability distribution, then the estimation process is called statistical learning, and the system is said to be a statistical model or a learning machine.

A lot of statistical models have hierarchical layers, hidden variables, a collection of modules, or grammatical structures. Such models are nonidentifiable and contain singularities in their parameter spaces. In fact, the map from a parameter to a statistical model is not one-to-one, and the Fisher information matrix is not positive definite. Such statistical models are called singular. It has been difficult to examine the learning process of singular models, because there has been no mathematical theory for such models.

In this book, we establish a mathematical foundation which enables us to understand the learning process of singular models. This chapter gives an overview of the book before a rigorous mathematical foundation is developed.

1.1 Basic concepts in statistical learning

To describe what statistical learning is, we need some basic concepts in probability theory. For the reader who is unfamiliar with probability theory, Section 1.6 summarizes the key results.

1.1.1 Random samples

Let N be a natural number and $\mathbb{R}^N$ be the N-dimensional real Euclidean space. We study a case when information data are represented by vectors in $\mathbb{R}^N$.

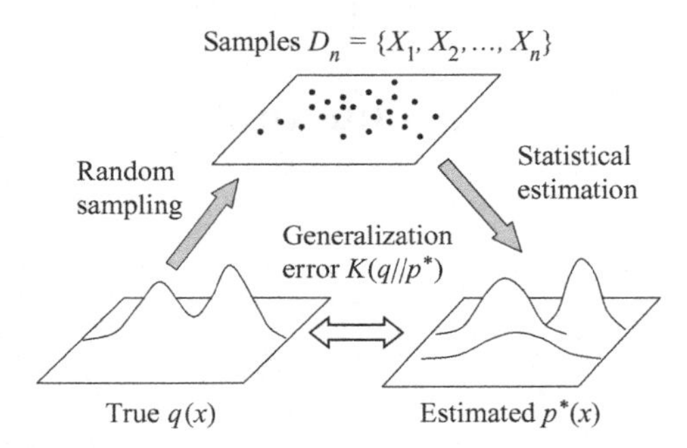

Fig. 1.1. Statistical learning

Firstly, using Figure 1.1, let us explain what statistical learning is. Let $(\Omega, \mathcal{B}, P)$ be a probability space and $X : \Omega \to \mathbb{R}^N$ be a random variable which is subject to a probability distribution $q(x)dx$. Here $q(x)$ is a probability density function and dx is the Lebesgue measure on $\mathbb{R}^N$.

We assume that random variables $X_1, X_2, \ldots, X_n$ are independently subject to the same probability distribution as X, where n is a natural number. In statistical learning theory, $q(x)$ is called a true probability density function, and random variables $X_1, X_2, \ldots, X_n$ are random samples, examples, random data, or training samples. The probability density function of the set of independent random samples is given by

$$q(x_1)q(x_2)\cdots q(x_n).$$

In practical applications, we obtain their realizations by observations. The natural number n is said to be the number of random samples. The set of random samples or random data is denoted by

$$D_n = \{X_1, X_2, \ldots, X_n\}.$$

The main purpose of statistical learning is to construct a method to estimate the true probability density function $q(x)$ from the set D_n.

In this book, we study a method which employs a parametric probability density function. A conditional probability density function $p(x|w)$ of $x \in \mathbb{R}^N$ for a given parameter $w \in W$ is called a learning machine or a statistical model, where W is the set of all parameters. Sometimes the notation $p(x|w) = p_w(x)$ is used. Then $w \mapsto p_w$ gives a map from the parameter to the probability density function. We mainly study the case when W is a subset of the d-dimensional real Euclidean space $\mathbb{R}^d$ or a d-dimensional real analytic manifold. An *a priori* probability density function $\varphi(w)$ is defined on W. We assume that, for any $w \in W$, the support of the probability distribution $p(x|w)$ is equal to that of

$q(x)$ and does not depend on w. That is to say, for any $w \in W$,

$$\overline{\{x \in \mathbb{R}^N; p(x|w) > 0\}} = \overline{\{x \in \mathbb{R}^N; q(x) > 0\}},$$

where $\overline{S}$ is the closure of a set S in $\mathbb{R}^N$.

In statistical learning or statistical inference, the main study concerns a method to produce a probability density function $p^*(x)$ on $\mathbb{R}^N$ based on the set of random samples D_n, using the parametric model $p(x|w)$. Such a function

$$D_n \mapsto p^*(x)$$

is called a statistical estimation method or a learning algorithm. Note that there are a lot of statistical estimation methods and learning algorithms. The probability density function $p^*(x)$, which depends on the set of random variables D_n, is referred to as the estimated or trained probability density function. Generally, it is expected that the estimated probability density function $p^*(x)$ is a good approximation of the true density function $q(x)$, and that it becomes better as the number of random samples increases.

1.1.2 Kullback–Leibler distance

In order to compare two probability density functions, we need a quantitative value which shows the difference between two probability density functions.

Definition 1.1 (Kullback–Leibler distance) For given probability density functions $q(x)$, $p(x) > 0$ on an open set $A \subset \mathbb{R}^N$, the Kullback–Leibler distance or relative entropy is defined by

$$K(q\|p) = \int_A q(x) \log \frac{q(x)}{p(x)} dx.$$

If the integral is not finite, $K(q\|p)$ is defined as $K(q\|p) = \infty$.

Theorem 1.1 *Assume that $q(x)$, $p(x) > 0$ are continuous probability density functions on an open set A. Then the following hold.*
(1) For arbitrary $q(x)$, $p(x)$, $K(q\|p) \geq 0$.
(2) $K(q\|p) = 0$ if and only if $q(x) = p(x)$ for any $x \in A$.

Proof of Theorem 1.1 Let us introduce a real function

$$S(t) = -\log t + t - 1 \quad (0 < t < \infty).$$

Then $S(t) \geq 0$, and $S(t) = 0$ if and only if $t = 1$. Since $\int q(x)dx = \int p(x)dx = 1$,

$$K(q \| p) = \int_A q(x)\, S\Big(\frac{p(x)}{q(x)}\Big)\, dx,$$

which shows (1). Assume $K(q \| p) = 0$. Since $S(p(x)/q(x))$ is a nonnegative and continuous function of x, $S(p(x)/q(x)) = 0$ for any $x \in A$, which is equivalent to $p(x) = q(x)$. □

Remark 1.1 The Kullback–Leibler distance is called the relative entropy in physics. In information theory and statistics, the Kullback–Leibler distance $K(q \| p)$ represents the loss of the system $p(x)$ for the information source $q(x)$. The fact that $K(q \| p)$ is not symmetric for $q(x)$ and $p(x)$ may originate from the difference of their roles. Historically, relative entropy was first defined by Boltzmann and Gibbs in statistical physics in the nineteenth century. In the twentieth century it was found that relative entropy plays a central role in information theory and statistical estimation.

We can measure the difference between the true density function $q(x)$ and the estimated one $p^*(x)$ by the Kullback–Leibler distance:

$$K(q \| p^*) = \int q(x) \log \frac{q(x)}{p^*(x)} dx.$$

In statistical learning theory, $K(q \| p^*)$ is called the generalization error of the method of statistical estimation $D_n \mapsto p^*$. In general, $K(q \| p^*)$ is a measurable function of the set of random samples D_n, hence it is also a real-valued random variable. The training error is defined by

$$K_n(q \| p^*) = \frac{1}{n} \sum_{i=1}^{n} \log \frac{q(X_i)}{p^*(X_i)},$$

which is also a random variable. One of the main purposes of statistical learning theory is to clarify the probability distributions of the generalization and training errors for a given method of statistical estimation. The expectation values $E[K(q \| p^*)]$ and $E[K_n(q \| p^*)]$ are respectively called the mean generalization error and the training error. If the mean generalization error is smaller, the statistical estimation method is more appropriate. The other purpose of statistical learning theory is to establish a mathematical relation between the generalization error and the training error. If the generalization error can be estimated from the training error, we can select the suitable model or hyperparameter among several statistical possible models.

Definition 1.2 (Likelihood function) For a given set of random samples D_n and a statistical model $p(x|w)$, the likelihood function $L_n(w)$ of $w \in W \subset \mathbb{R}^d$ is defined by

$$L_n(w) = \prod_{i=1}^{n} p(X_i|w).$$

If $p(x|w) = q(x)$, then $L_n(w)$ is equal to the probability density function of D_n.

Definition 1.3 (Log likelihood ratio function) For a given true distribution $q(x)$ and a parametric model $p(x|w)$, the log density ratio function $f(x, w)$, the Kullback–Leibler distance $K(w)$, and the log likelihood ratio function $K_n(w)$ are respectively defined by

$$f(x, w) = \log \frac{q(x)}{p(x|w)}, \tag{1.1}$$

$$K(w) = \int q(x) f(x, w) dx, \tag{1.2}$$

$$K_n(w) = \frac{1}{n} \sum_{i=1}^{n} f(X_i, w), \tag{1.3}$$

where $K_n(w)$ is sometimes referred to as an empirical Kullback–Leibler distance.

From the definition,

$$E[f(X, w)] = E[K_n(w)] = K(w).$$

By using the empirical entropy

$$S_n = -\frac{1}{n} \sum_{i=1}^{n} \log q(X_i), \tag{1.4}$$

the likelihood function satisfies

$$-\frac{1}{n} \log L_n(w) = K_n(w) + S_n. \tag{1.5}$$

The empirical entropy S_n does not depend on the parameter w, hence maximization of the likelihood function $L_n(w)$ is equivalent to minimization of $K_n(w)$.

Remark 1.2 If a function $S(t)$ satisfies $S''(t) > 0$ and $S(1) = 0$, then

$$\int_A q(x)\, S\Big(\frac{p(x)}{q(x)}\Big)\, dx$$

has the same property as the Kullback–Leibler distance in Theorem 1.1. For example, using $S(t) = (1 - t^a)/a$ for a given a, $0 < a < 1$, a generalized distance is defined by

$$K^{(a)}(q\|p) = \int q(x)\Big(\frac{1 - (p(x)/q(x))^a}{a}\Big)\, dx.$$

For example, if $a = 1/2$, Hellinger's distance is derived,

$$K^{(1/2)}(q\|p) = \int (\sqrt{q(x)} - \sqrt{p(x)})^2 dx.$$

In general Jensen's inequality claims that, for any measurable function $F(x)$,

$$\int q(x)S(F(x))dx \geq S\Big(\int q(x)F(x)dx\Big),$$

where the equality holds if and only if $F(x)$ is a constant function on $q(x) > 0$. Hence $K^{(a)}(q\|p) \geq 0$ and $K^{(a)}(q\|p) = 0$ if and only if $q(x) = p(x)$ for all x. Hence $K^{(a)}(q\|p)$ indicates a difference of $p(x)$ from $q(x)$. The Kullback–Leibler distance is formally obtained by $a \to +0$. For arbitrary probability density functions $q(x)$, $p(x)$, the Kullback–Leibler distance satisfies $K(q\|p) \geq K^{(a)}(q\|p)$, because

$$K(q\|p) - K^{(a)}(q\|p) = \int q(x)\Big(\frac{af(x,w) + e^{-af(x,w)} - 1}{a}\Big)dx \geq 0.$$

Moreover, if $K(q\|p) \neq 0$ then

$$\lim_{a\to+0} \frac{K_a(q\|p)}{K(q\|p)} = 1.$$

Therefore, from the learning theory of $K(q\|p)$, we can construct a learning theory of $K^{(a)}(q\|p)$.

Remark 1.3 If $E[K(w)] < \infty$ then, by the law of large numbers, the convergence in probability

$$K_n(w) \to K(w)$$

holds for each $w \in W$. Furthermore, if $E[K(w)^2] < \infty$ then, by the central limit theorem,

$$\sqrt{n}(K_n(w) - K(w))$$

converges in law to the normal distribution, for each $w \in W$. Therefore, for each w, the convergence in probability

$$-\frac{1}{n} \log L_n(w) \to K(w) - S$$

holds, where S is the entropy of the true distribution $q(x)$,

$$S = -\int q(x) \log q(x) dx.$$

It might seem that minimization of $K_n(w)$ is equivalent to minimization of $K(w)$. If these two minimization problems were equivalent, then maximization of $L_n(w)$ would be the best method in statistical estimation. However, minimization and expectation cannot be commutative.

$$E[\min_w K_n(w)] \neq \min_w E[K_n(w)] = \min_w K(w). \qquad (1.6)$$

Hence maximization of $L_n(w)$ does not mean minimization of $K(w)$. This is the basic reason why statistical learning does not result in a simple optimization problem. To clarify the difference between $K(w)$ and $K_n(w)$, we have to study the meaning of the convergence $K_n(w) \to K(w)$ in a functional space. There are many nonequivalent functional topologies. For example, sup-norm, L^p-norm, weak topology of Hilbert space L^2, Schwartz distribution topology, and so on. It strongly depends on the topology of the function space whether the convergence $K_n(w) \to K(w)$ holds or not. The Bayes estimation corresponds to the Schwartz distribution topology, whereas the maximum likelihood or *a posteriori* method corresponds to the sup-norm. This difference strongly affects the learning results in singular models.

1.1.3 Fisher information matrix

Definition 1.4 (Fisher information matrix) For a given statistical model or a learning machine $p(x|w)$, where $x \in \mathbb{R}^N$ and $w \in \mathbb{R}^d$, the Fisher information matrix

$$I(w) = \{I_{jk}(w)\} \quad (1 \le j, k \le d)$$

is defined by

$$I_{jk}(w) = \int \Big(\frac{\partial}{\partial w_j} \log p(x|w)\Big) \Big(\frac{\partial}{\partial w_k} \log p(x|w)\Big)\, p(x|w)\, dx$$

if the integral is finite.

By the definition, the Fisher information matrix is always symmetric and positive semi-definite. It is not positive definite in general. In some statistics textbooks, it is assumed that the Fisher information matrix is positive definite, and that the Cramer–Rao inequality is proven; however, there are a lot of statistical models and learning machines in which Fisher information matrices have zero eigenvalue. The Fisher information matrix is positive definite if and only if

$$\left\{\frac{\partial}{\partial w_j}\log p(x|w)\right\}_{j=1}^{d}$$

is linearly independent as a function of x on the support of $p(x|w)$. Since

$$\frac{\partial}{\partial w_j}\log p(x|w) = -\frac{\partial}{\partial w_j} f(x,w),$$

the Fisher information matrix is positive definite if and only if

$$\left\{\frac{\partial}{\partial w_j} f(x,w)\right\}_{j=1}^{d}$$

is linearly independent as a function of x. By using $\int p(x|w)dx = 1$ for an arbitrary w, it is easy to show that

$$I_{jk}(w) = -\int \Big(\frac{\partial^2}{\partial w_j \partial w_k}\log p(x|w)\Big)\, p(x|w)\, dx.$$

If $q(x) = p(x|w_0)$, then

$$I_{jk}(w_0) = \frac{\partial^2}{\partial w_j \partial w_k} K(w_0).$$

Therefore, the Fisher information matrix is equal to the Hessian matrix of the Kullback–Leibler distance at the true parameter.

Remark 1.4 If the Fisher information matrix is positive definite in a neighborhood of the true parameter w_0, $K(w) > 0$ $(w \neq w_0)$ holds, and the Kullback–Leibler distance can be approximated by the positive definite quadratic form, then

$$K(w) \approx \tfrac{1}{2}(w - w_0)\cdot I(w_0)(w - w_0),$$

where $u \cdot v$ shows the inner product of two vectors u, v. If the Fisher information matrix is not positive definite, then $K(w)$ cannot be approximated by any quadratic form in general. This book establishes the mathematical foundation for the case when the Fisher information matrix is not positive definite.

Remark 1.5 (Cramer–Rao inequality) Assume that random samples $\{X_i; i = 1, 2, \ldots, n\}$ are taken from the probability density function $\prod_{i=1}^{n} p(x_i|w)$, where $w = (w_1, w_2, \ldots, w_d) \in \mathbb{R}^d$. A function from random samples to the parameter space

$$\{u_j(x_1, x_2, \ldots, x_n); j = 1, 2, \ldots, d\} \in \mathbb{R}^d$$

is called an unbiased estimator if it satisfies

$$E[u_j(X_1, X_2, \ldots, X_n) - w_j] \equiv \int (u_j(x_1, x_2, \ldots, x_n) - w_j) \times \prod_{i=1}^{n} p(x_i|w)dx_i = 0$$

for arbitrary $w \in \mathbb{R}^d$. Under certain are conditions which ensure that $\int dx_j$ and $(\partial/\partial w_k)$ are commutative for arbitrary j, k,

$$\begin{aligned} 0 &= \frac{\partial}{\partial w_k} E[u_j(X_1, X_2, \ldots, X_n) - w_j] \\ &= E[(u_j - w_j) \sum_{i=1}^{n} \frac{\partial}{\partial w_k} \log p(X_i, w)] - \delta_{jk}. \end{aligned}$$

Therefore,

$$\delta_{jk} = E[(u_j - w_j) \sum_{i=1}^{n} \frac{\partial}{\partial w_k} \log p(X_i, w)].$$

For arbitrary d-dimensional vectors $\mathbf{a} = (a_j)$, $\mathbf{b} = (b_k)$,

$$(\mathbf{a} \cdot \mathbf{b}) = E\Big[\Big(\sum_{j=1}^{d} a_j(u_j - w_j)\Big)\Big(\sum_{k=1}^{d} \sum_{i=1}^{d} b_k \frac{\partial}{\partial w_k} \log p(X_i, w)\Big)\Big].$$

By applying the Cauchy–Schwarz inequality

$$(\mathbf{a} \cdot \mathbf{b})^2 \leq n\,(\mathbf{a} \cdot V\mathbf{a})(\mathbf{b} \cdot I(w)\mathbf{b}), \tag{1.7}$$

where $V = (V_{jk})$ is the covariance matrix of $u - w$,

$$V_{jk} = E[(u_j - w_j)(u_k - w_k)]$$

and $I(w)$ is the Fisher information matrix. If $I(w)$ is positive definite, by putting

$$\begin{aligned} \mathbf{a} &= I(w)^{1/2}\mathbf{c}, \\ \mathbf{b} &= I(w)^{-1/2}\mathbf{c}, \end{aligned}$$

it follows that

$$\|\mathbf{c}\|^2 \leq n\,(\mathbf{c} \cdot I(w)^{1/2} V I(w)^{1/2} \mathbf{c})$$

holds for arbitrary vector $\mathbf{c}$, hence

$$V \geq \frac{I(w)^{-1}}{n}. \tag{1.8}$$

This relation, the Cramer–Rao inequality, shows that the covariance matrix of any unbiased estimator cannot be made smaller than the inverse of the Fisher information matrix. If $I(w)$ has zero eigenvalue and $\mathbf{b}$ is an eigenvector for zero eigenvalue, eq.(1.7) shows that either V is not a finite matrix or no unbiased estimator exists. For statistical models which have a degenerate Fisher information matrix, we have no effective unbiased estimator in general.

1.2 Statistical models and learning machines

1.2.1 Singular models

Definition 1.5 (Identifiablity) A statistical model or a learning machine $p(x|w)$ ($x \in \mathbb{R}^N$, $w \in W \subset \mathbb{R}^d$) is called identifiable if the map

$$W \ni w \mapsto p(\ |w)$$

is one-to-one, in other words,

$$p(x|w_1) = p(x|w_2) \quad (\forall x \in \mathbb{R}^d) \Longrightarrow w_1 = w_2.$$

A model which is not identifiable is called nonidentifiable or unidentifiable.

Definition 1.6 (Positive definite metric) A statistical model or a learning machine $p(x|w)$ ($x \in \mathbb{R}^N$, $w \in W \subset \mathbb{R}^d$) is said to have a positive definite metric if its Fisher information matrix $I(w)$ is positive definite for arbitrary $w \in \mathbf{W}$. If a statistical model does not have a positive definite metric, it is said to have a degenerate metric.

Definition 1.7 (Singular statistical models) Assume that the support of the statistical model $p(x|w)$ is independent of w. A statistical model $p(x|w)$ is said to be regular if it is identifiable and has a positive definite metric. If a statistical model is not regular, then it is called strictly singular. The set of singular statistical models consists of both regular and strictly singular models.

Mathematically speaking, identifiability is neither a necessary nor a sufficient condition of positive definiteness of the Fisher information matrix. In fact, if $p(x|a)$ ($x, a \in \mathbb{R}^1$) is a regular statistical model, then $p(x|a^3)$ is identifiable but has a degenerate Fisher information matrix. Also $p(x|a^2)$ ($|a| > 1$) has a nondegenerate Fisher information matrix but is nonidentifiable. These are trivial examples in which an appropriate transform or restriction of a parameter makes models regular.

However, a lot of statistical models and learning machines used in information science have simultaneously nonidentifiability and a degenerate metric.

Moreover, they contain a lot of singularities which cannot be made regular by any transform or restriction.

In this book, we mainly study singular statistical models or singular learning machines. The following statistical models are singular statistical models.

(1) Layered neural networks
(2) Radial basis functions
(3) Normal mixtures
(4) Binomial and multinomial mixtures
(5) Mixtures of statistical models
(6) Reduced rank regressions
(7) Boltzmann machines
(8) Bayes networks
(9) Hidden Markov models
(10) Stochastic context-free grammar

These models play the central role of information processing systems in artificial intelligence, pattern recognition, robotic control, time series prediction, and bioinformatics. They determine the preciseness of the application systems. Singular models are characterized by the following features.

(1) They are made by superposition of parametric functions.
(2) They have hierarchical structures.
(3) They contain hidden variables.
(4) They consists of several information processing modules.
(5) They are designed to obtain hidden knowledge from random samples.
(6) They estimate the probabilistic grammars.

In singular statistical models, the knowledge or grammar to be discovered corresponds to singularities in general. Figure 1.2 shows an example of the correspondence between parameters and probability distributions in normal mixtures.

Remark 1.6 (Equivalence relation) The condition that $p(x|w_1) = p(x|w_2)$ for arbitrary x does not mean

$$\frac{\partial^k p(x|w_1)}{\partial w_1^k} = \frac{\partial^k p(x|w_2)}{\partial w_2^k} \quad (k = 1, 2, 3, \ldots).$$

Even if $p(x|w_1) \approx p(x|w_2)$, their derivatives are very different in general. The preciseness of statistical estimation is determined by the derivative of $p(x|w)$, hence results of statistical estimations are very different if $p(x|w_1) \approx p(x|w_2)$.

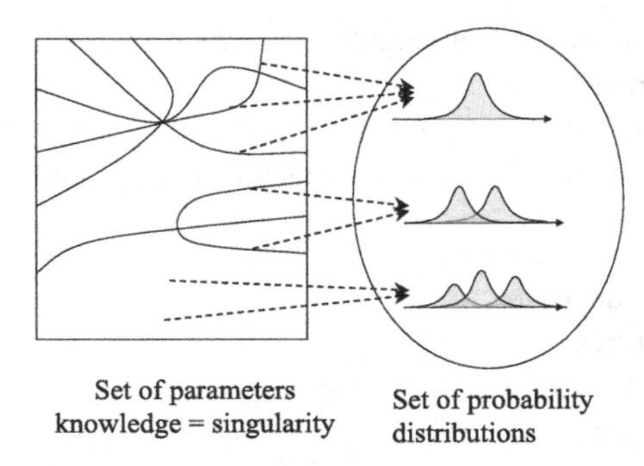

Fig. 1.2. Map from parameter to probability distribution

One can introduce an equivalence relation $\sim$ into the set of parameters W,

$$w_1 \sim w_2 \Longleftrightarrow p(x|w_1) = p(x|w_2) \quad (\forall x).$$

Then the map $(W/_{\sim}) \ni w \to p_w$ is one-to-one. However, the quotient set $(W/_{\sim})$ is neither the Euclidean space nor a manifold. Therefore, it is still difficult to construct statistical learning theory on $(W/_{\sim})$. In this book, we show that there is a birational map in algebraic geometry which enables us to establish singular learning theory.

Remark 1.7 (No asymptotic normality) If a model is regular, then the Bayes *a posteriori* distribution can be approximated by the normal distribution

$$\frac{1}{Z_n} \exp\Big(-\frac{n}{2}(w - w_0)I(w_0)(w - w_0)\Big),$$

where w_0 is the unique parameter such that $q(x) = p(x|w_0)$. Also the maximum likelihood estimator and the maximum *a posteriori* estimator are asymptotically subject to the normal distribution. Such a property is called asymptotic normality. However, singular statistical models do not have such a property, with the result that almost all statistical theories using asymptotic normality do not hold in singular statistical models.

Remark 1.8 (True generic condition) In a lot of statistical models and learning machines, the set of parameters at which the Fisher information matrices are degenerate

$$W_{(0)} = \{w \in W \; ; \; \det(I(w)) = 0\}$$

is a measure zero subset in $\mathbb{R}^d$. Hence one might suppose that, in generic cases, the true parameter w_0 is seldom contained in $W_{(0)}$, and that the learning theory assuming $\det(I(w_0)) > 0$ may be sufficient in practical applications. However, this consideration is wrong. On the contrary, in general cases, we have to optimize a statistical model or a learning machine by comparing several probable models and hyperparameters. In such cases, we always examine models under the condition that the optimal parameter lies in a neighborhood of $W_{(0)}$. Especially in model selection, hyperparameter optimization, or hypothesis testing, we need the theoretical results of the case $w_0 \in W_{(0)}$ because we have to determine whether $w_0 \in W_{(0)}$ or not. Therefore, the superficial generic condition $\det(I(w)) > 0$ does not have true generality.

Remark 1.9 (Singular theory contains regular theory) Statistical theory of regular models needs identifiability and a nondegenerate Fisher information matrix. In this book, singular learning theory is established on the assumption that neither identifiability nor a positive definite Fisher information matrix is necessary. Of course, even if a model is regular, the singular learning theory holds. In other words, a regular model is understood as a very special example to which singular learning theory can be applied. From the mathematical point of view, singular learning theory contains regular learning theory as a very special part. For example, the concepts AIC (Akaike's information criterion) and BIC (Bayes information criterion) in regular statistical theory are completely generalized in this book.

1.2.2 Density estimation

Let us introduce some examples of regular and singular statistical models.

Example 1.1 (Regular model) A parametric probability density function of $(x, y) \in \mathbb{R}^2$ for a given parameter $w = (a, b) \in \mathbb{R}^2$ defined by

$$p(x, y|a, b) = \frac{1}{2\pi} \exp\Big(-\frac{(x-a)^2 + (y-b)^2}{2}\Big)$$

is a regular statistical model, where the set of parameters is $W = \{(a, b) \in \mathbb{R}^2\}$. This is a two-dimensional normal distribution. For given random samples (X_i, Y_i), the likelihood function is

$$L_n(a, b) = \frac{1}{(2\pi)^n} \exp\Big(-\frac{1}{2}\sum_{i=1}^{n}\{(X_i - a)^2 + (Y_i - b)^2\}\Big).$$

If the true distribution is given by (a_0, b_0),

$$q(x, y) = p(x, y|a_0, b_0),$$

the log likelihood ratio function is

$$K_n(a,b) = \frac{a^2 - a_0^2 + b^2 - b_0^2}{2} - (a-a_0)\Big(\frac{1}{n}\sum_{i=1}^{n} X_i\Big) - (b-b_0)\Big(\frac{1}{n}\sum_{i=1}^{n} Y_i\Big).$$

The Kullback–Leibler distance is

$$K(a,b) = \tfrac{1}{2}\{(a-a_0)^2 + (b-b_0)^2\}.$$

Note that $K(a,b) = 0$ if and only if $a = a_0$ and $b = b_0$. The Fisher information matrix

$$I(a,b) = \begin{pmatrix} 1 & 0 \\ 0 & 1 \end{pmatrix}$$

is positive definite for an arbitrary (a,b).

Example 1.2 (Singular model) Let us introduce another parametric probability density function of $x \in \mathbb{R}^1$ defined by

$$p(x|a,b) = \frac{1}{\sqrt{2\pi}}\,\{(1-a)\,e^{-\frac{x^2}{2}} + a\,e^{-\frac{(x-b)^2}{2}}\}.$$

The set of parameters is

$$W = \{w = (a,b); 0 \le a \le 1,\ -\infty < b < \infty\}.$$

This model is called a normal mixture. If the true distribution is given by

$$q(x) = p(x|a_0, b_0),$$

then the log likelihood ratio function is

$$K_n(a,b) = \frac{1}{n}\sum_{i=1}^{n}\log\Big(\frac{1 + a_0\,(\exp(b_0 X_i - b_0^2/2) - 1)}{1 + a\,(\exp(bX_i - b^2/2) - 1)}\Big)$$

and

$$K(a,b) = \int \log\Big(\frac{1 + a_0\,(\exp(b_0 x - b_0^2/2) - 1)}{1 + a\,(\exp(bx - b^2/2) - 1)}\Big)\,q(x)\,dx.$$

If $a_0 b_0 \neq 0$, then $K(a,b) = 0$ is equivalent to $a = a_0$ and $b = b_0$. In such cases, the Fisher information matrix $I(a_0, b_0)$ is positive definite. However, if $a_0 b_0 = 0$, then $K(a,b) = 0$ is equivalent to $ab = 0$, and the Fisher information matrix $I(a_0, b_0) = 0$. The function $K(a,b)$ can be expanded as

$$K(a,b) = \tfrac{1}{2}a^2 b^2 + \cdots,$$

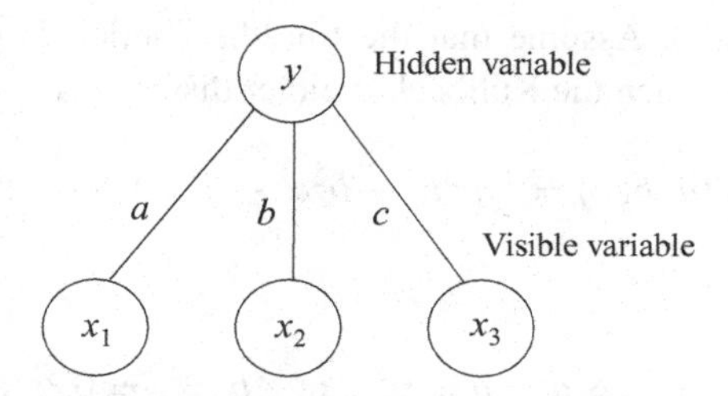

Fig. 1.3. Bayesian network with hidden unit

which shows that $K(a,b)$ cannot be approximated by any quadratic form. If we make a model selection algorithm or a hypothesis test procedure for this model, then we have to study the case $K(a,b) \cong 1/\sqrt{n}$ where n is the number of random samples. Hence we have to evaluate the effect of the singularity in the set $ab = 0$.

Example 1.3 (Bayesian network with a hidden unit) Let X_1, X_2, X_3 and Y are random variables which take values $\{-1, 1\}$. The Bayesian network shown in Figure 1.3 is defined by the probability distribution of $X = (X_1, X_2, X_3)$ and Y,

$$p(x, y|w) = \frac{1}{Z(a,b,c)} \exp(ax_1 y + bx_2 y + cx_3 y),$$

where $w = (a, b, c)$ and $Z(a,b,c)$ is a normalizing constant. Let X be a set of visible units and Y a hidden unit. The probability distribution of X is given by the marginal distribution,

$$\begin{aligned} p(x|w) &= \frac{1}{Z(a,b,c)} \sum_{y=\pm 1} \exp(ax_1 y + bx_2 y + cx_3 y) \\ &= \frac{1}{2\,Z(a,b,c)} \cosh(ax_1 + bx_2 + cx_3). \end{aligned}$$

By using $\tanh(ax_i) = \tanh(a)x_i$ for $x_i = \pm 1$, and

$$\begin{aligned} \cosh(u+v) &= \cosh(u)\cosh(v) + \sinh(u)\sinh(v), \\ \sinh(u+v) &= \sinh(u)\cosh(v) + \cosh(u)\sinh(v), \end{aligned}$$

we have

$$p(x|w) = \tfrac{1}{8}\{1 + t(a)t(b)x_1x_2 + t(b)t(c)x_2x_3 + t(c)t(a)x_3x_1\},$$

where $t(a) = \tanh(a)$. Assume that the true distribution is given by $q(x) = p(x|0, 0, 0) = 1/8$. Then the Kullback–Leibler distance is

$$K(a, b, c) = \tfrac{1}{2}(a^2b^2 + b^2c^2 + c^2a^2) + \cdots .$$

Therefore

$$q(x) = p(x|a, b, c) \Longleftrightarrow a = b = 0, \quad \text{or} \quad b = c = 0, \quad \text{or} \quad c = a = 0.$$

The Fisher information matrix is equal to zero at (0, 0, 0). If we want to judge whether the hidden variable Y is necessary to explain a given set of random samples, we should clarify the effect of the singularity of $K(a, b, c) = 0$.

1.2.3 Conditional probability density

Example 1.4 (Regular model) A probability density function of $(x, y) \in \mathbb{R}^2$,

$$p(x, y|a, b) = q_0(x) \frac{1}{\sqrt{2\pi}} \exp(-\tfrac{1}{2}(y - ax - b)^2), \tag{1.9}$$

is a statistical model, where the set of parameters is $W = \{w = (a, b) \in \mathbb{R}^2\}$ and $q_0(x)$ is a constant probability density function of x. This model is referred to as a line regression model. If the true distribution is $q(x, y)$, the true conditional probability density function

$$q(y|x) = \frac{q(x, y)}{\int q(x, y')\, dy'}$$

is estimated by the conditional probability density function

$$p(x|y, a, b) = \frac{1}{\sqrt{2\pi}} \exp(-\tfrac{1}{2}(y - ax - b)^2). \tag{1.10}$$

The two models eq.(1.9) and eq.(1.10) have the same log likelihood ratio function and the same Kullback–Leibler distance, hence the two models are equivalent from a statistical point of view. In other words, estimation of the conditional density function of y for a given x can be understood as the estimation of a joint probability density function of (x, y), if $q(x)$ is not estimated.

If the true distribution is given by $w_0 = (a_0, b_0)$,

$$q(x, y) = p(x, y|a_0, b_0).$$

The log likelihood ratio function is

$$K_n(a,b) = \frac{(a^2 - a_0^2)}{2}\Big(\frac{1}{n}\sum_{i=1}^{n} X_i^2\Big) + \frac{(b^2 - b_0^2)}{2} + (ab - a_0 b_0)\Big(\frac{1}{n}\sum_{i=1}^{n} X_i\Big)$$

$$- (a - a_0)\Big(\frac{1}{n}\sum_{i=1}^{n} X_i Y_i\Big) - (b - b_0)\Big(\frac{1}{n}\sum_{i=1}^{n} Y_i\Big).$$

The Kullback–Leibler distance is

$$K(a,b) = \frac{1}{2}\int (ax + b - a_0 x - b_0)^2 \, q_0(x)\, dx$$
$$= \tfrac{1}{2}(w - w_0)\cdot I(w - w_0),$$

where

$$I = \begin{pmatrix} m_2 & m_1 \\ m_1 & 1 \end{pmatrix},$$

and

$$m_i = \int x^i \, q_0(x)\, dx.$$

The Fisher information matrix is always equal to I, which does not depend on the true parameter w_0. It is positive definite if and only if $m_2 \neq m_1^2$. In other words, I is degenerate if and only if the variance of $q_0(x)$ is equal to zero.

Example 1.5 (Singular model) Another example of a statistical model of $y \in \mathbb{R}^1$ for $x \in \mathbb{R}^1$ is

$$p(x, y|a, b) = q_0(x)\,\frac{1}{\sqrt{2\pi}}\,\exp(-\tfrac{1}{2}(y - a\tanh(bx))^2),$$

where the set of parameters is $W = \{(a, b) \in \mathbb{R}^2\}$. This model is the simplest three-layer neural network. If the true distribution is given by $w = (a_0, b_0)$,

$$q(x, y) = p(x, y|a_0, b_0),$$

the log likelihood ratio function is

$$K_n(a,b) = \frac{1}{2n}\sum_{i=1}^{n}\{(Y_i - a\tanh(bX_i))^2 - (Y_i - a_0\tanh(b_0 X_i))^2\},$$

and the Kullback–Leibler distance is

$$K(a,b) = \frac{1}{2}\int (a\tanh(bx) - a_0\tanh(b_0 x))^2 \, q_0(x)\, dx.$$

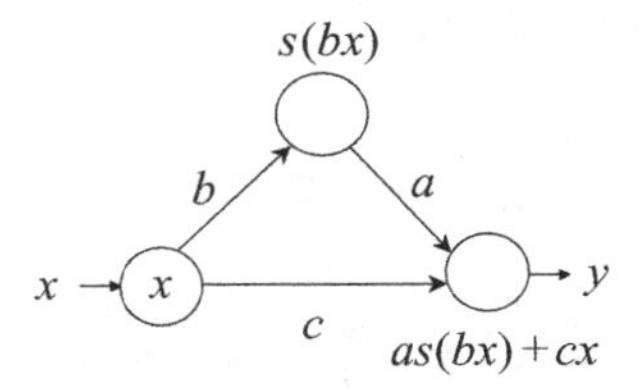

Fig. 1.4. Layered neural network

If $a_0 b_0 \neq 0$, then $K(a, b) = 0$ if and only if $a = a_0$ and $b = b_0$, and the Fisher information matrix $I(a_0, b_0)$ is positive definite. However, if $a_0 b_0 = 0$, then $K(a, b) = 0$ is equivalent to $ab = 0$, and the Fisher information matrix is degenerate. In practical applications of three-layer neural networks, we have to decide whether a three-layer neural network

$$\sum_{h=1}^{H} a_h \tanh(b_h x + c_h)$$

almost approximates the true regression function or not. In such cases, the more precisely the model approximates the true regression function, the more degenerate the Fisher information matrix is. Therefore, we cannot assume that the Fisher information matrix is positive definite in the model evaluation.

Example 1.6 (Layered neural network) Let $x, y \in \mathbb{R}^1$ and $w = (a, b, c) \in \mathbb{R}^3$. The statistical model shown in Figure 1.4,

$$p(y|x, w) = \frac{1}{\sqrt{2\pi}} \exp(-\tfrac{1}{2}(y - as(bx) - cx)^2),$$

where $s(t) = t + t^2$, is a layered statistical model. If the true distribution is $q(y|x) = p(y|x, 0, 0, 0)$ and if $q(x)$ is the standard normal distribution, then

$$K(a, b, c) = \tfrac{1}{2}(ab + c)^2 + \tfrac{3}{2}a^2 b^2.$$

Hence

$$q(y|x) = p(y|x, w) \Longleftrightarrow ab = c = 0.$$

The Fisher information matrix is equal to zero at (0, 0, 0).

1.3 Statistical estimation methods

In this section, let us introduce some statistical estimation methods, Bayes and Gibbs estimations, the maximum likelihood and *a posteriori* estimations.

1.3.1 Evidence

Let $D_n = \{X_1, X_2, \ldots, X_n\}$ be a set of random samples. For a given set of a statistical model $p(x|w)$ and an *a priori* probability density function $\varphi(w)$, the *a posteriori* probability density function $p(w|D_n)$ with the inverse temperature $\beta > 0$ is defined by

$$p(w|D_n) = \frac{1}{Z_n}\, \varphi(w) \prod_{i=1}^{n} p(X_i|w)^\beta,$$

where Z_n is the normalizing constant determined so that $p(w|D_n)$ is a probability density function of w,

$$Z_n = \int dw\, \varphi(w) \prod_{i=1}^{n} p(X_i|w)^\beta.$$

If $\beta = 1$, $p(w|D_n)$ is called a strict Bayes *a posteriori* density; if $\beta \neq 1$, it is a generalized version. When $\beta \to \infty$, it converges to $\delta(w - \hat{w})$, where $\hat{w}$ is the maximum likelihood estimator. Note that Z_n is a measurable function of D_n, hence it is also a random variable. The random variable Z_n is called the evidence, the marginal likelihood, or the partition function.

Remark 1.10 (Meaning of evidence) If $\beta = 1$, $Z_n = Z_n(X_1, X_2, \ldots, X_n)$ satisfies

$$\int dx_1 dx_2 \cdots dx_n Z_n(x_1, x_2, \ldots, x_n) = 1.$$

Therefore, Z_n with $\beta = 1$ defines a probability density function of D_n for a given pair of $p(x|w)$ and φ. In other words, Z_n can be understood as a likelihood function of the pair $(p(x|w), \varphi(w))$.

The predictive distribution $p(x|D_n)$ is defined by

$$p(x|D_n) = \int p(x|w)\, p(w|D_n)\, dw.$$

The Bayes estimation is defined by

$$p^*(x) = p(x|D_n),$$

in other words, the Bayes estimation is the map

$$D_n \mapsto p^*(x) = p(x|D_n).$$

The Bayes generalization error B_g is the Kullback–Leibler distance from $q(x)$ to $p^*(x)$,

$$B_g = \int q(x) \log \frac{q(x)}{p(x|D_n)} dx.$$

Here B_g is a measurable function of D_n, hence it is also a random variable. The Bayes training error B_t is defined by

$$B_t = \frac{1}{n} \sum_{i=1}^{n} \log \frac{q(X_i)}{p(X_i|D_n)}.$$

In Bayes learning theory, there are several important observables. The stochastic complexity, the minus log marginal likelihood, or the free energy is defined by

$$F_n = -\log Z_n.$$

Since Z_n with $\beta = 1$ can be understood as the likelihood of the pair $p(x|w)$ and $\varphi(w)$, F_n with $\beta = 1$ is the minus log likelihood of them.

In analysis of B_g, B_t, and F_n, we have some useful equations. The normalized evidence is defined by

$$Z_n^0 = \frac{Z_n}{\prod_{i=1}^{n} q(X_i)^{\beta}}. \tag{1.11}$$

Then, by using eq.(1.5), the *a posteriori* distribution is rewritten as

$$p(w|D_n) = \frac{1}{Z_n^0} \exp(-n\beta K_n(w)) \, \varphi(w),$$

where $K_n(w)$ is the log likelihood ratio function defined in eq.(1.3), and

$$Z_n^0 = \int dw \; \varphi(w) \; \exp(-n\beta K_n(w)).$$

In the same way, the normalized stochastic complexity is defined by

$$F_n^0 = -\log Z_n^0.$$

The empirical entropy is given by

$$S_n = -\frac{1}{n} \sum_{i=1}^{n} \log q(X_i).$$

Then

$$F_n = F_n^0 - n\beta S_n.$$

By the definition of the predictive distribution, it follows that

$$p(X_{n+1}|D_n) = \frac{\displaystyle\int dw\, \varphi(w)\, p(X_{n+1}|w) \prod_{i=1}^{n} p(X_i|w)^{\beta}}{\displaystyle\int dw\, \varphi(w) \prod_{i=1}^{n} p(X_i|w)^{\beta}}. \tag{1.12}$$

Theorem 1.2 *For an arbitrary natural number n, the Bayes generalization error with $\beta = 1$ and its mean satisfy the following equations.*

$$B_g = E_{X_{n+1}}\big[F_{n+1}^0\big] - F_n^0,$$
$$E[B_g] = E\big[F_{n+1}^0\big] - E[F_n^0].$$

Proof of Theorem 1.2 From eq.(1.12) with $\beta = 1$,

$$p(X_{n+1}|D_n) = \frac{Z_{n+1}}{Z_n}$$

holds for an arbitrary natural number n. The logarithm of this equation results in

$$-\log p(X_{n+1}|D_n) = F_{n+1} - F_n.$$

Also,

$$\frac{p(X_{n+1}|D_n)}{q(X_{n+1})} = \frac{\displaystyle\int dw\, \varphi(w)\, \exp(-(n+1)K_{n+1}(w))}{\displaystyle\int dw\, \varphi(w)\, \exp(-nK_n(w))}$$

shows

$$\frac{p(X_{n+1}|D_n)}{q(X_{n+1})} = \frac{Z_{n+1}^0}{Z_n^0}.$$

Therefore,

$$\log \frac{q(X_{n+1})}{p(X_{n+1}|D_n)} = F_{n+1}^0 - F_n^0. \tag{1.13}$$

Based on eq.(1.13), the two equations in the theorem are respectively given by the expectations of X_{n+1} and D_{n+1}. □

This theorem shows that the Bayes generalization error with $\beta = 1$ is equal to the increase of the normalized stochastic complexity.

1.3.2 Bayes and Gibbs estimations

Let $E_w[\,\cdot\,]$ be the expectation value using the *a posteriori* distribution $p(w|D_n)$. In Bayes estimation, the true distribution is estimated by the predictive distribution $E_w[p(x|w)]$. In the other method of statistical estimation, Gibbs estimation a parameter w is randomly chosen from $p(w|D_n)$, then the true distribution is estimated by $p(x|w)$. Gibbs estimation depends on a random choice of the parameter w. Hence, to study its generalization error, we need the expectation value over random choices of w. Bayes and Gibbs estimations respectively have generalization and training errors. The set of four errors is referred to as the Bayes quartet.

Definition 1.8 (Bayes quartet) For the generalized *a posteriori* distribution $p(w|D_n)$, the four errors are defined as follows.
(1) The Bayes generalization error,

$$B_{\mathrm{g}} = E_X\Big[\log \frac{q(X)}{E_w[p(X|w)]}\Big],$$

is the Kullback–Leibler distance from $q(x)$ to the predictive distribution.
(2) The Bayes training error,

$$B_{\mathrm{t}} = \frac{1}{n}\sum_{i=1}^{n} \log \frac{q(X_i)}{E_w[p(X_i|w)]},$$

is the empirical Kullback–Leibler distance from $q(x)$ to the predictive distribution.
(3) The Gibbs generalization error,

$$G_{\mathrm{g}} = E_w\Big[E_X\Big[\log \frac{q(X)}{p(X|w)}\Big]\Big],$$

is the mean Kullback–Leibler distance from $q(x)$ to $p(x|w)$.
(4) The Gibbs training error,

$$G_{\mathrm{t}} = E_w\Big[\frac{1}{n}\sum_{i=1}^{n} \log \frac{q(X_i)}{p(X_i|w)}\Big],$$

is the mean empirical Kullback–Leibler distance from $q(x)$ to $p(x|w)$.

Remark 1.11 The Bayes *a posteriori* distribution $p(w|D_n)$ depends on the set of random samples, D_n. Hence the Bayes quartet is a set of random variables. The most important variable among them is the Bayes generalization error because it is used in practical applications; however, the other variables have important information about statistical estimation. In fact, we prove that there are mathematical relations among them.

Theorem 1.3 (Representation of Bayes quartet) *By using the log density ratio function $f(x, w) = \log(q(x)/p(x|w))$, the four errors are rewritten as*

$$B_{\rm g} = E_X\Big[-\log E_w[e^{-f(X,w)}]\Big],$$

$$B_{\rm t} = \frac{1}{n}\sum_{i=1}^{n} - \log E_w[e^{-f(X_i,w)}],$$

$$G_{\rm g} = E_w[K(w)],$$

$$G_{\rm t} = E_w[K_n(w)].$$

Proof of Theorem 1.3 The first and the second equations are derived from

$$\log \frac{q(X)}{E_w[p(X|w)]} = -\log E_w[e^{-f(X,w)}].$$

The third and the fourth equations are derived from the definitions of the Kullback–Leibler distance $K(w)$ and the empirical one $K_n(w)$. □

Remark 1.12 (Generalization errors and square error) Let $p(y|x, w)$ be a conditional probability density of $y \in \mathbb{R}^N$ for a given $x \in \mathbb{R}^M$ defined by

$$p(y|x, w) = \frac{1}{\sqrt{2\pi\sigma^2}^N} \exp\Big(-\frac{1}{2\sigma^2}|y - h(x, w)|^2\Big),$$

where $h(x, w)$ is a function from $\mathbb{R}^M \times \mathbb{R}^d$ to $\mathbb{R}^N$, $|\cdot|$ is the norm of $\mathbb{R}^N$, and $\sigma > 0$ is a constant. Let us compare generalization errors with the square error. If the true conditional distribution is $p(y|x, w_0)$, then the log density ratio function is

$$\begin{aligned} f(x, y, w) &= \frac{1}{2\sigma^2}\{|y - h(x, w)|^2 - |y - h(x, w_0)|^2\} \\ &= \frac{1}{2\sigma^2}\{2(y - h(x, w_0)) \cdot (h(x, w_0) - h(x, w)) \\ &\quad + |h(x, w_0) - h(x, w)|^2\}. \end{aligned} \tag{1.14}$$

The Kullback–Leibler distance is

$$K(w) = \frac{1}{2\sigma^2} E_X[\ |h(X, w_0) - h(X, w)|^2].$$

The Gibbs generalization error is

$$G_{\rm g} = \frac{1}{2\sigma^2} E_X[E_w[\ |h(X, w_0) - h(X, w)|^2]].$$

The Bayes generalization error is

$$B_{\rm g} = E_X E_Y[-\log E_w[e^{-f}]], \tag{1.15}$$

where $f = f(X, Y, w)$. On the other hand, the regression function of the estimated distribution $E_w[p(y|x, w)]$ is equal to

$$\int y\, E_w[p(y|x, w)]\, dy = E_w\Big[\int yp(y|x, w)dy\Big]$$
$$= E_w[h(x, w)].$$

Let us define the square error of the estimated and true regression functions by

$$E_{\rm g} = \frac{1}{2\sigma^2} E_X[|E_w[h(x, w)] - h(X, w_0)|^2]. \tag{1.16}$$

In general, $B_{\rm g} \neq E_{\rm g}$. However, asymptotically, $B_{\rm g} \cong E_{\rm g}$. In fact, on a natural assumption, the *a posteriori* distribution $p(w|X_n)$ converges so that $f \to 0$, hence

$$B_{\rm g} = E_X E_Y\Big[-\log E_w\Big[1 - f + \frac{f^2}{2} + o(f^2)\Big]\Big]$$
$$= E_X E_Y\Big(E_w\Big[f - \frac{f^2}{2}\Big] + \tfrac{1}{2}E_w[f]^2 + o(f^2)\Big).$$

By $E_X E_Y E_w(f - f^2/2) = o(f^2)$ using eq.(1.14),

$$B_{\rm g} = \tfrac{1}{2} E_X E_Y[E_w[f(X, Y, w)]^2] + o(f^2)]],$$

hence $B_{\rm g} \cong E_{\rm g}$.

1.3.3 Maximum likelihood and *a posteriori*

Let $q(x)$, $p(x|w)$, and $\varphi(w)$ be the true distribution, a statistical model, and an *a priori* probability density function, respectively. The generalized log likelihood function is given by

$$R_n(w) = -\sum_{i=1}^{n} \log p(X_i|w) - a_n \log \varphi(w),$$

where $\{a_n\}$ is a sequence of nonnegative real values. By using a log density ratio function

$$f(x, w) = \log(q(x)/p(x|w)),$$

the generalized log likelihood function can be rewritten as

$$R_n(w) = R_n^0(w) + nS_n,$$

where $R_n^0(w)$ is given by

$$R_n^0(w) = \sum_{i=1}^{n} f(X_i, w) - a_n \log \varphi(w). \tag{1.17}$$

Note that, in singular statistical models, sometimes

$$\inf_w R_n(w) = -\infty,$$

which means that there is no parameter that minimizes $R_n(w)$. If a parameter $\hat{w}$ that minimizes $R_n(w)$ exists, then a statistical estimation method

$$D_n \mapsto p(x|\hat{w})$$

is defined. The generalization error $R_{\rm g}$ and the training error $R_{\rm t}$ of this method are respectively defined by

$$R_{\rm g} = \int q(x) \log \frac{q(x)}{p(x|\hat{w})} dx,$$

$$R_{\rm t} = \frac{1}{n} \sum_{i=1}^{n} \log \frac{q(X_i)}{p(X_i|\hat{w})}.$$

By using $K(w)$ and $K_n(w)$ in equations (1.2) and (1.3) respectively, they can be rewritten as

$$R_{\rm g} = K(\hat{w}),$$

$$R_{\rm t} = K_n(\hat{w}).$$

Definition 1.9 (Maximum likelihood and maximum *a posteriori*)
(1) If $a_n = 0$ for arbitrary n, then $\hat{w}$ is called the maximum likelihood (or ML) estimator and the statistical estimation method is called the maximum likelihood (or ML) method.
(2) If $a_n = 1$ for arbitrary n, then $\hat{w}$ is called the maximum *a posteriori* estimator (or MAP) and the method is called the maximum *a posteriori* (or MAP) method.
(3) If a_n is an increasing function of n, then $\hat{w}$ is the generalized maximum *a posteriori* estimator and the method is called the generalized maximum *a posteriori* method.

Remark 1.13 (Formal relation between Bayes and ML)
(1) If $\beta \to \infty$, both Bayes and Gibbs estimations formally result in the maximum likelihood estimation.
(2) In regular statistical models in which the maximum likelihood estimator has asymptotic normality, the leading terms of the asymptotic generalization

errors of Bayes, ML, and MAP are equal to each other. However, in singular statistical models, they are quite different.

Example 1.7 (Divergence of MLE) Let $g(x|a,\sigma)$ be the normal distribution on $\mathbb{R}^1$,

$$g(x|a,\sigma) = \frac{1}{\sqrt{2\pi\sigma^2}} \exp\Big(-\frac{(x-a)^2}{2\sigma^2}\Big).$$

Let us study a normal mixture,

$$p(x|a,b,c,\sigma,\rho) = a\ g(x|b,\sigma) + (1-a)\ g(x|c,\rho),$$

where the set of parameters is

$$W = \{(a,b,c,\sigma,\rho)\ ;\ 0 \le a \le 1, |b|, |c| < \infty, \sigma, \rho > 0\}.$$

Then the likelihood function for a given D_n

$$L_n(a,b,c,\sigma,\rho) = \prod_{i=1}^{n} p(X_i|a,b,c,\sigma,\rho)$$

is an unbounded function, because

$$\lim_{\rho\to 0} L_n(a,b,X_1,\sigma,\rho) = \infty.$$

Therefore the normal mixture $p(x|a,b,c,\sigma,\rho)$ does not have a maximum likelihood estimator for arbitrary true distribution. To avoid this problem, we should restrict the parameter set or adopt the generalized maximum *a posteriori* method. In singular statistical models, the maximum likelihood estimator often diverges.

1.4 Four main formulas

In this section, we give an outline of singular learning theory. Because singular learning theory is quite different from regular statistical theory, the reader is advised to read this overview of the results of the book in advance. The equations and explanations in this section are intuitively described, because rigorous definitions and proofs are given in subsequent chapters.

1.4.1 Standard form of log likelihood ratio function

To evaluate how appropriate the statistical models $p(x|w)$ and $\varphi(w)$ are for a given data set $D_n = \{X_1, X_2, \ldots, X_n\}$, we have to study the case when the set

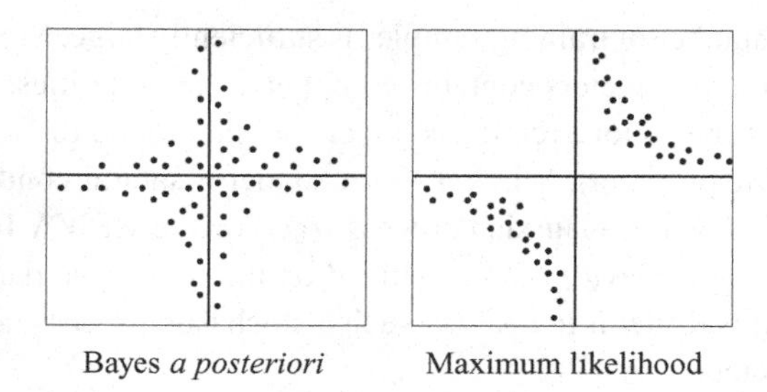

Fig. 1.5. Maximum likelihood and Bayes *a posteriori*

of true parameters

$$\begin{aligned} W_0 &= \{w \in W \ ; \ q(x) = p(x|w) \ (\forall x)\} \\ &= \{w \in W; K(w) = 0\} \end{aligned}$$

consists of not one point but a union of several manifolds. If $K(w)$ is a polynomial, then W_0 is called an algebraic set; if $K(w)$ is an analytic function, then W_0 is called an analytic set. If W_0 is not one point, neither the Bayes *a posteriori* distribution nor the distribution of the maximum likelihood estimators converges to the normal distribution. For example, the left-hand side of Figure 1.5 shows a Bayes *a posteriori* distribution when the set of true parameters is $\{(a, b); ab = 0\}$. The right-hand side shows the probability distribution of the maximum likelihood estimator. We need a method to analyze such a singular distribution.

The basic term in statistical learning is the empirical Kullback–Leibler distance,

$$K_n(w) = \frac{1}{n} \sum_{i=1}^{n} f(X_i, w),$$

which is a function of $w \in W \subset \mathbb{R}^d$. For $w \in W \setminus W_0$, a random process

$$\psi_n(w) = \sum_{i=1}^{n} \frac{K(w) - f(X_i, w)}{\sqrt{n\,K(w)}}$$

is well-defined. The log likelihood ratio function is rewritten as

$$nK_n(w) = nK(w) - \sqrt{nK(w)}\ \psi_n(w).$$

This representation has two mathematical problems.

(1) (Geometrical problem). In a singular model, W_0 is not one point but a real analytic set hence the log likelihood ratio function cannot be treated locally

even if the number of training samples is sufficiently large. Moreover, since the set of true parameters contains complicated singularities, it is difficult to analyze its behavior even in each local neighborhood of W_0.

(2) (Probabilistic problem). When $n \to \infty$, under a natural condition, $\psi_n(w)$ converges in law to a Gaussian process $\psi(w)$ on the set $W \setminus W_0$. However, neither $\psi_n(w)$ nor $\psi(w)$ is well-defined on the set of true parameters W_0. Therefore it is difficult to analyze such a stochastic process near the set of true parameters.

In this book, we propose an algebraic geometrical transform that is powerful enough to overcome these two problems. For a real analytic function $K(w)$, the fundamental theorem in algebraic geometry ensures that there exists a real d-dimensional manifold $\mathcal{M}$ and a real analytic map

$$g : \mathcal{M} \ni u \mapsto w \in W$$

such that, for each coordinate $\mathcal{M}_\alpha$ of $\mathcal{M}$, $K(g(u))$ is a direct product,

$$K(g(u)) = u_1^{2k_1} \, u_2^{2k_2} \cdots u_d^{2k_d},$$

where $k_1, k_2, \ldots, k_d$ are nonnegative integers. Morevoer, there exists a function $\phi(u) > 0$ and nonnegative integers $h_1, h_2, \ldots, h_d$ such that

$$\varphi(g(u))|g'(u)| = \phi(u)\big|u_1^{h_1} \, u_2^{h_2} \; \cdots \; u_d^{h_d}\big|,$$

where $|g'(u)|$ is Jacobian determinant of $w = g(u)$. Note that $k_1, k_2, \ldots, k_d$ and $h_1, h_2, \ldots, h_d$ depend on a local coordinate. By using the notation

$$\begin{aligned} u &= (u_1, u_2, \ldots, u_d), \\ k &= (k_1, k_2, \ldots, k_d), \\ h &= (h_1, h_2, \ldots, h_d), \end{aligned}$$

the function $K(g(u))$ and the *a priori* distribution $\varphi(g(u))|g'(u)|$ are respectively expressed as

$$\begin{aligned} K(g(u)) &= u^{2k}, \\ \varphi(g(u))|g'(u)| &= \phi(u)|u^h|. \end{aligned}$$

The theorem that ensures the existence of such a real analytic manifold $\mathcal{M}$ and a real analytic map $w = g(u)$ is called Hironaka's theorem or resolution of singularities. The function $w = g(u)$ is called a resolution map. In Chapters 2 and 3, we give a rigorous statement of the theorem and a method to find the set $(\mathcal{M}, g)$, respectively. Then by using

$$K(g(u)) = 0 \Longrightarrow f(x, g(u)) = 0,$$

we can prove that there exists a real analytic function $a(x, u)$ such that

$$f(x, g(u)) = a(x, u)\, u^k \quad (\forall x).$$

From the definition of the Kullback–Leibler distance,

$$\int f(x, g(u))q(x)dx = K(g(u)) = u^{2k}.$$

It follows that

$$\int a(x, u)q(x)dx = u^k.$$

Moreover, by $f(x, g(u)) = \log(q(x)/p(x|g(u)))$,

$$K(g(u)) = \int (f(x, g(u)) + e^{-f(x,g(u))} - 1)q(x)dx.$$

It is easy to show

$$\lim_{t \to 0} \frac{t + e^{-t} - 1}{t^2} \to \frac{1}{2}.$$

Therefore, if $u^{2k} = 0$, then

$$\int a(x, u)^2 q(x)dx = \lim_{u^{2k} \to 0} \frac{2K(g(u))}{u^{2k}} = 2.$$

Here we can introduce a well-defined stochastic process on $\mathcal{M}$,

$$\xi_n(u) = \frac{1}{\sqrt{n}} \sum_{i=1}^{n} \{u^k - a(X_i, u)\},$$

from which we obtain a representation,

$$nK_n(g(u)) = nu^{2k} - \sqrt{n}u^k \xi_n(u). \tag{1.18}$$

By definition, $\xi_n(u)$ satisfies

$$E[\xi_n(u)] = 0 \quad (\forall u \in \mathcal{M}),$$
$$E[\xi_n(u)\xi_n(v)] = E_X[a(X, u)a(X, v)] - u^k v^k \quad (\forall u, v \in \mathcal{M}).$$

If $K(g(u)) = K(g(v)) = 0$, then

$$E[\xi_n(u)\xi_n(v)] = E_X[a(X, u)a(X, v)],$$

and $E[\xi_n(u)^2] = 2$. By the central limit theorem, for each $u \in \mathcal{M}$, $\xi_n(u)$ converges in law to a Gaussian distribution with mean zero and variance 2. In Chapter 5, we prove the convergence in law $\xi_n \to \xi$ as a random variable on the space of bounded and continuous functions on $\mathcal{M}$. Then the Gaussian process $\xi(u)$ is uniquely determined by its mean and covariance. Here we attain the first main formula.

Main Formula I (Standard form of log likelihood ratio function)
Under natural conditions, for an arbitrary singular statistical model, there exist a real analytic manifold $\mathcal{M}$ and a real analytic map $g : \mathcal{M} \to W$ such that the log likelihood ratio function is represented by

$$K_n(g(u)) = u^{2k} - \frac{1}{\sqrt{n}} u^k \xi_n(u), \tag{1.19}$$

where $\xi_n(u)$ converges in law to the Gaussian process $\xi(u)$. Also $w = g(u)$ gives the relation

$$\varphi(g(u))|g'(u)| = \phi(u)|u^h|, \tag{1.20}$$

where $\phi(u) > 0$ is a positive real analytic function.

Remark 1.14 (1) Note that the log likelihood ratio function of any singular statistical model can be changed to the standard form by algebraic geometrical transform, which allows $|g'(u)| = 0$.
(2) The integration over the manifold $\mathcal{M}$ can be written as the finite sum of the integrations over local coordinates. There exists a set of functions $\{\sigma_\alpha(u)\}$ such that $\sigma_\alpha(u) \geq 0$, $\sum_\alpha \sigma_\alpha(u) = 1$, and the support of $\sigma_\alpha(u)$ is contained in $\mathcal{M}_\alpha$. By using a function $\phi^*(u) = \phi(u)\sigma_\alpha(u) \geq 0$, where dependence of α in ϕ^* is omitted, for an arbitrary integrable function $F(w)$,

$$\begin{aligned}\int_W F(w)\varphi(w)dw &= \int_{\mathcal{M}} F(g(u))\varphi(g(u))|g'(u)|du \\ &= \sum_\alpha \int F(g(u))\phi^*(u)|u^h|du.\end{aligned} \tag{1.21}$$

(3) In regular statistical models, the set of true parameters consists of one point, $W_0 = \{w_0\}$. By the transform $w = g_0(u) = w_0 + I(w_0)^{1/2}u$

$$\begin{aligned}K(g_0(u)) &\cong \tfrac{1}{2}|u|^2, \\ K_n(g_0(u)) &\cong \tfrac{1}{2}|u|^2 - \tfrac{\xi_n}{\sqrt{n}} \cdot u,\end{aligned}$$

where $I(w_0)$ is the Fisher information matrix and $\xi_n = (\xi_n(1), \xi_n(2), \ldots, \xi_n(d))$ is defined by

$$\xi_n(k) = \frac{1}{\sqrt{n}} \sum_{i=1}^{n} \frac{\partial}{\partial u_k} \log p(X_i|g_0(u))\Big|_{u=0}.$$

Here each $\xi_n(k)$ converges in law to the standard normal distribution. This property is called asymptotic normality. If a statistical model has asymptotic normality, Bayes generalization and training errors, MAP, and ML estimations are obtained by using the normal distribution. However, singular statistical models do not have asymptotic normality. The standard form of the log likelihood ratio function, eq.(1.19), is the universal base for singular statistical models.

1.4.2 Evidence of singular model

In singular learning theory, the zeta function of a statistical model plays an important role.

Definition 1.10 (Zeta function of a statistical model) For a given set (p, q, φ), where $p(x|w)$ is a statistical model, $q(x)$ is a true probability distribution, and $\varphi(w)$ is an *a priori* probability density function with compact support, the zeta function $\zeta(z)$ $(z \in \mathbb{C})$ of a statistical model is defined by

$$\zeta(z) = \int K(w)^z \, \varphi(w) \, dw,$$

where $K(w)$ is the Kullback–Leibler distance from $q(x)$ to $p(x|w)$.

By the definition, the zeta function is holomorphic in $Re(z) > 0$. It can be rewritten by using resolution map $w = g(u)$,

$$\zeta(z) = \int_{\mathcal{M}} K(g(u))^z \, \varphi(g(u)) \, |g'(u)| \, du.$$

By using $K(g(u)) = u^{2k}$ and eq.(1.20),(1.21), in each local coordinate,

$$\zeta(z) = \sum_{\alpha} \int_{\mathcal{M}_\alpha} u^{2kz+h} \phi^*(u) du.$$

It is easy to show that

$$\int_0^b u_1^{2k_1 z + h_1} du_1 = \frac{b^{2k_1 z + h_1}}{2k_1 z + h_1 + 1}.$$

Therefore, the Taylor expansion of $\phi^*(u)$ around the origin in arbitrary order shows that $\zeta(z)$ ($\mathrm{Re}(z) > 0$) can be analytically continued to the meromorphic function on the entire complex plane $\mathbb{C}$, whose poles are all real, negative, and rational numbers. They are ordered from the larger to the smaller,

$$0 > -\lambda_1 > -\lambda_2 > -\lambda_3 > \cdots .$$

The largest pole $(-\lambda_1)$ is determined by

$$\lambda_1 = \min_{\alpha} \min_{1 \le j \le d} \Big(\frac{h_j + 1}{2k_j}\Big). \tag{1.22}$$

Let m_k be the order of the pole $(-\lambda_k)$. The order m_1 is the maximum number of the elements of the set $\{j\}$ that attain the minimum of eq.(1.22). Therefore, the zeta function has the Laurent expansion,

$$\zeta(z) = \zeta_0(z) + \sum_{k=1}^{\infty} \sum_{m=1}^{m_k} \frac{c_{km}}{(z + \lambda_k)^{m_k}}, \tag{1.23}$$

where $\zeta_0(z)$ is a holomorphic function and c_{km} is a coefficient. Let the state density function of $t > 0$ be

$$\begin{aligned} v(t) &= \int \delta(t - K(w))\, \varphi(w)\, dw \\ &= \sum_{\alpha} \int \delta(t - u^{2k})|u^h|\phi^*(u)du. \end{aligned}$$

The zeta function is equal to its Mellin transform,

$$\zeta(z) = \int_0^{\infty} v(t)\, t^z\, dt.$$

Conversely, $v(t)$ is uniquely determined as the inverse Mellin transform of $\zeta(z)$. The inverse Mellin transform of

$$F(z) = \frac{(m-1)!}{(z + \lambda)^m}$$

is equal to

$$f(t) = \begin{cases} t^{\lambda-1}\,(-\log t)^{m-1} & (0 < t < 1) \\ 0 & \text{otherwise} \end{cases}.$$

By eq.(1.23), we obtain the asymptotic expansion of $v(t)$ for $t \to 0$,

$$v(t) = \sum_{k=1}^{\infty} \sum_{m=1}^{m_k} c'_{km}\, t^{\lambda_k - 1}\, (-\log t)^{m-1}.$$

This expansion holds for arbitrary $\phi^*(u)$, and c'_{km} is a linear transform of $\phi^*(u)$, therefore there exists a set of Schwartz distributions $\{D_{km}(u)\}$ whose supports are contained in $\mathcal{M}_0 = g^{-1}(W_0)$ such that the asymptotic expansion

$$\delta(t - u^{2k})u^h\phi^*(u) = \sum_{k=1}^{\infty} \sum_{m=1}^{m_k} D_{km}(u)\, t^{\lambda_k - 1}\, (-\log t)^{m-1}$$

holds for $t \to 0$. Let $Y_n(w)dw$ be a measure defined by

$$Y_n(w)dw \equiv \exp(-n\beta K_n(w))\, \varphi(w)\, dw,$$

then we have an asymptotic expansion,

$$\begin{aligned}
Y_n(w)dw &= Y_n(g(u))\, |g'(u)|\, du \\
&= \sum_{\alpha} e^{-n\beta u^{2k}+\sqrt{n\beta}u^k \xi_n(u)}\, \phi^*(u)|u^h|du \\
&= \sum_{\alpha} \int_0^\infty dt\, \delta(t-u^{2k}) \\
&\quad \times \phi^*(u)|u^h|e^{-n\beta t+\sqrt{nt\beta}\,\xi_n(u)}\, du \\
&= \sum_{\alpha} \sum_{k=1}^{\infty} \sum_{r=0}^{m_k-1} D_{km}(u)du \\
&\quad \times \int_0^\infty \frac{dt}{n} \left(\frac{t}{n}\right)^{\lambda_k-1} \left(\log\frac{n}{t}\right)^r e^{-\beta t+\sqrt{t\beta}\,\xi_n(u)}.
\end{aligned}$$

For simplicity we use the notation $\lambda = \lambda_1$, $m = m_1$ and

$$du^* = \sum_{\alpha^*} D_{1\, m_1}(u)du, \tag{1.24}$$

where $\sum_{\alpha^*}$ shows the sum of local coordinates that attain the minimum λ and the maximum m in eq.(1.22). Such local coordinates are called essential coordinates in this book. By using the convergence in law $\xi_n(u) \to \xi(u)$, the largest term of the asymptotic expansion of the *a posteriori* distribution is given by

$$Y_n(w)\, dw \cong \frac{(\log n)^{m-1}}{n^\lambda}\, du^* \int_0^\infty dt\, t^{\lambda-1}\, e^{-\beta t+\sqrt{t\beta}\,\xi(u)}. \tag{1.25}$$

The normalized evidence is

$$Z_n^0 = \int Y_n(w)\, dw.$$

It follows that

$$\begin{aligned}
F_n^0 &= -\log Z_n^0 \\
&\cong \lambda\, \log n - (m-1)\log\log n + F^R(\xi),
\end{aligned} \tag{1.26}$$

where $F^R(\xi)$ is a random variable

$$F^R(\xi) = -\log\Big(\int du^* \int_0^\infty dt\ t^{\lambda-1}\ e^{-\beta t + \sqrt{t\beta}\ \xi(u)}\Big).$$

We obtain the second main result.

Main Formula II (Convergence of stochastic complexity)
Let $(-\lambda)$ and m be respectively the largest pole and its order of the zeta function

$$\zeta(z) = \int K(w)^z \varphi(w) dw$$

of a statistical model. The normalized stochastic complexity has the following asymptotic expansion,

$$F_n^0 = \lambda \log n - (m-1)\log\log n + F^R(\xi) + o_p(1),$$

where $F^R(\xi)$ is a random variable and $o_p(1)$ is a random variable which satisfies the convergence in probability $o_p(1) \to 0$. Therefore the stochastic complexity F_n has the asymptotic expansion

$$F_n = n\beta S_n + \lambda \log n - (m-1)\log\log n + F^R(\xi) + o_p(1),$$

where S_n is the empirical entropy defined by eq.(1.4).

Remark 1.15 If a model is regular then $K(w)$ is equivalent to $|w|^2$, hence $\lambda = d/2$ and $m = 1$ where d is the dimension of the parameter space. The asymptotic expansion of F_n with $\beta = 1$ in a regular statistical model is well known as the Bayes information criterion (BIC) or the minimum description length (MDL). Hence Main Formula II contains BIC and MDL as a special case. If a model is singular, then $\lambda \neq d/2$ in general. In Chapter 3, we give a method to calculate λ and m, and in Chapter 7, we show examples in several statistical models. The constant λ is an important birational invariant, which is equal to the real log canonical threshold if $\varphi(w) > 0$ at singularities. Therefore Main Formula II claims that the stochastic complexity is asymptotically determined by the algebraic geometrical birational invariant.

1.4.3 Bayes and Gibbs theory

In real-world problems, the true distribution is unknown in general. The third formula is useful because it holds independently of the true distribution $q(x)$.

The expectation value of an arbitrary function $F(w)$ over the *a posteriori* distribution is defined by

$$E_w[F(w)] = \frac{\int F(w)Y_n(w)dw}{\int Y_n(w)dw},$$

where $Y_n(w) = \exp(-nK_n(w))\varphi(w)$. When the number of training samples goes to infinity, this distribution concentrates on the union of neighborhoods of $K(w) = 0$. In such neighborhoods, the renormalized *a posteriori* distribution $E_{u,t}[\ ]$ is defined for an arbitrary function $A(u, t)$,

$$E_{u,t}[A(u,t)] = \frac{\displaystyle\int du^* \int_0^\infty dt\ A(u,t)\, t^{\lambda-1}\, e^{-\beta t+\beta\sqrt{t}\,\xi(u)}}{\displaystyle\int du^* \int_0^\infty dt\ t^{\lambda-1}\, e^{-\beta t+\beta\sqrt{t}\,\xi(u)}},$$

where du^* is defined in eq.(1.24). Then eq.(1.25) shows the convergence in law

$$E_w[(\sqrt{n} f(x,w))^s] \to E_{u,t}[(\sqrt{t}\ a(x,u))^s]$$

for $s > 0$, where the relations of the paramaters are

$$\begin{aligned} w &= g(u), \\ t &= nK(w) = nu^{2k}, \\ f(x,w) &= a(x,u)u^k. \end{aligned}$$

Based on these properties, we can derive the asymptotic behavior of the Bayes quartet from Theorem 1.3. Firstly, Gibbs generalization error is

$$G_g = E_w[K(w)] = \frac{1}{n} E_{u,t}[t].$$

Secondly, Gibbs training error is

$$G_t = \frac{1}{n}\sum_{i=1}^{n} E_w[f(X_i,w)] = \frac{1}{n} E_{u,t}[\xi(u)t^{1/2}] + o_p\Big(\frac{1}{n}\Big),$$

where $o_p(1/n)$ is a random variable which satisfies the convergence in probability, $n\ o_p(1/n) \to 0$. Thirdly, the Bayes generalization error is

$$\begin{aligned} B_g &= E_X[-\log E_w[1 - f(X,w) + \tfrac{1}{2} f(X,w)^2]] + o_p\Big(\frac{1}{n}\Big) \\ &= E_X[-\log(1 - E_w[f(X,w)] + \tfrac{1}{2} E_w[f(X,w)^2])] + o_p(1/n) \end{aligned}$$

Then by using $-\log(1-\epsilon) = \epsilon + \epsilon^2/2 + o(\epsilon^2)$ and

$$E_w[f(X,w)] = \frac{1}{\sqrt{n}} E_{u,t}[a(X,u)t^{1/2}],$$

$$E_X[E_w[f(X,w)]] = E_w[K(w)] = \frac{1}{n} E_{u,t}[t],$$

$$E_X[E_w[f(X,w)^2]] = E_X[E_{u,t}[a(X,u)^2 t]] = \frac{2}{n} E_{u,t}[t] + o_p(1/n),$$

where we used $E_X[a(X,u)^2] = 2$, it follows that

$$B_{\mathrm{g}} = \frac{1}{2n} E_X[E_{u,t}[a(X,u)t^{1/2}]^2] + o_p(1/n). \tag{1.27}$$

And, lastly, the Bayes training error is

$$\begin{aligned}
B_{\mathrm{t}} &= \frac{1}{n}\sum_{i=1}^{n}[-\log E_w[1 - f(X_i,w) + \tfrac{1}{2} f(X_i,w)^2]] + o_p\Big(\frac{1}{n}\Big) \\
&= \frac{1}{n}\sum_{i=1}^{n}[-\log(1 - E_w[f(X_i,w)] + \tfrac{1}{2}E_w[f(X_i,w)^2])] + o_p\Big(\frac{1}{n}\Big) \\
&= \frac{1}{n^2}\sum_{i=1}^{n}\big\{E_{u,t}[a(X_i,u)t^{1/2}] - \tfrac{1}{2}E_{u,t}[a(X_i,u)^2 t] \\
&\quad + \tfrac{1}{2}E_{u,t}[a(X_i,u)t^{1/2}]^2\big\} + o_p\Big(\frac{1}{n}\Big) \\
&= G_{\mathrm{t}} - G_{\mathrm{g}} + B_{\mathrm{g}} + o_p(1/n),
\end{aligned}$$

where we used the law of large numbers

$$\frac{1}{n}\sum_{i=1}^{n} a(X_i,u)a(X_i,v) = E_X[a(X,u)a(X,v)] + o_p(1)$$

in the last equation. By using convergence in law $\xi_n(u) \to \xi(u)$, we prove the convergences in law of the Bayes quartet,

$$nB_{\mathrm{g}} \to B_{\mathrm{g}}^*, \quad nB_{\mathrm{t}} \to B_{\mathrm{t}}^*,$$

$$nG_{\mathrm{g}} \to G_{\mathrm{g}}^*, \quad nG_{\mathrm{t}} \to G_{\mathrm{t}}^*,$$

where $B^*_{\rm g}$, $B^*_{\rm t}$, $G^*_{\rm g}$, $G^*_{\rm t}$ are random variables represented by the random process $\xi(u)$. Let us introduce the notation ($a \in \mathbb{R}$),

$$S_\lambda(a) = \int_0^\infty dt\ t^{\lambda-1}\ e^{-\beta t + a\beta\sqrt{t}},$$

$$Z(\xi) = \int du^*\ S_\lambda(\xi(u)).$$

Then

$$S'_\lambda(a) = \beta \int_0^\infty dt\ t^{\lambda-1/2}\ e^{-\beta t + a\beta\sqrt{t}},$$

$$S''_\lambda(a) = \beta^2 \int_0^\infty dt\ t^{\lambda}\ e^{-\beta t + a\beta\sqrt{t}}.$$

Finally we obtain

$$E[B^*_{\rm g}] = \frac{1}{2\beta^2} E\left[E_X\left[\left(\frac{\int du^* a(X,u) S'_\lambda(\xi(u))}{Z(\xi)} \right)^2 \right] \right],$$

$$E[B^*_{\rm t}] = E[B^*_{\rm g}] + E[G^*_{\rm t}] - E[G^*_{\rm g}],$$

$$E[G^*_{\rm g}] = \frac{1}{\beta^2} E\left[\frac{\int du^* S''_\lambda(\xi(u))}{Z(\xi)} \right],$$

$$E[G^*_{\rm t}] = \frac{1}{\beta^2} E\left[\frac{\int du^* S''_\lambda(\xi(u))}{Z(\xi)} \right] - \frac{1}{\beta} E\left[\frac{\int du^*\ \xi(u) S'_\lambda(\xi(u))}{Z(\xi)} \right].$$

These equations show that the expectations of the Bayes quartet are represented by linear sums of three expectation values over the random process $\xi(u)$. On the other hand, $\xi(u)$ is a Gaussian process which is represented by

$$\xi(u) = \sum_{i=1}^{\infty} b_k(u) g_k,$$

where $\{g_k\}$ is a set of random variables that are independently subject to the standard normal distribution and $b_k(u) = E[\xi(u) g_k]$. By using the partial integration $E[g_k F(g_k)] = E[(\partial/\partial g_k) F(g_k)]$ for an arbitrary integrable function $F(\ )$, we can prove

$$E[B^*_{\rm g}] = \frac{1}{\beta^2} E\Big[\frac{\int du^*\ S''_\lambda(\xi(u))}{Z(\xi)} \Big] - \frac{1}{2\beta^2} E\Big[\frac{\int du^*\ \xi(u)\ S'_\lambda(\xi(u))}{Z(\xi)} \Big].$$

Therefore four errors are given by the linear sums of two expectations of $S'_\lambda(\xi(u))$ and $S''_\lambda(\xi(u))$. By eliminating two expectations from four equations, we obtain two equations which hold for the Bayes quartet.

Main Formula III (Equations of states in statistical estimation) There are two universal relations in Bayes quartet.

$$E[B_{\mathrm{g}}^*] - E[B_{\mathrm{t}}^*] = 2\beta(E[G_{\mathrm{t}}^*] - E[B_{\mathrm{t}}^*]), \tag{1.28}$$

$$E[G_{\mathrm{g}}^*] - E[G_{\mathrm{t}}^*] = 2\beta(E[G_{\mathrm{t}}^*] - E[B_{\mathrm{t}}^*]). \tag{1.29}$$

These equations hold for an arbitrary true distribution, an arbitrary statistical model, an arbitrary *a priori* distribution, and arbitrary singularities.

Remark 1.16 (1) Main Formula III holds in both regular and singular models. Although the four errors themselves strongly depend on $q(x)$, $p(x|w)$, and $\varphi(w)$, these two equations do not. By this formula, we can estimate the Bayes and Gibbs generalization errors from the Bayes and Gibbs training errors without any knowledge of the true distributions. The constant

$$\nu(\beta) = \beta(E[G_{\mathrm{t}}^*] - E[B_{\mathrm{t}}^*]) \tag{1.30}$$

is the important birational invariant called a singular fluctuation. Then Main Formula III claims that

$$E[B_{\mathrm{g}}^*] = E[B_{\mathrm{t}}^*] + 2\nu(\beta), \tag{1.31}$$

$$E[G_{\mathrm{g}}^*] = E[G_{\mathrm{t}}^*] + 2\nu(\beta). \tag{1.32}$$

We can estimate $\nu(\beta)$ from samples. In fact, by defining two random variables,

$$V_0 = \sum_{i=1}^{n} (\log E_w[p(X_i|w)] - E_w[\log p(X_i|w)]),$$

$$V = \sum_{i=1}^{n} (E_w[(\log p(X_i|w))^2] - E_w[\log p(X_i|w)]^2),$$

we have $V_0 = nG_{\mathrm{t}} - nB_{\mathrm{t}}$ and $E[V/2]$ is asymptotically equal to $E[nG_{\mathrm{t}}] - E[nB_{\mathrm{t}}]$,

$$\nu(\beta) = \beta E[V_0] + o(1) = (\beta/2)E[V] + o(1).$$

(2) If a model is regular then, for any $\beta > 0$,

$$\nu(\beta) = d/2, \tag{1.33}$$

where d is the dimension of the parameter space. If a model is regular, both Bayes and Gibbs estimation converge to the maximum likelihood estimation, when $\beta \to \infty$. Then two equations of states result in one equation,

$$E[nR_{\mathrm{g}}] = E[nR_{\mathrm{t}}] + d, \tag{1.34}$$

where R_g and R_t are the generalization and training errors of the maximum likelihood estimator. The equation (1.34) is well known as the Akaike information criterion (AIC) of a regular statistical model, hence Main Formula III contains AIC as a very special case. In singular learning machines, eq.(1.33) does not hold in general, hence AIC cannot be applied. Moreover, Main Formula III holds even if the trae distribution is not contained in the model [120].

1.4.4 ML and MAP theory

The last formula concerns the maximum likelihood or *a posteriori* method. Let W be a compact set, and $f(x, w)$ and $\varphi(w)$ be respectively analytic and C^2-class functions of $w \in W$. Then there exists a parameter $\hat{w} \in W$ that minimizes the generalized log likelihood ratio function,

$$R_n^0(w) = \sum_{i=1}^{n} f(X_i, w) - a_n \log \varphi(w),$$

where a_n is a nondecreasing sequence. Note that, if W is not compact, the parameter that minimizes $R_n^0(w)$ does not exist in general. By applying the standard form of the log likelihood ratio function and a simple notation $\sigma(u) = -\log \varphi(g(u))$, in each coordinate, the function $R_n^0(g(u))$ is represented by

$$\frac{1}{n} R_n^0(g(u)) = u^{2k} - \frac{1}{\sqrt{n}}\, u^k\, \xi_n(u) + \frac{a_n}{n}\sigma(u),$$

where $\xi_n(u) \to \xi(u)$ in law. For an arbitrary u, a new parameterization (t, v) is defined by

$$t = u^k,$$
$$v = \mathrm{Proj}(u),$$

where the function Proj( ) maps u to v on the set $\{v; v^{2k} = 0\}$ along the ordinary differential equation $u(T)$ for $T \geq 0$,

$$\frac{d}{dT} u(T) = -\nabla(u(T)^{2k}). \tag{1.35}$$

Here $v = \mathrm{Proj}(u)$ is determined by $v = u(T = \infty)$ for the initial condition $u = u(T = 0)$. More precisely, see Chapter 6 and Figure 6.3. In each local coordinate,

$$\frac{1}{n} R_n^0(g(t, v)) = t^2 - \frac{1}{\sqrt{n}}\, t\, \xi_n(t, v) + \frac{a_n}{n}\sigma(t, v).$$

Then we can prove that, for arbitrary C^1-class function $f(u)$ on a compact set, there exist constants $C, \delta > 0$ such that

$$|f(t, v) - f(0, v)| \leq t^{\delta} \|\nabla f\|,$$

where

$$\|\nabla f\| \equiv \sup_j \sup_u \left| \frac{\partial f}{\partial u_j} \right|.$$

Let $\hat{t}$ be the parameter that minimizes $R_n^0(t, v)$, then $\hat{t}$ should be in proportion to $1/\sqrt{n}$, hence

$$\frac{1}{n} R_n^0(g(\hat{t}, v)) = \hat{t}^2 - \frac{1}{\sqrt{n}}\, \hat{t}\, \xi(0, v) + \frac{a_n}{n} \sigma(0, v) + o_p\Big(\frac{1}{n}\Big).$$

We can prove that $o_p(1/n)$ does not affect the main terms. Let $\hat{v}$ be the parameter that minimizes $R_n^0(g(\hat{t}, v))$. Then

$$\hat{t} = \frac{1}{2\sqrt{n}} \max_{\alpha} \max\{0, \xi(0, \hat{v})\},$$

where α shows the local coordinate. If $a_n \equiv 0$, then $\hat{v}$ is determined by minimizing

$$-\max_{\alpha} \max\{0, \xi(\hat{v})\}^2.$$

Hence the generalization and training errors are given by

$$R_{\mathrm{g}} = \frac{1}{4n} \Big(\max_{u \in \mathcal{M}_0} \{0, \xi(u)\}^2 \Big),$$

$$R_{\mathrm{t}} = -\frac{1}{4n} \Big(\max_{u \in \mathcal{M}_0} \{0, \xi(u)\}^2 \Big),$$

where $\mathcal{M}_0 = g^{-1}(W_0)$ is the set of true parameters. The symmetry of generalization and training errors holds if $a_n/n^p \to \infty$ for arbitrary $p > 0$. Therefore,

$$E[nR_{\mathrm{g}}] = -E[nR_{\mathrm{t}}] + o(1).$$

For the other sequence a_n, the same result is obtained.

Main Formula IV (Symmetry of generalization and training errors)
If the maximum likelihood or generalized maximum *a posteriori* method is applied, the symmetry of generalization and training errors holds,

$$\lim_{n\to\infty} E[nR_{\mathrm{g}}] = -\lim_{n\to\infty} E[nR_{\mathrm{t}}].$$

Remark 1.17 (1) In regular statistical models, $E[nR_g] = d/2$ where d is the dimension of the parameter space. In singular statistical models $E[nR_g] >> d/2$ in general, because it is the mean of the maximum value of a Gaussian process. If the parameter space is not compact, then the maximum likelihood estimator sometimes does not exist. Even if it exists, it often diverges for $n \to \infty$, which means that

$$E[nR_t] \to -\infty.$$

In such a case, the behavior of the generalization error is still unknown. It is expected that the symmetry still holds, in which case

$$E[nR_g] \to +\infty.$$

Hence the maximum likelihood method is not appropriate for singular statistical models. Even if the set of parameters is compact, it is still difficult to estimate the generalization error from the training error without knowledge of the true distribution. From a statistical point of view, the maximum likelihood estimator is asymptotically the sufficient statistic in regular models. However, it is not in singular models, because the likelihood function does not converge to the normal distribution.
(2) In singular statistical models, two limiting procedures $n \to \infty$ and $\beta \to \infty$ are not commutative in general. In other words,

$$\lim_{\beta\to\infty} \lim_{n\to\infty} E[nB_g] \neq \lim_{n\to\infty} E[nR_g].$$

In fact, the Bayes generalization error is determined by the sum of the essential local coordinates $\sum_{\alpha^*}$, whereas the maximum likelihood generalization error is determined by the set of all coordinates $\sum_{\alpha}$.

1.5 Overview of this book

The main purpose of this book is to establish the mathematical foundation on which the four main formulas are proved.

In Chapter 2, we introduce singularity theory and explain the resolution theorem which claims that, for an arbitrary analytic function $K(w)$, there exist a manifold and an analytic function $w = g(u)$ such that $K(g(u)) = u^{2k}$.

In Chapter 3, elemental algebraic geometry is explained. The relation between algebra and geometry, Hilbert's basis theorem, projective space, and blow-ups are defined and illustrated. We show how to find the resolution map using recursive blow-ups for a given statistical model.

In Chapter 4, the mathematical relation between the zeta function and the singular integral is clarified. We need Schwartz distribution theory to connect these two concepts. Several inequalities which are used in the following sections are proved.

In Chapter 5, we study the convergence in law of the empirical process to a Gaussian process, $\xi_n(u) \to \xi(u)$. This is the central limit theorem on the functional space. Also we introduce the partial integral on the function space.

Based on mathematical foundations in chapters 2, 3, 4, and 5, the four main formulas are rigorously proved in Chapter 6. These are generalizations of the conventional statistical theory of regular models to singular models. We find two birational invariants, the maximum pole of the zeta function and the singular fluctuation, which determine the statistical learning process.

Chapters 7 and 8 are devoted to applications of this book to statistics and information science.

1.6 Probability theory

In this section, fundamental points of probability theory are summarized. Readers who are familiar with probability theory can skip this section.

Definition 1.11 (Metric space) Let Ω be a set. A function D

$$D : \Omega \times \Omega \ni (x, y) \mapsto D(x, y) \in \mathbb{R}$$

is called a metric if it satisfies the following three conditions.
(1) For arbitrary $x, y \in \Omega$, $D(x, y) = D(y, x) \geq 0$.
(2) $D(x, y) = 0$ if and only if $x = y$.
(3) For arbitrary $x, y, z \in \Omega$, $D(x, y) + D(y, z) \geq D(x, z)$.
A set Ω with a metric is called a metric space. The set of open neighborhoods of a point $x \in \Omega$ is defined by $\{U_\epsilon(x); \epsilon > 0\}$ where

$$U_\epsilon(x) = \{y \in \Omega \; ; \; D(x, y) < \epsilon\}.$$

The topology of the metric space is determined by all open neighborhoods. A metric space Ω is called separable if there exists a countable and dense subset. A set $\{x_n; n = 1, 2, 3, \ldots\}$ is said to be a Cauchy sequence if, for arbitrary $\delta > 0$, there exists M such that

$$m, n > M \Longrightarrow D(x_m, x_n) < \delta.$$

If any Cauchy sequence in a metric space Ω converges in Ω, then Ω is called a complete metric space. A complete and separable metric space is called a Polish space.

Example 1.8 In this book, we need the following metric spaces.
(1) The finite-dimensional real Euclidean space $\mathbb{R}^d$ is a metric space with the metric

$$D(x, y) = |x - y| \equiv \Big(\sum_{i=1}^{d}(x_i - y_i)^2\Big)^{1/2},$$

where $x = (x_i)$, $y = (y_i)$, and $|\cdot|$ is a norm of $\mathbb{R}^d$. The real Euclidean space $\mathbb{R}^d$ is a complete and separable metric space.
(2) A subset of $\mathbb{R}^d$ is a metric space with the same metric. Sometimes a finite or countable subset in $\mathbb{R}^d$ is studied.
(3) Let K be a compact subset in $\mathbb{R}^d$. The set of all continuous function from K to $\mathbb{R}^{d'}$

$$\Omega = \{f\ ;\ f : K \to \mathbb{R}^{d'}\}$$

is a metric space with the metric

$$D(f, g) = \|f - g\| \equiv \max_{x \in K} |f(x) - g(x)|,$$

where $|\cdot|$ is the norm of $\mathbb{R}^{d'}$. By the compactness of K in $\mathbb{R}^d$, it is proved that Ω is a complete and separable metric space.

Definition 1.12 (Probability space) Let Ω be a metric space. A set $\mathcal{B}$ composed of subsets contained in Ω is called a sigma algebra or a completely additive set if it satisfies the following conditions. ($\mathcal{B}$ contains the empty set.)
(1) If $A_1, A_2 \in \mathcal{B}$ then $A_1 \cap A_2 \in \mathcal{B}$.
(2) If $A \in \mathcal{B}$ then $A^c \in \mathcal{B}$ (A^c is the complementary set of A).
(3) If $A_1, A_2, A_3 \ldots, \in \mathcal{B}$ then the countable union $\cup_{k=1}^{\infty} A_k \in \mathcal{B}$.
The smallest sigma algebra that contains all open sets of Ω is said to be a Borel field. A pair of a metric space and a sigma algebra $(\Omega, \mathcal{B})$ is called a measurable space. A function P,

$$P : \mathcal{B} \ni A \mapsto 0 \leq P(A) \leq 1,$$

is called a probability measure if it satisfies
(1) $P(\Omega) = 1$.
(2) For $\{B_k\}$ which satisfies $B_k \cap B_{k'} = \emptyset$ $(k \neq k')$, $P(\cup_{k=1}^{\infty} B_k) = \sum_{k=1}^{\infty} P(B_k)$.

A triple of a metric space, a sigma algebra, and a probability measure $(\Omega, \mathcal{B}, P)$ is called a probability space.

Remark 1.18 Let $(\mathbb{R}^N, \mathcal{B}, P)$ be a probability space, where $\mathbb{R}^N$ is the N-dimensional real Euclidean space, $\mathcal{B}$ the Borel field, and P a probability distribution. If P is defined by a function $p(x) \geq 0$,

$$P(A) = \int_A p(x)dx \quad (A \in \mathcal{B}),$$

then $p(x)$ is called a probability density function.

Definition 1.13 (Random variable) Let $(\Omega, \mathcal{B}, P)$ be a probability space and $(\Omega_1, \mathcal{B}_1)$ a measurable space. A function

$$X : \Omega \ni \omega \mapsto X(\omega) \in \Omega_1$$

is said to be measurable if $X^{-1}(B_1) \in \mathcal{B}$ for arbitrary $B_1 \in \mathcal{B}_1$. A measurable function X on a probability space is called a random variable. Sometimes X is said to be an Ω_1-valued random variable. By the definition

$$\mu(B_1) = P(X^{-1}(B_1)), \tag{1.36}$$

μ is a probability measure on $(\Omega_1, \mathcal{B}_1)$, hence $(\Omega_1, \mathcal{B}_1, \mu)$ is a probability space. The probability measure μ is called a probability distribution of the random variable X. Then X is said to be subject to μ. Note that μ is the probability distribution on the image space of a function of X. Equation (1.36) can be rewritten as

$$\int_{B_1} \mu(dx) = \int_{X^{-1}(B_1)} P(da).$$

Remark 1.19 (1) In probability theory, the simplified notation

$$P(f(X) > 0) \equiv P(\{\omega \in \Omega; f(X(\omega)) > 0\})$$

is often used. Then by definition, $P(f(X) > 0) = \mu(\{x \in \Omega_1; f(x) > 0\})$.
(2) The probability measure μ to which a random variable X is subject is often denoted by P_X. The map $X \mapsto P_X$ is not one-to-one in general. For example, on a probability space $(\Omega, 2^\Omega, P)$ where $\Omega = \{1, 2, 3, 4\}$ and $P(\{i\}) = 1/4$ $(i = 0, 1, 2, 3)$, two different random variables

$$X(i) = \begin{cases} 0 & (i = 0, 1) \\ 1 & (i = 2, 3) \end{cases}$$

$$Y(i) = \begin{cases} 0 & (i = 0, 2) \\ 1 & (i = 1, 3) \end{cases}$$

are subject to the same probability distribution. Therefore, in general, even if X and Y are subject to the same probability distribution, we cannot predict the realization of Y from a realization of X.

(3) In descriptions of definitions and theorems, sometimes we need only the information of the image space of a random variable X and the probability distribution P_X. In other words, there are some definitions and theorems in which the explicit statement of the probability space $(\Omega, \mathcal{B}, P)$ is not needed. In such cases, the explicit definition of the probability space is omitted, resulting in a statement such as "for Ω_1-valued random variable X which is subject to a probability distribution P_X satisfies the following equality . . ."

Definition 1.14 (Expectation) Let X be a random variable from the probability space $(\Omega, \mathcal{B}, P)$ to $(\Omega_1, \mathcal{B}_1)$ which is subject to the probability distribution P_X. If the integration

$$E[X] = \int X(\omega)P(d\omega) = \int x\ P_X(dx)$$

is well defined and finite in Ω_1, $E[X] \in \Omega_1$ is called the expectation or the mean of X. Let S be a subset of Ω_1. The partial expectation is defined by

$$E[X]_S = \int_{X(\omega)\in S} X(\omega)P(d\omega) = \int_S x\ P_X(dx).$$

Remark 1.20 These are fundamental remarks.
(1) Let $(\Omega_1, \mathcal{B}_1)$ and X be same as Definition 1.14 and $(\Omega_2, \mathcal{B}_2)$ be a measurable space. If $f : \Omega_1 \to \Omega_2$ is a measurable function then $f(X)$ is a random variable on $(\Omega, \mathcal{B}, P)$. The expectation of $f(X)$ is equal to

$$E[f(X)] = \int f(X(\omega))P(d\omega) = \int f(x)\ P_X(dx).$$

This expectation is often denoted by $E_X[f(X)]$.
(2) Two random variables which have the same probability distribution have the same expectation value. Hence if X and Y have the same probability distribution, we can predict $E[Y]$ based on the information of $E[X]$.
(3) In statistical learning theory, it is important to predict the expectation value of the generalization error from the training error.
(4) If $E[|X|] = C$ then, for arbitrary $M > 0$,

$$\begin{aligned} C = E[|X|] &\geq E[|X|]_{\{|X|>M\}} \\ &\geq M E[1]_{\{|X|>M\}} = M P(|X| > M). \end{aligned}$$

Hence

$$P(|X| > M) \leq \frac{C}{M},$$

which is well known as Chebyshev's inequality. The same derivation is often effective in probability theory.

(5) The following conditions are equivalent.

$$E[|X|] < \infty \Longleftrightarrow \lim_{M\to\infty} E[|X|]_{\{|X|\geq M\}} = 0.$$

(6) If there exist constants $\delta > 0$ and $M_0 > 0$ such that for an arbitrary $M > M_0$

$$P(|X| \geq M) \leq \frac{1}{M^{1+\delta}},$$

then $E[|X|] < \infty$.

Definition 1.15 (Convergence of random variables) Let $\{X_n\}$ and X be a sequence of random variables and a random variable on a probability space $(\Omega, \mathcal{B}, P)$, respectively.
(1) It is said that X_n converges to X almost surely (almost everywhere), if

$$P\Big(\{\omega \in \Omega\,;\ \lim_{n\to\infty} X_n(\omega) = X(\omega)\}\Big) = 1.$$

(2) It is said that X_n converges to X in the mean of order $p > 0$, if

$$\lim_{n\to\infty} E[(X_n - X)^p] = 0.$$

(3) It is said that X_n converges to X in probability, if

$$\lim_{n\to\infty} P(D(X_n, X) > \epsilon) = 0$$

for arbitrary $\epsilon > 0$, where $D(\cdot,\cdot)$ is the metric of the image space of X.

Remark 1.21 There are well-known properties of random variables.
(1) If X_n converges to X almost surely or in the mean of order $p > 0$, then it does in probability.
(2) If X_n converges to X in probability, then it does in law. For the definition of convergence in law, see chapter 5.
(3) In general, "almost surely" is neither sufficient nor necessary condition of "in the means of order $p > 0$."

Remark 1.22 (Limit theorem) From the viewpoint of probability theory, in this book we obtain the limit theorem of the random variables,

$$F_n = -\log \int p(X_1|w)^\beta p(X_2|w)^\beta \cdots p(X_n|w)^\beta \varphi(w)dw,$$

and

$$B_g = E_X\left[-\log \frac{\int p(X|w)p(X_1|w)^\beta \cdots p(X_n|w)^\beta \varphi(w)dw}{\int q(X)p(X_1|w)^\beta \cdots p(X_n|w)^\beta \varphi(w)dw}\right],$$

where $X_1, \ldots, X_n$ are independently subject to the same distribution as X. As the central limit theorem is characterized by the mean and the variance of the random variables, the statistical learning theory is characterized by the largest pole of the zeta function and the singular fluctuation. The large deviation theory indicates that $F_n/n \to pS$, where S is the entropy of X. Main Formulas II and III show more precise results than the large deviation theory.

2

Singularity theory

A lot of statistical models and learning machines contain singularities in their parameter spaces. Singularities determine the behavior of the learning process, hence it is not until we understand singularities that we obtain statistical learning theory. In this chapter, the definition of singularities and the basic theorem for resolution of singularities are introduced. To explain the resolution of singularities, the definition of a manifold is necessary, which is included in Section 2.6.

2.1 Polynomials and analytic functions

Let d be a natural number. Let $\mathbb{R}$ and $\mathbb{C}$ be the set of all real numbers and the set of all complex numbers respectively. A d-dimensional multi-index α is defined by

$$\alpha = (\alpha_1, \alpha_2, \ldots, \alpha_d),$$

where $\alpha_1, \alpha_2, \ldots, \alpha_d$ are nonnegative integers. For given $x, b \in \mathbb{R}^d$

$$x = (x_1, x_2, \ldots, x_d),$$
$$b = (b_1, b_2, \ldots, b_d),$$

and $a_\alpha \in \mathbb{R}$, we define

$$a_\alpha (x - b)^\alpha = a_{\alpha_1 \alpha_2 \cdots \alpha_d} (x_1 - b_1)^{\alpha_1} (x_2 - b_2)^{\alpha_2} \cdots (x_d - b_d)^{\alpha_d}.$$

A sum of such terms

$$f(x) = \sum_{\alpha_1=0}^{\infty} \cdots \sum_{\alpha_d=0}^{\infty} a_{\alpha_1 \alpha_2 \cdots \alpha_d} (x_1 - b_1)^{\alpha_1} \cdots (x_d - b_d)^{\alpha_d} \tag{2.1}$$

is said to be a power series, which is written by

$$f(x) = \sum_{\alpha} a_\alpha (x - b)^\alpha. \tag{2.2}$$

If the number of the nonzero terms in the sum of eq.(2.2) is finite, then $f(x)$ is called a polynomial. If $f(x)$ is a polynomial, then it uniquely determines a function $f : \mathbb{R}^d \to \mathbb{R}$. The set of all polynomials with real coefficients is written by $\mathbb{R}[x_1, x_2, \ldots, x_d]$.

If there exists an open set $U \subset \mathbb{R}^d$ which contains b such that, for arbitrary $x \in U$,

$$\sum_{\alpha} |a_\alpha| \, |x - b|^\alpha < \infty,$$

then $f(x)$ is called an absolutely convergent power series. If $f(x)$ is an absolutely convergent power series, then the sum of eq.(2.1) converges independently of the order of sums $\sum_{\alpha_1}, \sum_{\alpha_2}, \ldots, \sum_{\alpha_d}$, and uniquely determines a function $f : U \to \mathbb{R}$. This function is called a real analytic function. If $f(x)$ is a real analytic function then the coefficient of the Taylor expansion around b satisfies

$$a_\alpha = \frac{1}{\alpha!} \frac{\partial^\alpha f}{\partial x^\alpha}(b),$$

where

$$\alpha! = \prod_{i=1}^{d} \alpha_i !$$

$$\frac{\partial^\alpha}{\partial x^\alpha} = \prod_{i=1}^{d} \frac{\partial^{\alpha_i}}{\partial x_i^{\alpha_i}}.$$

A real analytic function $f(x)$ has the expansion around any $b' \in U$

$$f(x) = \sum_{\alpha} a'_\alpha \, (x - b')^\alpha,$$

which absolutely converges in some open set in U. If $f(x)$ also absolutely converges in U', the domain of the analytic function $f(x)$ can be extended to $U \cup U'$. Such a method of extending the domain of an analytic function is called an analytic continuation. A complex analytic function $f(x)$ of $x \in U \subset \mathbb{C}^d$ is also defined by using $a_\alpha \in \mathbb{C}$. Absolute convergence and analytic continuation are defined in the same way.

Definition 2.1 (Function of class C^r) Let U be an open set in d-dimensional real Euclidean space $\mathbb{R}^d$. A function $f : U \to \mathbb{R}^{d'}$ is said to be of class C^r in U if partial derivatives

$$\frac{\partial^{n_1+n_2+\cdots+n_d} f}{\partial x^{n_1} \partial x^{n_2} \cdots \partial x^{n_d}}(x)$$

are well defined and continuous for all nonnegative integers $n_1, n_2, \ldots, n_d$ such that

$$n_1 + n_2 + \cdots + n_d \leq r.$$

If $f(x)$ is a function of class C^r for $r \geq 1$, then it is of class $C^{r'}$ for all $0 \leq r' \leq r$. If $f(x)$ is a function of class C^r for all natural numbers r, it is said to be of class C^∞. If a function $f(x)$ is real analytic in U, it is said to be of class C^ω.

Example 2.1 A finite sum

$$f(x, y, z) = x^3 y^5 z^2 + xy^6 + z^5 + 2$$

is a polynomial. A power series

$$f(x, y) = \sum_{m=0}^{\infty} \sum_{n=0}^{\infty} \frac{x^m y^n}{m!n!}$$

absolutely converges if $|x|, |y| < \infty$, which defines a real analytic function

$$f(x, y) = e^{x+y}.$$

A function

$$g(x, y) = \begin{cases} \exp\left(-\dfrac{1}{x^2 y^2}\right) & (xy \neq 0) \\ 0 & (xy = 0) \end{cases}$$

is not a real analytic function but of class C^∞, by defining that the arbitrary times derivative of $g(x, y)$ at $xy = 0$ is zero. Although

$$g(x, y) = 0 \Longleftrightarrow xy = 0$$

holds, there exists no pair of integers (k_1, k_2) which satisfies

$$g(x, y) = a(x, y) x^{k_1} y^{k_2}$$

for some function $a(x, y) > 0$.

2.2 Algebraic set and analytic set

Definition 2.2 (Real algebraic set) Let $f : \mathbb{R}^d \to \mathbb{R}$ be a function defined by a polynomial. The set of all points that make $f(x) = 0$,

$$\mathbf{V}(f) = \{x \in \mathbb{R}^d; f(x) = 0\},$$

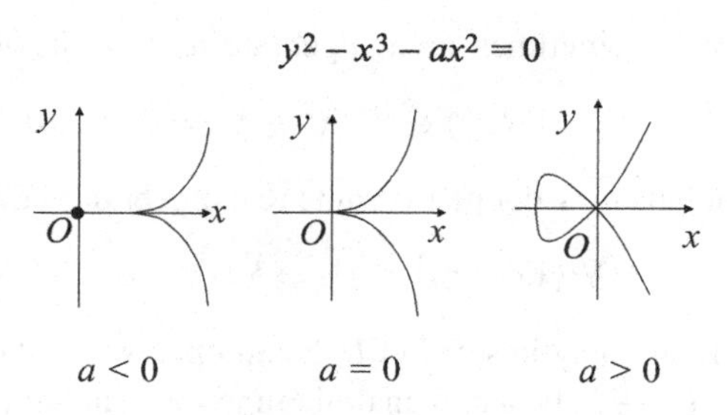

Fig. 2.1. Examples of real algebraic sets

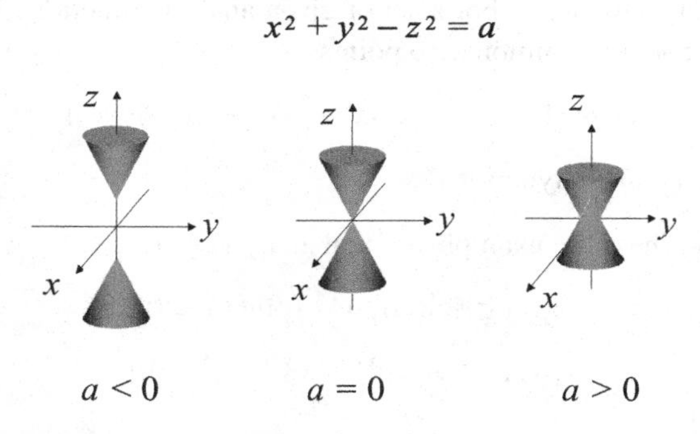

Fig. 2.2. Example of a real algebraic set

is called a real algebraic set. Let $f_1(x), f_2(x), \ldots, f_k(x)$ be polynomials. The set of all points that make all functions zero

$$\mathbf{V}(f_1, f_2, \ldots, f_k) = \{x \in \mathbb{R}^d;\ f_1(x) = f_2(x) = \cdots = f_k(x) = 0\}$$

is also called a real algebraic set.

Example 2.2 Three real algebraic sets

$$\mathbf{V}(y^2 - x^3 - ax^2) = \{(x, y) \in \mathbb{R}^2;\ y^2 - x^3 - ax^2 = 0\}$$

for $a < 0$, $a = 0$, and $a > 0$ are illustrated in Figure 2.1. For different a, they have very different shapes. For the case $a < 0$, the origin is isolated from other lines. For the case $a = 0$, the shape of the origin is called a cusp. In all cases, the origin is a singularity, where the singularity is defined in Definition 2.6. Figure 2.2 shows real algebraic sets

$$\mathbf{V}(x^2 + y^2 - z^2 - a) = \{(x, y, z) \in \mathbb{R}^3;\ x^2 + y^2 - z^2 - a = 0\}$$

for three different a. Sometimes a real algebraic set consists of one point

$$\mathbf{V}(x^2 + y^2) = \{(x, y) \in \mathbb{R}^2; x^2 + y^2 = 0\} = \{(0, 0)\}.$$

Different polynomials may define the same real algebraic set.

$$\mathbf{V}((xy + z)^2 + z^4) = \mathbf{V}(xy, z).$$

Definition 2.3 (Real analytic set) Let U be an open set in the real Euclidean space $\mathbb{R}^d$ and $f : U \to \mathbb{R}$ be a real analytic function. The set of all zero points

$$\{x \in U; f(x) = 0\}$$

is called a real analytic set. For a set of given analytic functions $f_1(x)$, $f_2(x)$, $\ldots$, $f_k(x)$, the set of common zero points

$$\{x \in U; f_1(x) = f_2(x) = \cdots = f_k(x) = 0\}$$

is also called a real analytic set.

Example 2.3 These are examples of real analytic sets.

$$\{(x, y) \in \mathbb{R}^2; \cos(x) - \sin(y) = 0\},$$
$$\{(x, y, z) \in \mathbb{R}^3; e^{xy} + e^{yz} + z^3 = 0\}.$$

Another example is

$$\{(x, y, z) \in U; x^2 - y \log z = 0\}$$

where

$$U = \{(x, y, z); x, y \in \mathbb{R}, z > 0\}.$$

A function $f : \mathbb{R}^2 \to \mathbb{R}^1$ defined by

$$f(x, y) = \begin{cases} xy \sin(1/(xy)) & (x \neq 0) \\ 0 & (x = 0) \end{cases}$$

is not a real analytic function, because it is not represented by any absolutely convergent power series at the origin. Note that the set

$$\{(x, y) \in \mathbb{R}^2; f(x, y) = 0\}$$

is not a real analytic set, because this set contains a sequence $(x_n, y_n) = (\pi/n, 1)$ which converges to (0, 1). However, in $U = \{(x, y) \in \mathbb{R}^2; 0 < x, y < 1\}$,

$$\{(x, y) \in U; f(x, y) = 0\} = \{(x, y) \in U; xy = 1/(n\pi), n = 1, 2, \ldots\}$$

is a real analytic set.

2.3 Singularity

Let U be an open set in $\mathbb{R}^d$ and $f : U \to \mathbb{R}^1$ be a function of C^1 class. The d-dimensional vector $\nabla f(x) \in \mathbb{R}^d$ defined by

$$\nabla f(x) = \left(\frac{\partial f}{\partial x_1}(x), \frac{\partial f}{\partial x_2}(x), \ldots, \frac{\partial f}{\partial x_d}(x) \right)$$

is said to be the gradient vector of $f(x)$.

Definition 2.4 (Critical point of a function) Let U be an open set of $\mathbb{R}^d$, and $f : U \to \mathbb{R}^1$ be a function of C^1 class.
(1) A point $x^* \in U$ is called a critical point of f if it satisfies

$$\nabla f(x^*) = 0.$$

If x^* is a critical point of f, then $f(x^*)$ is called a critical value.
(2) If there exists an open set $U' \subset U$ such that $x^* \in U'$ and

$$f(x) \leq f(x^*) \quad (\forall x \in U'),$$

then x^* is called a local maximum point of f. If x^* is a local maximum point, then $f(x^*)$ is called a local maximum value.
(3) If there exists an open set $U' \subset U$ such that $x^* \in U'$ and

$$f(x) \geq f(x^*) \quad (\forall x \in U'),$$

then x^* is called a local minimum point of f. If x^* is a local minimum point, then $f(x^*)$ is called a local minimum value.

If f is a function of C^1 class, then a local maximum or minimum point is a critical point of f. However, a critical point is not always a local maximum or minimum point.

Example 2.4 (1) A function on $\mathbb{R}^2$

$$f(x, y) = x^2 + y^4 + 3$$

has a unique local minimum point $(0, 0)$.
(2) For a function on $\mathbb{R}^3$

$$f(x, y, z) = (x + y + z)^4 + 1,$$

all points (x^*, y^*, z^*) which satisfy $x^* + y^* + z^* = 0$ are local minimum points.
(3) For a function on $\mathbb{R}^2$

$$f(x, y) = x^2 - y^2,$$

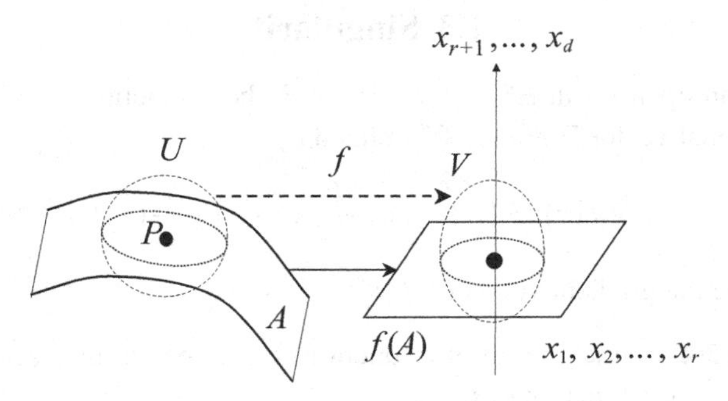

Fig. 2.3. Definition of a nonsingular point

there is no local maximum or minimum point. The origin $(0, 0)$ is a critical point of f. If $|x| \geq |y|$, then $f(x, y) \geq 0$, and if $|x| \leq |y|$, then $f(x, y) \leq 0$. Such a critical point is said to be a saddle point.

Definition 2.5 (C^r Isomorphism) Let U, V be open sets of the real Euclidean space $\mathbb{R}^d$. If there exists a one-to-one map $f : U \to V$ such that both f and f^{-1} are functions of C^r class, then U is said to be C^r isomorphic to V, and f is called a C^r isomorphism. If both f and f^{-1} are analytic functions, then U is said to be analytically isomorphic to V and f is called an analytic isomorphism.

Example 2.5 Two open sets

$$U = \{(x, y)\ ;\ x^2 + y^2 < 1\},$$
$$V = \{(x', y')\ ;\ x'^2 + y'^2 + 2y'e^{x'} + e^{2x'} < 1\}$$

are analytically isomorphic because

$$x' = x,$$
$$y' = y - e^x$$

is an analytic isomorphism.

Definition 2.6 (Singularities of a set) Let A be a nonempty set contained in the real Euclidean space $\mathbb{R}^d$.

(1) A point P in A is said to be nonsingular if there exist open sets $U, V \subset \mathbb{R}^d$ and an analytic isomorphism $f : U \to V$ such that

$$f(A \cap U) = \{(x_1, x_2, \ldots, x_r, 0, 0, \ldots, 0); x_i \in \mathbb{R}\} \cap V, \qquad (2.3)$$

where r is a nonnegative integer (Figure 2.3). If all points of A are nonsingular, then A is called a nonsingular set.

(2) If a point P in A is not nonsingular, it is called a singularity or a singular point of the set A. The set of all singularities in A is called the singular locus of A, which is denoted by

$$\mathrm{Sing}(A) = \{P \in A\ ;\ P \text{ is a singularity of A}\}.$$

Example 2.6 (1) The set $A = \{(x, y); y - x^3 = 0\}$ is nonsingular, $\mathrm{Sing}(A) = \emptyset$. Here the origin $(0, 0)$ is a nonsingular point of A because we can choose $U = V = \{(x, y); |x| < 1\}$ and

$$x' = x,$$
$$y' = y - x^3$$

is an analytic isomorphism. The origin is not a critical point of the function $f(x, y) = y - x^3$.
(2) The set $A = \{(x, y, z); (xy + z)^2 = 0\}$ is nonsingular. The origin is a nonsingular point of A because there exist $U = V = \{(x, y, z); |x| < 1, |y| < 1\}$ and an analytic isomorphism,

$$x' = x,$$
$$y' = y,$$
$$z' = z - xy.$$

The origin is a critical point of the function $f(x, y, z) = (xy + z)^2$.
(3) In the set $A = \{(x, y); xy = 0\}$, the origin is a singularity of A, which is a critical point of the function $f(x, y) = xy$. The singular locus is $\mathrm{Sing}(A) = \{(0, 0)\}$.
(4) In the set $A = \{(x, y); y^2 - x^3 = 0\}$, the origin is a singularity of A, which is a critical point of the function $f(x, y) = y^3 - x^2$.
(5) In the set $A = \{(x, y); x^5 - y^3 = 0\}$, the origin is a singularity of A, which is a critical point of $f(x, y) = x^5 - y^3$. The set A has a tangent line $y = 0$.
(6) In the set $A = \{(x, y, z); xyz = 0\}$,

$$\mathrm{Sing}(A) = \{(x, y, z); x = y = 0, \text{ or } y = z = 0, \text{ or } z = x = 0\}.$$

The set $B = \{(x, y, z); x = y = 0\}$ is a nonsingular set contained in $\mathrm{Sing}(A)$. Such a set is called a nonsingular set contained in the singular locus of A.

Remark 2.1 (1) A nonsingular analytic set is a real analytic manifold, because the neighborhood of a nonsingular point is analytically isomorphic to an r-dimensional open set in real Euclidean space, where r is equal to the number in eq.(2.3).
(2) At a nonsingular point, we can define a tangent plane. At a singularity, in general a tangent plane cannot be defined.

(3) We can check whether a point P in an algebraic set is a singularity or not by a condition of the Jacobian matrix (see Theorem 2.2).
(4) A critical point of a function f may not be a singularity of a real analytic set $\{x; f(x) = 0\}$.

Let U be an open set in $\mathbb{R}^d$ and $f : U \to \mathbb{R}^d$,

$$f(x) = (f_1(x), f_2(x), \ldots, f_d(x)),$$

is a function of class C^1. The Jacobian matrix at $x \in U$ is a $d \times d$ matrix defined by

$$J(x) = \begin{pmatrix} \frac{\partial f_1}{\partial x_1}(x) & \cdots & \frac{\partial f_1}{\partial x_d}(x) \\ \vdots & \ddots & \vdots \\ \frac{\partial f_d}{\partial x_1}(x) & \cdots & \frac{\partial f_d}{\partial x_d}(x) \end{pmatrix}. \tag{2.4}$$

The Jacobian determinant is defined by $\det J(x)$. The absolute value of the Jacobian determinant is denoted by

$$|f'(x)| = |\det J(x)|.$$

Theorem 2.1 (Inverse function theorem) *Let U be an open set in $\mathbb{R}^d$ and $f : U \to \mathbb{R}^d$ be a function of C^r class ($1 \le r \le \omega$). If the Jacobian matrix at $x_0 \in U$ is invertible, then there exists an open set $U' \subset U$ such that f is C^r isomorphism of U' and $f(U')$.*

(Explanation of Theorem 2.1) This theorem is the well-known inverse function theorem. See, for example, [95].

Theorem 2.2 (A sufficient condition for a nonsingular point). *Let U be an open set in the real Euclidian space $\mathbb{R}^d$, and $f_1(x), f_2(x), \ldots, f_k(x)$ be a set of analytic functions ($1 \le k \le d$). We define a real analytic set by*

$$A = \{x \in U; f_1(x) = f_2(x) = \cdots = f_k(x) = 0\}.$$

If a point $x_0 \in A$ satisfies

$$\det \begin{pmatrix} \frac{\partial f_1}{\partial x_1}(x_0) & \cdots & \frac{\partial f_1}{\partial x_k}(x_0) \\ \vdots & \ddots & \vdots \\ \frac{\partial f_k}{\partial x_1}(x_0) & \cdots & \frac{\partial f_k}{\partial x_k}(x_0) \end{pmatrix} \neq 0, \tag{2.5}$$

then x_0 is a nonsingular point of A.

Proof of Theorem 2.2 Since $k \le d$, we add $(d - k)$ independent functions,

$$f_i(x) = x_i \quad (k < i \le d).$$

Then $f(x) = (f_1(x), \ldots, f_d(x))$ satisfies the condition of Theorem 2.1. Therefore, there exists an open set V, which contains x_0, such that $f : V \to f(V)$ is an analytic isomorphism. If $x = (x_1, x_2, \ldots, x_d) \in A \cap V$, then $f_1(x) = \cdots = f_k(x_0) = 0$, hence

$$f(x) = (0, 0, \ldots, 0, x_{k+1}, \ldots, x_d) \in f(V).$$

By Definition 2.6, x_0 is not a singularity. □

Remark 2.2 (Implicit function theorem) From the proof of Theorem 2.2, the function

$$f^{-1} : (0, 0, \ldots, 0, x_{k+1}, \ldots, x_d) \mapsto (x_1, x_2, \ldots, x_d) \in A \cap V$$

can be understood as a function from $\hat{x} = (x_{k+1}, \ldots, x_d) \in \mathbb{R}^{d-k}$ into $x = (x_1, x_2, \ldots, x_d) \in \mathbb{R}^d$, which is denoted by $g(\hat{x})$. If a function $\pi : \mathbb{R}^d \to \mathbb{R}^k$ is defined by

$$\pi(x_1, \ldots, x_d) = (x_1, \ldots, x_k),$$

then $\varphi(\hat{x}) \equiv \pi(g(\hat{x}))$ satisfies

$$\begin{aligned} f_1(\varphi(\hat{x}), \hat{x}) &= 0 \\ &\vdots \\ f_r(\varphi(\hat{x}), \hat{x}) &= 0. \end{aligned}$$

That is to say, under the condition of eq.(2.5), such a function $\varphi(\hat{x})$ exists. This result is called the implicit function theorem.

Remark 2.3 (1) Let the generalized Jacobian matrix $(k \times d)$ $(k \leq d)$ be

$$J(x_0) = \begin{pmatrix} \frac{\partial f_1}{\partial x_1}(x_0) & \cdots & \frac{\partial f_1}{\partial x_d}(x_0) \\ \vdots & \ddots & \vdots \\ \frac{\partial f_k}{\partial x_1}(x_0) & \cdots & \frac{\partial f_k}{\partial x_d}(x_0) \end{pmatrix}. \tag{2.6}$$

If the rank of this matrix is full, that is to say, $\text{rank} J(x_0) = k$, then there exists a set of analytic functions $\{g_i(x); i = 1, 2, \ldots, k\}$ made of a linear combination of $\{f_i(x); i = 1, 2, \ldots, d\}$ which satisfies the same condition as eq.(2.5) instead of $\{f_k\}$, hence x_0 is a nonsingular point in A.
(2) Even if x_0 is a nonsingular point, $\text{rank} J(x_0)$ is not equal to k, in general. However, there is a set of functions $f_1(x), \ldots, f_k(x)$ which ensures that

$$x_0 \text{ is nonsingular} \iff \text{rank} J(x_0) = k.$$

The condition of such a set of functions is shown in chapter 3.

Corollary 2.1 *For a real analytic function f, a singularity of the real analytic set $A = \{x \in U : f(x) = 0\}$ is a critical point of the function f. However, in general, a critical point of the function f may not be a singularity of the set A.*

Proof of Corollary 2.1 From Theorem 2.2, if x_0 is not a critical point of f, then x_0 is a nonsingular point. On the other hand, if $f(x, y) = (x + y)^2$, then the origin is a critical point of f, but a nonsingular point in $\{(x, y); x + y = 0\}$. □

Remark 2.4 (1) (Sard's theorem) Let f be a function of C^∞ class from an open set in $\mathbb{R}^d$ to $\mathbb{R}^d$. Then the Lebesgue measure of the set of all critical values is zero in $\mathbb{R}^d$.
(2) If f is a real analytic function whose domain is restricted in a compact set then the set of all critical values is a finite set (Theorem 2.9).

2.4 Resolution of singularities

Let $U \subset \mathbb{R}^d$ be an open set which contains x_0 and $f : U \to \mathbb{R}^1$ a real analytic function that satisfies $f(x_0) = 0$. If x_0 is not a critical point of f, in other words, $\nabla f(x_0) \neq 0$, then x_0 is not a singularity of the real analytic set $\{x \in U; f(x) = 0\}$. On the other hand, if $\nabla f(x_0) = 0$, then the point x_0 may be a singularity and the real analytic set may have a very complex shape at x_0. Therefore, if $\nabla f(x_0) = 0$, it seems to be very difficult to analyze the function $y = f(x)$ in the neighborhood of x_0. However, the following theorem shows that any neighborhood of a real analytic set can be understood as an image of normal crossing singularities. This theorem plays a very important role in this book.

Theorem 2.3 (Hironaka's theorem, resolution of singularities) *Let $f(x)$ be a real analytic function from a neighborhood of the origin in the real Euclidean space $\mathbb{R}^d$ to $\mathbb{R}^1$, which satisfies $f(0) = 0$ and $f(x)$ is not a constant function. Then, there exists a triple (W, U, g) where*

(a) W is an open set in $\mathbb{R}^d$ which contains 0,
(b) U is a d-dimensional real analytic manifold,
(c) $g : U \to W$ is a real analytic map,

which satisfies the following conditions.
(1) The map g is proper, in other words, for any compact set $C \subset W$, $g^{-1}(C)$ is a compact set in U.
(2) We use the notation $W_0 = \{x \in W; f(x) = 0\}$ and $U_0 = \{u \in U; f(g(u)) = 0\}$. The real analytic function g is a real analytic isomorphism of $U \setminus U_0$ and

$W \setminus W_0$, in other words, it is a one-to-one and onto real analytic function from $U \setminus U_0$ to $W \setminus W_0$.
(3) For an arbitrary point $P \in U_0$, there is a local coordinate $u = (u_1, u_2, \ldots, u_d)$ of U in which P is the origin and

$$f(g(u)) = S\, u_1^{k_1}\, u_2^{k_2} \cdots u_d^{k_d}, \tag{2.7}$$

where $S = 1$ or $S = -1$ is a constant, $k_1, k_2, \ldots, k_d$ are nonnegative integers, and the Jacobian determinant of $x = g(u)$ is

$$g'(u) = b(u) u_1^{h_1}\, u_2^{h_2} \cdots u_d^{h_d}, \tag{2.8}$$

where $b(u) \neq 0$ is a real analytic function, and $h_1, h_2, \ldots, h_d$ are nonnegative integers.

Remark 2.5 (1) This fundamental theorem was proved by Hironaka in 1964 [40], for which he received the Fields Medal. For the proof of this theorem, see [40]. Application of this theorem to Schwartz distribution theory and differential equations was pointed out by Atiyah in 1970. In the paper [14], the resolution theorem is introduced as "for an analytic function $f(x)$ there exists an invertible analytic function $a(u)$ such that $f(g(u)) = a(u)u^k$". Here "invertible" means not that $a^{-1}(u)$ exists but that $1/a(u)$ exists, in other words, the inverse Schwartz distribution $1/a$ exists. This paper also shows that the resolution theorem of real analytic functions has complexification. A mathematical relation between this theorem and Bernstein–Sato's b-function was shown by Kashiwara in 1976 [46], where the more direct expression "$f(g(u)) = u^k$" was applied. It was proposed in 1999 that this theorem is the foundation for statistical learning theory [99, 100].
(2) As is well known in linear algebra, any linear transform L on a finite-dimensional vector space can be represented by a matrix of Jordan form by choosing an appropriate coordinate. The resolution of singularities claims that any analytic function can be represented by a normal crossing function by choosing an appropriate manifold.

Remark 2.6 These are remarks on Theorem 2.3.
(1) Figure 2.4 illustrates this theorem.
(2) By using multi-indices, eq.(2.7) and eq.(2.8) are respectively expressed by

$$f(g(u)) = S\, u^k,$$
$$g'(u) = b(u)\, u^h.$$

(3) The constant S and the multi-indeces k and h depend on local coordinates in general.

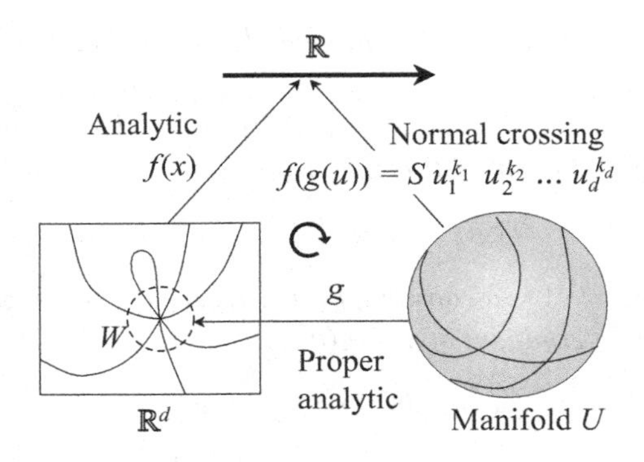

Fig. 2.4. Resolution of singularities

(4) In this theorem, we used the notation,

$$g^{-1}(C) = \{u \in U; g(u) \in C\},$$

and

$$W \setminus W_0 = \{x \in W; x \notin W_0\},$$
$$U \setminus U_0 = \{u \in U; u \notin U_0\}.$$

(5) Although g is a real analytic morphism of $U \setminus U_0$ and $W \setminus W_0$, it is not of U and W in general. It is not a one-to-one map from U_0 to W_0 in general.
(6) This theorem holds for any analytic function f such that $f(0) = 0$, even if 0 is not a critical point of f.
(7) The triple (W, U, g) is not unique. There is an algebraic procedure by which we can find a triple, which is shown in Chapter 3. The manifold U is not orientable in general.
(8) If $f(x) \geq 0$ in the neighborhood of the origin, then all of $k_1, k_2, \ldots, k_d$ in eq.(2.7) should be even integers and $S = 1$, hence eq.(2.7) can be replaced by

$$f(g(u)) = u_1^{2k_1}\, u_2^{2k_2} \cdots u_d^{2k_d}. \tag{2.9}$$

(9) The theorem shows resolution of singularities in the neighborhood of the origin. For the other point x_0, if $f(x_0) = 0$, then the theorem can be applied to $x_0 \in \mathbb{R}^d$, which implies that there exists another triple (W, U, g) such that

$$f(g(u) - x_0) = S\, u^k,$$
$$g'(u) = b(u)\, u^h,$$

where S, k, h are different from those of the origin.
(10) Let K be a compact set in an open domain of the real analytic function $f(x)$. By collecting and gluing triples $\{W, U, g\}$ for all points of K, we obtain

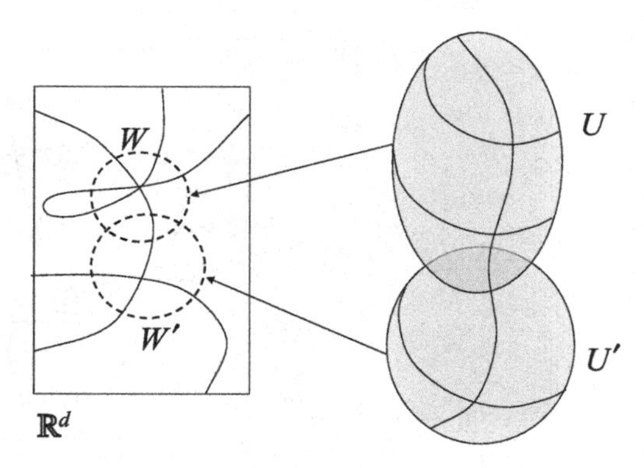

Fig. 2.5. Collection of local resolutions

a global resolution of $\{x \in K; f(x) = 0\}$ (Figure 2.5). Since K is compact, the number of local coordinates is finite.

Remark 2.7 If there exists a real analytic function $v = t(u)$,

$$v = t(u) = (t_1(u), t_2(u), \ldots, t_d(u)),$$

whose Jacobian matrix is invertible, and if

$$f(g(u)) = t_1(u)^{k_1}\, t_2(u)^{k_2}\, \cdots\, t_d(u)^{k_d}, \tag{2.10}$$

then the same result as Theorem 2.3 is obtained, because

$$f(g(t^{-1}(v))) = v_1^{k_1}\, v_2^{k_2}\, \cdots\, v_d^{k_d}.$$

Specifically, if we find a real analytic function g such that

$$f(g(u)) = a(u) u_1^{k_1}\, u_2^{k_2}\, \cdots\, u_d^{k_d}, \tag{2.11}$$

where $a(u) \neq 0$, then by using an analytic morphism $v = t(u)$ defined by

$$v_1 = |a(u)|^{1/k_1}\, u_1,$$
$$v_i = u_i \quad (2 \leq i \leq d),$$

we have

$$f(g(t^{-1}(v))) = v_1^{k_1}\, v_2^{k_2}\, \cdots\, v_d^{k_d}.$$

Therefore if we find a function g such that eq.(2.10) or eq.(2.11) holds, then we obtain the same result as Theorem 2.3.

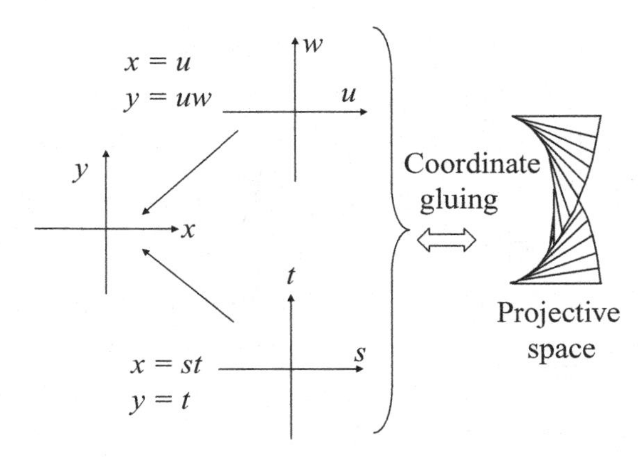

Fig. 2.6. Example of desingularization

Example 2.7 (1) Let us study $f(x, y) = x^2 + y^2$ in $\mathbb{R}^2$. The triple (W, U, g) is defined as follows. Firstly, $W = \mathbb{R}^2$. Secondly, two local open sets are defined by

$$U_1 = \{(x_1, y_1); -\infty < x_1, y_1 < \infty\},$$

$$U_2 = \{(x_2, y_2); -\infty < x_2, y_2 < \infty\}.$$

The manifold U is made by gluing two local coordinates $U_1 \cup U_2$ as in Figure 2.6. Here (x_1, y_1) and (x_2, y_2) are identified as a point in the manifold U if and only if

$$x_1 y_1 = x_2,$$

$$y_1 = x_2 y_2.$$

Thirdly, the map $g : U \to W$ $(U = U_1 \cup U_2)$ is defined on U_1 by

$$x = x_1 y_1,$$

$$y = y_1,$$

and on U_2 by

$$x = x_2,$$

$$y = x_2 y_2.$$

Then the map g is well defined as a function from U to W. The manifold U is a two-dimensional projective space introduced in Chapter 3, which is not orientable. Then $f(g(\cdot))$ is a well-defined function on each coordinate,

$$f(g(x_1, y_1)) = y_1^2(1 + x_1^2),$$
$$f(g(x_2, y_2)) = x_2^2(1 + y_2^2).$$

Here $1 + x_1^2 > 0$ and $1 + y_2^2 > 0$ are positive real analytic functions. The Jacobian matrix of $f(g(u))$ on U_1 is

$$J = \begin{pmatrix} y_1 & x_1 \\ 0 & 1 \end{pmatrix},$$

hence the Jacobian determinant is $g'(u) = y_1$. Note that g is not proper as a function $g : U_1 \to W$, because the inverse of a compact set

$$g^{-1}(\{0, 0\}) = \{(x_1, y_1); y_1 = 0\}$$

is not compact. However, it is proper as a function $g : U \to W$.
(2) The function introduced in Example 1.3 is

$$f(x, y, z) = x^2y^2 + y^2z^2 + z^2x^2,$$

which is defined on $W = \{(x, y, z) \in \mathbb{R}^3\}$. We prepare three open sets defined by

$$U_1 = \{(x_1, y_1, z_1); -\infty < x_1, y_1, z_1 < \infty\},$$
$$U_2 = \{(x_2, y_2, z_2); -\infty < x_2, y_2, z_2 < \infty\},$$
$$U_3 = \{(x_3, y_3, z_3); -\infty < x_3, y_3, z_3 < \infty\}.$$

A map from $g : U_1 \cup U_2 \cup U_3 \to W$ is defined by

$$x = x_1y_1z_1,$$
$$y = y_1z_1,$$
$$z = z_1,$$

on U_1. On U_2 and U_3, g is defined by

$$x = x_2y_2 = x_3,$$
$$y = y_2 = y_3z_3x_3,$$
$$z = z_2x_2y_2 = z_3x_3.$$

Then U is a manifold of local coordinates $U_1 \cup U_2 \cup U_3$ and $g : U \to W$ is a well-defined function. On U_1,

$$g(x_1, y_1, z_1) = y_1^2z_1^4(1 + x_1^2 + x_1^2y_1^2z_1^2),$$

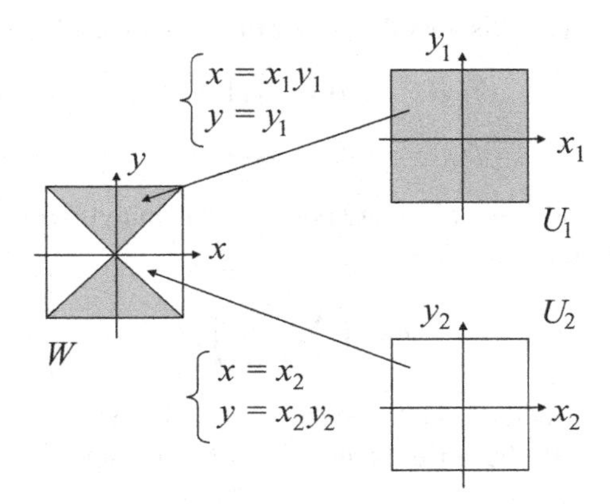

Fig. 2.7. Desingularization and integral

and the Jacobian determinant on U_1 is

$$g'(x_1, y_1, z_1) = y_1 z_1^2.$$

Therefore (W, U, g) gives the resolution of singularities of $f(x, y, z)$.
(3) The above two examples are typical cases of desingularization. For general functions and an algorithm to find a triple (W, U, g), see Chapter 3.

Example 2.8 Let $W = \{(x, y); |x|, |y| \leq 1\}$ in $\mathbb{R}^2$ and

$$U_1 = \{(x_1, y_1); |x_1|, |y_1| \leq 1\},$$
$$U_2 = \{(x_2, y_2); |x_2|, |y_2| \leq 1\}.$$

The map $g : U_1 \cup U_2 \to W$ is defined on U_1 by

$$x = x_1 y_1,$$
$$y = y_1,$$

and on U_2 by

$$x = x_2,$$
$$y = x_2 y_2,$$

which means that the integration over W is the sum of the integrations over U_1 and U_2 as in Figure 2.7,

$$\int_W f(x, y) dx dy = \int_{U_1} f(x_1 y_1, y_1) |y_1| dx_1 dy_1 + \int_{U_2} f(x_2, x_2 y_2) |x_2| dx_2 dy_2.$$

If $f(x, y)$ has a singularity at the origin, then the integral over W is sometimes made easier on $U_1 \cup U_2$ by resolution of singularities.

Definition 2.7 (Real log canonical threshold) Let $f(x)$ be a real analytic function defined on an open set $O \subset \mathbb{R}^d$. Let C be a compact set which is contained in O. For each point $P \in C$ such that $f(p) = 0$, there exists a triple (W, U, g) as in Theorem 2.3

$$f(g(u) - P) = S\, u_1^{k_1}\, u_2^{k_2} \cdots u_d^{k_d},$$

$$g'(u) = b(u) u_1^{h_1}\, u_2^{h_2} \cdots u_d^{h_d},$$

where $(k_1, k_2, \ldots, k_d)$ and $(h_1, h_2, \ldots, h_d)$ depend on the point P and triple (W, U, g). The real log canonical threshold for a given compact set C is defined by

$$\Lambda(C) = \inf_{P \in C} \min_{1 \le j \le d} \left(\frac{h_j + 1}{k_j} \right),$$

where, if $k_j = 0$, we define $(h_j + 1)/k_j = \infty$.

Theorem 2.4 *The real log canonical threshold does not depend on the triple* (W, U, g).

Proof of Theorem 2.4 Let us introduce a zeta function of $z \in \mathbb{C}$

$$\zeta(z) = \int_C |f(x)|^z dx.$$

Then this function is holomorphic in $\mathrm{Re}(z) > 0$ and

$$\begin{aligned} \zeta(z) &= \int_{g^{-1}(C)} |f(g(u))|^z |g'(u)| du \\ &= \sum_\alpha \int_{U_\alpha \cap g^{-1}(C)} u^{kz+h} b(u) du, \end{aligned}$$

where U_α is the local coordinate. By using the Taylor expansion of $g'(u)$ for an arbitrary order at the origin of each local coordinate, it follows that the zeta function can be analytically continued to a meromorphic function on the entire complex plane whose poles are all real, negative, and rational numbers. The log canonical threshold $\Lambda(C)$ corresponds the real largest pole $(-\min(h_j + 1)/k_j)$ of the zeta function, which does not depend on the triple. □

Remark 2.8 (1) A number which does not depend on the triple is called a birational invariant. The real log canonical threshold is a birational invariant [50, 62, 78]. In Chapter 4, we study the generalized concept of the real log canonical threshold.

(2) In this book, we mainly study real algebraic geometry. In complex algebraic geometry, there is the same concept, the log canonical threshold, but they are not equal to each other in general.

2.5 Normal crossing singularities

By resolution of singularities, any singularities can be understood as the image of normal crossing singularities. Normal crossing singularities have very good properties which enable us to construct statistical learning theory.

Definition 2.8 Let U be an open set in $\mathbb{R}^d$. A real analytic function $f(x)$ on U is said to be normal crossing at $x^* = (x_1^*, x_2^*, \ldots, x_d^*)$, if there exists an open set $U' \subset U$ such that

$$f(x) = a(x) \prod_{j=1}^{d} (x_j - x_j^*)^{k_j} \quad (x \in U'),$$

where $a(x)$ is a real analytic function ($|a(x)| > 0$), and $k_1, k_2, \ldots, k_d$ are non-negative integers.

Theorem 2.5 *Let r be a natural number ($1 \leq r \leq d$). If a real analytic function $f(x)$ defined on an open set $U \subset \mathbb{R}^d$ satisfies*

$$(\forall x \in U), \quad \text{“}x_1 x_2 \cdots x_r = 0 \Longleftrightarrow f(x) = 0,\text{”} \tag{2.12}$$

where U contains the origin O, then there exist an open set U' ($O \in U' \subset U$) and a real analytic function $a(x)$ such that

$$f(x) = a(x)\, x_1\, x_2\, \cdots\, x_r.$$

Proof of Theorem 2.5 Let W be an open set in which the Taylor expansion of $f(x)$ absolutely converges. By using the notation $\hat{x} = (x_2, x_3, \ldots, x_d)$,

$$f(x) = \sum_{i=0}^{\infty} a_i(\hat{x}) x_1^i.$$

By assumption,

$$a_0(\hat{x}) = 0 \quad ((0, \hat{x}) \in W).$$

Thus $f^*(x) = f(x)/x_1$ is analytic in W and

$$f(x) = x_1 f^*(x).$$

By the recursive procedure, we obtain the theorem. □

Remark 2.9 (Factor theorem) (1) If a real analytic function $f(x)$ of a single variable x satisfies $f(a) = 0$ then $f(x)/(x - a)$ is a real analytic function, which is the well-known factor theorem.
(2) If the dimension of x is bigger than 1, then such a relation does not hold in general. For example, even if

$$x^2 + y^2 = 0 \Longleftrightarrow f(x, y) = 0,$$

a real analytic function $a(x, y)$ which satisfies

$$f(x, y) = a(x, y)(x^2 + y^2)$$

does not exist in general. In fact, there is an example, $f(x, y) = x^2 + y^4$. Division by a normal crossing function can be understood as the generalization of the factor theorem.

Theorem 2.6 *Let r be a natural number $(1 \leq r \leq d)$. Assume that a real analytic function $f(x)$ on an open set $U \subset \mathbb{R}^d$ satisfies*

$$|f(x)| \leq A\left|x_1^{k_1} x_2^{k_2} \cdots x_r^{k_r}\right| \quad (x \in U), \tag{2.13}$$

where $A > 0$ is a constant and $k_1, k_2, \ldots, k_r$ are natural numbers. Then there exist an open set $W \subset U$ and a real analytic function $g(x)$ on W such that

$$f(x) = g(x) x_1^{k_1} x_2^{k_2} \cdots x_r^{k_r} \quad (x \in W).$$

Proof of Theorem 2.6 By Theorem 2.5, there exists a real analytic function $a(x)$ such that

$$f(x) = a(x) x_1 x_2 \cdots x_r,$$

hence by eq.(2.13)

$$|a(x)| \leq A\left|x_1^{k_1 - 1} \cdots x_r^{k_r - 1}\right|.$$

By the recursive procedure, we obtain the theorem. □

Remark 2.10 (1) Let U be an open set. Even if

$$|f(x)| \leq |g(x)| \quad (x \in U)$$

holds for real analytic functions, $f(x)/g(x)$ is not a real analytic function in general. For example, the Cauchy–Schwarz inequality

$$(xy + zw)^2 \leq (x^2 + z^2)(y^2 + w^2)$$

holds; however,

$$h(x, y, z, w) = \frac{(xy + zw)^2}{(x^2 + z^2)(y^2 + w^2)} \tag{2.14}$$

is not continuous at the origin. The limit value

$$\lim_{\mathbf{x}\to 0} h(\mathbf{x})$$

does not exist, hence $\mathbf{x} = 0$ is not a removable singularity. Theorem 2.6 claims that, if $g(x)$ is normal crossing and if $|f(x)| \leq |g(x)|$ holds, then the origin is a removable singularity, consequently $f(x)/g(x)$ can be made a real analytic function.

(2) Although the origin of eq.(2.14) is not a removable singularity, $h(x, y, z, w)$ can be made a real analytic function by resolution of singularities. In fact, by using

$$\begin{aligned} x &= x_1 = x_2 z_2, \\ y &= y_1 = y_2 w_2, \\ z &= x_1 z_1 = z_2, \\ w &= y_1 w_1 = w_2, \end{aligned}$$

the origin $\mathbf{x} = 0$ is a removable singularity of the function h on a manifold, and

$$h = \frac{(1 + z_1 w_1)^2}{(1 + z_1^2)(1 + w_1^2)} = \frac{(x_2 y_2 + 1)^2}{(x_2^2 + 1)(w_2^2 + 1)}.$$

The function that has the same property is sometimes called a blow-analytic function.

(3) In statistical learning theory, we have to study the log likelihood ratio function $\log(q(x)/p(x|w))$ divided by the Kullback–Leibler distance $K(w)$. In singular statistical models, such a function is ill-defined at singularities; however, it can be made well-defined by resolution of singularities.

Theorem 2.7 *Let $U \subset \mathbb{R}^d$ be an open set which contains the origin and r be a natural number ($1 \leq r \leq d$). Assume that two real analytic functions $f_1(x)$ and $f_2(x)$ on an open set U satisfy*

$$f_1(x) f_2(x) = x_1^{k_1} x_2^{k_2} \cdots x_r^{k_r},$$

where $k_1, k_2, \ldots, k_r$ are natural numbers. Then there exist an open set $W \subset U$ and real analytic functions $a_1(w)$, $a_2(w)$, such that

$$\begin{aligned} f_1(x) &= a_1(x) x_1^{j_1} x_2^{j_2} \cdots x_r^{j_r}, \\ f_2(x) &= a_2(x) x_1^{h_1} x_2^{h_2} \cdots x_r^{h_r}, \end{aligned}$$

where $a_1(x) a_2(x) = 1 \quad (x \in W) \quad$ (hence $a_1(x) \neq 0, a_2(x) \neq 0$) and $j_1, \ldots, j_r, h_1, \ldots, h_r$ are nonnegative integers.

Proof of Theorem 2.7 By the assumption,

$$\{x \in U; f_1(x) = 0\} \subset \{x \in U; x_1 x_2 \cdots x_r = 0\}.$$

Let I_1 be the set of i $(1 \le i \le r)$ which is defined by

$$f_1(x_1, x_2, \ldots, x_i = 0, \ldots, x_d) \equiv 0.$$

Then

$$\{x \in U; f_1(x) = 0\} = \{x \in U; \prod_{i \in I_1} x_i = 0\}.$$

Also we define I_2 for $f_2(x)$ in the same way. By Theorem 2.5, there exist real analytic functions $g_1(x)$ and $g_2(x)$ such that

$$f_1(x) = g_1(x) \prod_{i=1}^{r} x_i^{j_i},$$

$$f_2(x) = g_2(x) \prod_{i=1}^{r} x_i^{h_i},$$

where j_i and h_i are defined as follows. If $i \in I_1$ then $j_i = 1$; otherwise $j_i = 0$. If $i \in I_2$ then $h_i = 1$; otherwise $h_i = 0$. Therefore

$$g_1(x) g_2(x) = \prod_{1 \le i \le r} x_i^{k_i - j_i - h_i}.$$

By applying a recursive procedure, we obtain the theorem. □

Remark 2.11 If both $f_1(x)$ and $f_2(x)$ are polynomials, this theorem is equivalent to the uniqueness of factorization.

Theorem 2.8 (Simultaneous resolution of singularities) *Let k be an integer and $f_0(x), f_1(x), \ldots, f_k(x)$ be real analytic functions on an open set in $\mathbb{R}^d$ which contains the origin $x = 0$. Assume that, for all $0 \le i \le k$, $f_i(0) = 0$ and $f_i(x)$ is not a constant function. Then there exists a triple (W, U, g),*

(a) W is an open set in $\mathbb{R}^d$,
(b) U is a real analyic manifold,
(c) $g : U \to W$ is a real analytic map,

which satisfies the following conditions:
(1) g is proper.
(2) g is an analytic isomorphism of $U \setminus U_0$ and $W \setminus W_0$ where $U_0 = \cup_i \{x \in W; f_i(x) = 0\}$ and $W_0 = \cup_i \{u \in U; f_i(g(u)) = 0\}$.

(3) For an arbitrary point $P \in W_0$, there exists a local coordinate $u = (u_1, u_2, \ldots, u_d)$ such that

$$
\begin{aligned}
f_0(g(u)) &= u_1^{k_{01}}\, u_2^{k_{02}} \cdots u_d^{k_{0d}} \\
f_1(g(u)) &= a_1(u)\, u_1^{k_{11}}\, u_2^{k_{12}} \cdots u_d^{k_{1d}} \\
f_2(g(u)) &= a_2(u)\, u_1^{k_{21}}\, u_2^{k_{22}} \cdots u_d^{k_{2d}} \\
&\vdots \\
f_k(g(u)) &= a_k(u)\, u_1^{k_{k1}}\, u_2^{k_{k2}} \cdots u_d^{k_{kd}} \\
g'(u) &= b(w)\, u_1^{h_1}\, u_2^{h_2} \cdots u_d^{h_d},
\end{aligned}
$$

where k_{ij} and h_i are nonnegative integers, and $a_i(w) \neq 0$ $(1 \leq i \leq k)$ and $b(w) \neq 0$ are real analytic functions.

Proof of Theorem 2.8 Applying resolution of singularities to

$$f(x) = f_0(x) f_1(x) f_2(x) \cdots f_k(x),$$

there exists a triple (W, U, g) such that

$$f(g(u)) = u_1^{k_1} \cdots u_d^{k_d}.$$

By Theorem 2.7, we obtain the theorem. □

Remark 2.12 If a compact set $K \subset \mathbb{R}^d$ is defined by

$$K = \{x \in \mathbb{R}^d;\ f_1(x) \geq 0,\ f_2(x) \geq 0, \ldots, f_k(x) \geq 0\}$$

using real analytic functions $f_1(x), f_2(x), \ldots, f_k(x)$, then by applying simultaneous resolution of singularities, there exists a triple (W, U, g) by which all functions can be made normal crossing. Then the boundary of the set $g^{-1}(K)$ is contained in $u_1 u_2 \cdots u_d = 0$ in every local coordinate.

Theorem 2.9 (Finite critical values) *Let U be an open set in $\mathbb{R}^d$ and K be a compact set in U. Assume that $f(x) : U \to \mathbb{R}$ is a real analytic function. Then the set of critical values of a restricted function $f : K \to \mathbb{R}$*

$$\{y\ ;\ y = f(x), \nabla f(x) = 0,\ x \in K\}$$

has finite elements.

Proof of Theorem 2.9 Since $f(x)$ is a continuous function on a compact set, the set

$$f(K) = \{f(x) \in \mathbb{R}^1; x \in K\}$$

is compact in $\mathbb{R}$. Let y be a critical value in $f(K)$. The closed set

$$K_y \equiv \{x \in K;\ f(x) - y = 0\} \subset K$$

is also compact. For an arbitrary $x_0 \in K_y$, by applying the resolution theorem to $f(x) - y$, there exist an open set $W(x_0)$ that contains x_0, a manifold $M(x_0)$, and a real analytic map $g : M(x_0) \to W(x_0)$ such that $f(g(u)) - y$ is a normal crossing function. Because

$$\frac{\partial}{\partial u_i} f(g(u)) = \sum_{j=1}^{d} \frac{\partial f}{\partial x_j} \frac{\partial x_j}{\partial u_i},$$

the inverse of the critical points of $f(x) - y$ is contained in the set of critical points of $f(g(u)) - y$.

$$\begin{aligned} & g^{-1}(\{x \in W(x_0); f(x) - y = 0, \nabla f(x) = 0\}) \\ & \subset \{u \in M(x_0); f(g(u)) - y = 0, \nabla f(g(u)) = 0\}. \end{aligned}$$

Here $f(g(u)) - y$ is a normal crossing function, hence any critical point of $u \in M(x_0)$ is contained in the set of zero points of $f(g(u)) - y = 0$. Therefore any critical point of $f(x)$ in $W(x_0)$ satisfies $f(x) - y = 0$. In other words, in each $x_0 \in K_y$, we can choose $W(x_0)$ such that all critical points in $W(x_0)$ are contained in K_y. The set K_y can be covered by the union of neighborhoods

$$K_y \subset \cup_{x_0 \in K_y} W(x_0). \tag{2.15}$$

By the compactness, K_y is covered by an open set W_y that is a finite union of $W(x_0)$. There exists no critical point in $W_y \setminus K_y$. The set of all critical points is covered by the union of open sets $\{W_y\}$.

$$\{x \in K;\ \nabla f(x) = 0\} \subset \cup_y W_y. \tag{2.16}$$

The set of all critical points is compact because it is a set of all zero points of a continuous function ∇f. Hence $\cup_y$ in eq.(2.16) can be chosen to be a finite union. Therefore the set of critical values has finite elements. □

Remark 2.13 Let $f : \mathbb{R}^d \to \mathbb{R}^1$ be a real analytic function and t be a real number. In the following sections, we study a Schwartz distribution

$$\delta(t - f(x))$$

of $x \in \mathbb{R}^d$. If t is not a critical value of $f(x)$, $\delta(t - f(x))$ is a well-defined function, whereas, if t is a critical value of $f(x)$, such a distribution cannot be defined. By Theorem 2.9, $\delta(t - f(x))\varphi(x)$ is well-defined except for finite set of t if a C^∞ class function $\varphi(x)$ is equal to zero outside of a compact set.

2.6 Manifold

To study resolution of singularities, we need a mathematical preparation of a manifold.

Definition 2.9 (Topological space) Let M be a set. A set $\mathcal{U}$ consisting of subsets of M is called a family of open sets if it satisfies the following conditions.
(1) Both the set M and the empty set $\emptyset$ are contained in $\mathcal{U}$.
(2) If $\{U_\lambda\}$ are subsets of $\mathcal{U}$, then the union $\cup_\lambda U_\lambda$ is contained in $\mathcal{U}$.
(3) If U_1 and U_2 are contained in $\mathcal{U}$, then the intersection $U_1 \cap U_2$ is contained in $\mathcal{U}$.
The set M is called a topological space if a family of open sets is determined. It is called a Haussdorff space if, for arbitrary $x, y \in M$ $(x \neq y)$, there exist open sets U, V such that $x \in U$, $y \in V$, and $U \cap V = \emptyset$.

Definition 2.10 (System of local coordinates) Let M be a Haussdorff space and $\{U_\alpha\}$ be a family of open sets in M whose union covers M. Assume that, for each U_α, a map ϕ_α is defined from U_α to d-dimensional real Euclidean space $\mathbb{R}^d$. The pair $\{U_\alpha, \phi_\alpha\}$ is said to be a system of local coordinates if, for any α and any open set U in U_α, ϕ_α is a continuous and one-to-one map from U to $\phi_\alpha(U)$ whose inverse is also continuous. A Haussdorff space M that has a system of local coordinates is called a manifold. For each $r = 0, 1, 2, \ldots, \infty, \omega$, a manifold M is said to be of class C^r if it has a system of local coordinates such that, if $U^* \equiv U_1 \cap U_2 \neq \emptyset$, both of the maps

$$\phi_1(U^*) \ni x \mapsto \phi_2\big(\phi_1^{-1}(x)\big) \in \phi_2(U^*),$$

$$\phi_2(U^*) \ni x \mapsto \phi_1\big(\phi_2^{-1}(x)\big) \in \phi_1(U^*)$$

are of class C^r (Figure 2.8). If M is a manifold of class C^ω, it is called a real analytic manifold.

Example 2.9 Here are examples and counter examples.
(1) An open set that is not the empty set in $\mathbb{R}^d$ is a real analytic manifold.
(2) The d-dimensional sphere $\{x \in \mathbb{R}^{d+1}; \|x\|^2 = 1\}$ is a manifold.
(3) Let U be an open set in $\mathbb{R}^d$. If a real analytic function $f : U \to \mathbb{R}^1$ satisfies

$$\nabla f(x) \neq 0$$

on the real analytic set $U_0 = \{x \in U; f(x) = 0\}$, then U_0 is a manifold.

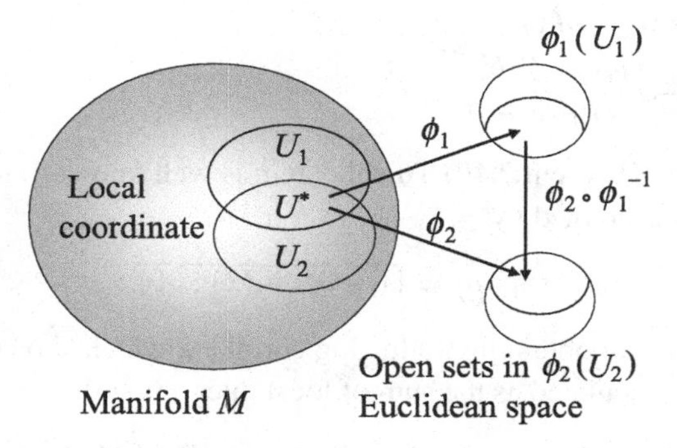

Fig. 2.8. Definition of manifold

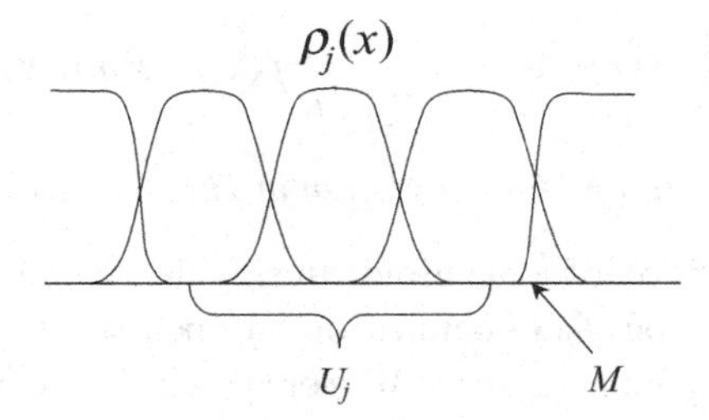

Fig. 2.9. Partition of unity

(4) A projective space, as defined in Chapter 3, is a manifold.
(5) Let $f(x)$ be a real analytic function which is represented as an absolutely convergent power series on an open set $U_1 \subset \mathbb{R}^d$. If it can be analytically continued to $U_1 \cup U_2$, where U_2 is another open set, then $U_1 \cup U_2$ is a manifold and f is a real analytic function on the real analytic manifold $U_1 \cup U_2$.
(6) The set $\{(x, y) \in \mathbb{R}^2; xy = 0\}$ is not a manifold because we cannot define a local coordinate in the neighborhood of the origin.
(7) The set $\{(x, y) \in \mathbb{R}^2; xy = 0, (x, y) \neq (0, 0)\}$ is a manifold.

Theorem 2.10 (Partition of a unity) *Let K be a compact set in a manifold M. Assume that the finite set of open sets $\{U_j, j = 1, 2, \ldots, J\}$ covers K,*

$$K \subset \bigcup_{j=1}^{J} U_j.$$

Then there exists a set of functions $\{\rho_j(x)\}$ of class C_0^∞ which satisfies the following conditions as in Figure 2.9.

(1) For each j, $0 \leq \rho_j(x) \leq 1$.
(2) For each j, *supp* $\rho_j \subset U_j$.
(3) If $x \in K$, $\sum_{j=1}^{J} \rho_j(x) = 1$.

(Explanation of Theorem 2.10) This theorem is well known. The support of a function $\rho_j(x)$ is defined by

$$\mathrm{supp}\rho_j = \overline{\{x \in M; \rho_j(x) > 0\}}.$$

By this theorem we obtain the following corollary, which shows that the integration can be calculated as the sum of local integrations.

Corollary 2.2 *Let* K *be a compact set of a manifold* M *and* μ *be a measure on* M. *Then an integral of a function* $f(x)$ *on* K *can be represented by*

$$\int_K f(x)\mu(dx) = \sum_{j=1}^{J} \int_K f(x)\rho_j(x)\mu(dx),$$

where $\{\rho_j\}$ *is the partition of a unity given in Theorem 2.10.*

Remark 2.14 (Partition of a manifold) In singular learning theory, we study a real analytic function $f_0(x)$ defined on an open set in $\mathbb{R}^d$. To analyze an integration over the parameter space W more precisely, we prepare the division of integration. Let $\pi_1(x)$, $\pi_2(x)$, $\ldots$, $\pi_k(x)$ be real analytic functions defined on the open set in $\mathbb{R}^d$. Let us study a compact set W defined by

$$W = \{x \in \mathbb{R}^d; \pi_1(x) \geq 0, \ \pi_2(x) \geq 0, \ldots, \pi_k(x) \geq 0\}.$$

By applying the simultaneous resolution theorem, Theorem 2.8, there exist a compact set $\mathcal{M}$ of a real analytic manifold and a real analytic map $g : M \to W$ such that $\pi_0(g(u)), \pi_1(g(u)), \ldots, \pi_k(g(u))$ are simultaneously normal crossing. For each point $p \in \mathcal{M}$, an open set of a local coordinate whose origin is p

$$(-b, b)^d \equiv \{u = (u_1, u_2, \ldots, u_d); |u_i| < b, i = 1, 2, \ldots, d\}$$

is denoted by $O_p(b)$, where $b > 0$ is a positive constant. Then $\mathcal{M}$ is covered by the union of $\{O_p\}$

$$\mathcal{M} \subset \cup_{p \in M} O_p(b).$$

Since $\mathcal{M}$ is a compact set, it is covered by a finite union of $\{O_p(b)\}$. Moreover, the set $(-b, b)^d$ is covered by 2^d sets which are respectively analytically isomorphic to $[0, b)^d$. Each set in $\mathcal{M}$ that is analytically isomorphic to $[0, b)^d$ is denoted by M_α. Then the set M is covered by a finite union of $\{M_\alpha\}$.

$$\mathcal{M} \subset \cup_\alpha M_\alpha.$$

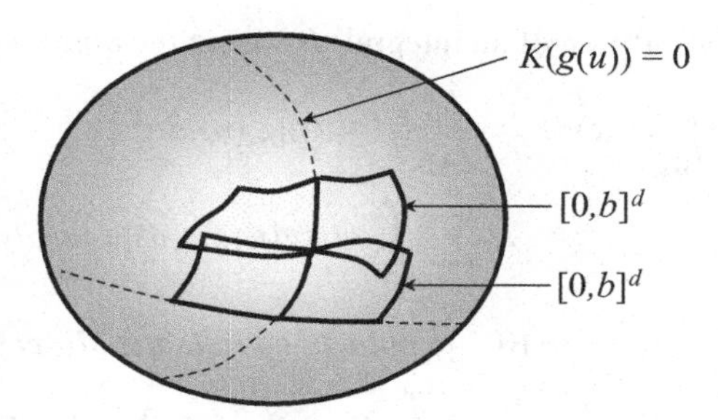

Fig. 2.10. Local coordinate made of $[0, b]^d$

Let us define a function $\sigma_\alpha(u)$ whose support is contained in M_α by

$$\sigma_\alpha^{(0)}(u) = \begin{cases} \prod_{i=1}^d \exp\Big(-\dfrac{1}{1-u_i}\Big) & (0 \leq u_i < 1\ (\forall i)) \\ 0 & \text{otherwise} \end{cases}$$

and $\sigma_\alpha(u)$ by

$$\sigma_\alpha(u) = \frac{\sigma_\alpha^{(0)}(u)}{\sum_\alpha \sigma_\alpha^{(0)}(u)}.$$

Then $\sigma_\alpha(u)$ is a function of class C^w in $(0, b)^d$, and

$$\sigma_\alpha(u) > 0 \quad (u \in [0, b)^d).$$

Moreover,

$$\sum_\alpha \sigma_\alpha(u) = 1 \quad (u \in \mathcal{M}).$$

For arbitrary integrable function $H(\cdot)$,

$$\begin{aligned} \int_W H(x)dx &= \int_{\mathcal{M}} H(g(u))|g'(u)|du \\ &= \sum_\alpha \int_{M_\alpha} H((g(u))\sigma_\alpha(u)|g'(u)|du. \end{aligned}$$

Moreover, since the number of elements $\{M_\alpha\}$ is finite if $H(x) \geq 0$,

$$C_1 \sum_\alpha \int_{M_\alpha} H(g(u))|g'(u)|du \leq \int_W H(x)dx \leq C_2 \sum_\alpha \int_{M_\alpha} H(g(u))|g'(u)|du$$

where $C_1, C_2 > 0$ are constants. This method is used in later chapters.

Theorem 2.11 (Partition of an integral) *By using the above notation,*

$$\int_W H(x)dx = \int_{\mathcal{M}} H(g(u))|g'(u)|du$$
$$= \sum_\alpha \int_{M_\alpha} H(g(u))\sigma_\alpha(u)|g'(u)|du.$$

where each M_α is equal to $[0, b]^d$ in local coordinate. Moreover, if $H(x) \geq 0$,

$$C_1 \sum_\alpha \int_{M_\alpha} H(g(u))|g'(u)|du \leq \int_W H(x)dx \leq C_2 \sum_\alpha \int_{M_\alpha} H(g(u))|g'(u)|du$$

where $C_1, C_2 > 0$ are constants.

3
Algebraic geometry

In this chapter, we introduce basic concepts in algebraic geometry: ring and ideal, the relation between algebra and geometry, projective spaces, and blow-ups. Based on this foundation, the method of how to find a resolution map is introduced. In this book, we mainly study real algebraic geometry.

3.1 Ring and ideal

In order to analyze a real algebraic set, we need an algebraic structure of polynomials. Let $\mathcal{R} = \mathbb{R}[x_1, x_2, \ldots, x_d]$ be the set of all polynomials of $x_1, x_2, \ldots, x_d$ with real coefficients. An element $f(x) \in \mathcal{R}$ can be written as a finite sum

$$f(x) = \sum_{\alpha} a_\alpha x^\alpha,$$

where α is a multi-index and a_α is a real number. For $f(x), g(x) \in \mathcal{R}$

$$f(x) = \sum a_\alpha x^\alpha,$$
$$g(x) = \sum b_\alpha x^\alpha,$$

and $s \in \mathbb{R}$, we define

$$f(x) + g(x) = \sum_{\alpha} (a_\alpha + b_\alpha) x^\alpha,$$
$$s f(x) = \sum_{\alpha} (s a_\alpha) x^\alpha,$$

by which $\mathcal{R}$ is an infinite-dimensional vector space over a field $\mathbb{R}$. Moreover, $\mathcal{R}$ is a ring by defining

$$f(x) g(x) = \sum_{\alpha} \sum_{\beta} a_\alpha b_\beta x^{\alpha+\beta}.$$

The ring $\mathcal{R}$ is called a polynomial ring.

Definition 3.1 (Ideal) A subset I in $\mathcal{R} = \mathbb{R}[x_1, x_2, \ldots, x_d]$ is said to be an ideal if

$$0 \in I,$$
$$f(x), g(x) \in I \Longrightarrow f(x) + g(x) \in I,$$
$$f(x) \in I, g(x) \in \mathcal{R} \Longrightarrow f(x)g(x) \in I.$$

For a given set of polynomials $f_1, f_2, \ldots, f_k$, we define a subset of $\mathcal{R}$ by

$$\langle f_1, f_2, \ldots, f_k \rangle = \left\{ \sum_{i=1}^{k} g_i(x) f_i(x) \; ; \; g_i \in \mathcal{R} \right\},$$

which is the minimum ideal that contains $f_1, f_2, \ldots, f_k$. This ideal is called the ideal generated by $f_1, f_2, \ldots, f_k$.

For a subset $S \subset \mathbb{R}[x_1, x_2, \ldots, x_d]$, the linear hull $L.H.(S)$ is defined by the smallest vector space that contains S,

$$L.H.(S) = \left\{ \sum_{i} a_i f_i \; ; \; a_i \in \mathbb{R}, \; f_i \in S \right\},$$

where $\sum_i$ shows the finite sum.

Example 3.1 The following are ideals in $\mathbb{R}[x, y]$:

$$\langle x^2 \rangle = L.H.(\{x^2, x^3, \ldots, x^2 y, x^2 y^2, \ldots, \}),$$
$$\langle x^3, y+1 \rangle = L.H.(\{x^3, x^4, \ldots, x^3 y, x^4 y, \ldots, y + 1, x(y+1)^2, \ldots, \}),$$
$$\langle x + y, x^2 - y^2 \rangle = \langle x + y \rangle,$$
$$\langle x + y^2, x - y^2, y^3 \rangle = \langle x, y^2 \rangle,$$
$$\langle x^2 \rangle \subset \langle x \rangle,$$
$$\langle x^2 + y^2 \rangle \subset \langle x, y \rangle.$$

If $x^m y^n$ is contained in I, then $x^{m+k} y^{n+l} \in I$ for arbitrary nonnegative integers k, l.

From the definition, the following lemma is immediately derived.

Lemma 3.1 *(1) An ideal in $\mathcal{R}$ is a ring. It is a subring of $\mathcal{R}$. A subring of $\mathcal{R}$ is not an ideal in general.*
(2) If an ideal in $\mathcal{R}$ is not $\langle 0 \rangle$, then it is an infinite-dimensional vector space.
(3) If an ideal contains 1*, then it is equal to $\mathcal{R}$.*

Theorem 3.1 (Hilbert's basis theorem) *For an arbitrary ideal I in $\mathcal{R} \equiv \mathbb{R}[x_1, x_2, \ldots, x_d]$, there exists a finite set $f_1, f_2, \ldots, f_k \in \mathbb{R}[x_1, x_2, \ldots, x_d]$ such that*

$$I = \langle f_1, f_2, \ldots, f_k \rangle.$$

(Explanation of Theorem 3.1) This is the fundamental theorem of ideal theory. For the proof, see [21]. This theorem holds in real and complex polynomial rings.

Remark 3.1 (1) A ring is said to be Noetherian if Hilbert's basis theorem holds for any ideal in the ring.
(2) If a ring A is Noetherian, then $A[x]$ is also Noetherian.
(3) A ring is Noetherian if and only if, for any nondecreasing sequence of ideals,

$$I_1 \subset I_2 \subset \cdots \subset I_n \subset \cdots,$$

there exists N such that for all $n \geq N$

$$I_n = I_{n+1} = I_{n+2} = \cdots = .$$

Example 3.2 In statistical learning theory, Hilbert's basis theorem plays an important role. Let us study a parametric function on $x \in [0, 1]$,

$$f(x, a, b, c, d) = a \sin(bx) + c \sin(dx).$$

Let S be a set of parameters which make $f(x, a, b, c, d) \equiv 0$, that is to say,

$$S = \{(a, b, c, d) \in \mathbb{R}^4; f(x, a, b, c, d) = 0 \quad (\forall x \in [0, 1])\}.$$

By using the defintion g_k,

$$g_k(a, b, c, d) = ab^{2k+1} + cd^{2k+1} \quad (k = 0, 1, 2, \ldots,),$$

the Taylor expansion shows that

$$f(x, a, b, c, d) = \sum_{k=0}^{\infty} \frac{(-1)^k x^{2k+1}}{(2k+1)!} g_k(a, b, c, d).$$

Since the set of functions $\{x^{2k+1}\}$ is linearly independent on $[0, 1]$, $f(x, a, b, c, d) \equiv 0$ if and only if

$$g_k(a, b, c, d) = 0 \ (\forall k).$$

Hence the set S is represented by

$$S = \{(a, b, c, d) \in \mathbb{R}^4; g_0(a, b, c, d) = g_1(a, b, c, d) = \cdots = 0\},$$

which shows S is a common locus defined by infinite polynomials. Let us define an ideal I_k by

$$I_k = \langle g_0, g_1, g_2, \dots, g_k \rangle.$$

Then it defines a nondecreasing sequence of ideals,

$$I_0 \subset I_1 \subset I_2 \subset I_3 \subset \cdots.$$

By Hilbert's basis theorem, there exists k such that

$$S = \{(a, b, c, d) \in \mathbb{R}^4; g_0(a, b, c, d) = \cdots = g_k(a, b, c, d) = 0\},$$

which shows S is a real algebraic set. To obtain k, we need concrete calculation. In this case, it is easy to show

$$\begin{aligned} g_{k+1}(a, b, c, d) &= g_1(a, b, c, d)(b^{2k} + d^{2k}) - g_0(a, b, c, d)(b^2 d^{2k} + b^{2k} d^2) \\ &\quad + b^2 d^2 g_{k-1}(a, b, c, d), \end{aligned}$$

hence $k = 1$. Therefore,

$$S = \{(a, b, c, d) \in \mathbb{R}^4; ab + cd = ab^3 + cd^3 = 0\}.$$

This relation can be generalized. A polynomial

$$p_n(a_1, \dots a_d, b_1, \dots, b_d) = \sum_{k=1}^{d} a_k b_k^n,$$

is contained in the ideal

$$p_n \in \langle p_1, p_2, \dots, p_d \rangle$$

for any natural number n.

3.2 Real algebraic set

A real algebraic set is a geometric object and an ideal is an algebraic object. There are mathematical relations between geometry and algebra.

In algebraic geometry, the set made of d times direct product $\mathbb{R} \times \mathbb{R} \times \cdots \times \mathbb{R}$ is called a d-dimensional affine space, which is denoted by $\mathbb{A}^d$. Since this set is equal to the Euclidean space $\mathbb{R}^d$ as a set, in this book we identify the affine space with the Euclidean space.

Definition 3.2 For an ideal I of $\mathbb{R}[x_1, x_2, \dots, x_d]$, a set $\mathbb{V}(I)$ is defined by

$$\mathbb{V}(I) = \{x \in \mathbb{R}^d \; ; \; f(x) = 0 \quad (\forall f \in I)\}.$$

A subset V of $\mathbb{R}^d$ is said to be a real algebraic set if there exists an ideal I such that $V = \mathbb{V}(I)$. The map $I \mapsto \mathbb{V}(I)$ is defined from an ideal to a real algebraic set.

Example 3.3 (1) The set of zero locus of polynomials $f_1(x), f_2(x), \ldots, f_k(x)$ is equal to $\mathbb{V}(\langle f_1, f_2, \ldots, f_k \rangle)$. For example the intersection of $x^2 + y^2 + z^2 = 10$ and $xyz = 1$ is $\mathbb{V}(\langle x^2 + y^2 + z^2 - 10, xyz - 1 \rangle)$.
(2) By definition, the map $I \mapsto \mathbb{V}(I)$ from the set of all ideals to the set of all real algebraic sets is surjective.
(3) The map $I \mapsto \mathbb{V}(I)$ is not one-to-one. For example,

$$I_1 = \langle x, y \rangle, \quad I_2 = \langle x^2, y^2 \rangle, \quad I_3 = \langle x^2 + y^2 \rangle.$$

Although three ideals I_i $(i = 1, 2, 3)$ are different from each other, the corresponding algebraic sets are equal to each other,

$$\mathbb{V}(I_1) = \mathbb{V}(I_2) = \mathbb{V}(I_3) = \{(0, 0)\}.$$

(4) For $I = \langle x^2 + 1 \rangle$, $\mathbb{V}(I)$ is the empty set.

Definition 3.3 (Defining ideal) For a given real algebraic set V, an ideal $\mathbb{I}(V)$ is defined by the set of all polynomials which are equal to zero on V,

$$\mathbb{I}(V) = \{f(x) \in \mathbb{R}[x_1, x_2, \ldots, x_d] \ ; \ f(x) = 0 \ (\forall x \in V)\}.$$

This set is an ideal of $\mathbb{R}[x_1, x_2, \ldots, x_d]$ because it satisfies the definition for an ideal. The ideal $\mathbb{I}(V)$ is called a defining ideal of a real algebraic set V.

Example 3.4 (1) For $V = \{(x, y); x + y + 1 = 0\}$, $\mathbb{I}(V) = \langle x + y + 1 \rangle$.
(2) For $V = \{(x, y); x^2y^3 = 0\}$, $\mathbb{I}(V) = \langle xy \rangle$.
(3) For $V = \{(x, y); x^3 + y^3 = 0\}$, $\mathbb{I}(V) = \langle x + y \rangle$.
(4) For $V = \{(x, y); x^2 + y^2 = 0\}$, $\mathbb{I}(V) = \langle x, y \rangle$.
(5) For the empty set $\emptyset$, $\mathbb{I}(\emptyset) = \langle 1 \rangle = \mathbb{R}[x_1, x_2, \ldots, x_d]$.

Theorem 3.2 (Real algebraic set and ideal) *Let V_1 and V_2 be real algebraic sets and I_1 and I_2 be ideals.*
(1) $V_1 \subset V_2 \Longleftrightarrow \mathbb{I}(V_1) \supset \mathbb{I}(V_2)$.
(2) $I_1 \subset I_2 \Longrightarrow \mathbb{V}(I_1) \supset \mathbb{V}(I_2)$.

Proof of Theorem 3.2 (1) ($\Longrightarrow$) Let $f \in \mathbb{I}(V_2)$. Then $f(x) = 0$ for all $x \in V_2$. Hence $f(x) = 0$ for all $x \in V_1$, giving $f \in \mathbb{I}(V_1)$. ($\Longleftarrow$) Let $x \in V_1$. Then $f(x) = 0$ for all $f \in \mathbb{I}(V_1)$. Hence $f(x) = 0$ for all $f \in \mathbb{I}(V_2)$, giving $x \in V_2$.
(2) ($\Longrightarrow$) Let $x \in \mathbb{V}(I_2)$. Then $f(x) = 0$ for all $f \in I_2$. Hence $f(x) = 0$ for all $f \in I_1$, giving $x \in \mathbb{V}(I_1)$. □

Remark 3.2 In Theorem 3.2 (2), $\Longleftarrow$ does not hold. For example, $I_1 = \langle x \rangle$, $I_2 = \langle x^2 + y^2 \rangle$. Then $\mathbb{V}(I_1) \supset \mathbb{V}(I_2)$, but $I_1 \not\subset I_2$.

Definition 3.4 (Radical of an ideal). For a given ideal I in $\mathbb{R}[x_1, x_2, \ldots, x_d]$, the radical of the ideal I is defined by

$$\sqrt{I} = \{f(x) \in \mathbb{R}[x_1, x_2, \ldots, x_d] \; ; \;\; f(x)^m \in I \;\; (\exists m : \text{natural number})\},$$

which is also an ideal.

Remark 3.3 (1) By definition, $I \subset \sqrt{I}$.
(2) If an ideal I satisfies $I = \sqrt{I}$, then I is called a radical ideal. Since $\sqrt{\sqrt{I}} = \sqrt{I}$, the radical $\sqrt{I}$ is a radical ideal.
(3) For any real algebraic set V, $\mathbb{I}(V)$ is a radical ideal.
(4) The map $V \mapsto \mathbb{I}(V)$ can be understood as a function from a real algebraic set to a radical ideal. The map $V \mapsto \mathbb{I}(V)$ is not surjective. For example, for a radical ideal $I = \langle x^2 + 1 \rangle$, there does not exist a real algebraic set V such that $\mathbb{I}(V) = I$.
(5) The map $V \mapsto \mathbb{I}(V)$ is one-to-one by Theorem 3.2 (1).

Theorem 3.3 *Let V be a real algebraic set and I be an ideal.*
(1) $\mathbb{V}(\mathbb{I}(V)) = V$.
(2) $I \subset \sqrt{I} \subset \mathbb{I}(\mathbb{V}(I))$.

Proof of Theorem 3.3 (1) Firstly, we prove $V \subset \mathbb{V}(\mathbb{I}(V))$. Let $x \in V$. Then $f(x) = 0$ for all $f \in \mathbb{I}(V)$, which is equivalent to $x \in \mathbb{V}(\mathbb{I}(V))$. Secondly, we prove $V \supset \mathbb{V}(\mathbb{I}(V))$. Since V is a real algebraic set, there exist polynomials $f_1, f_2, \ldots, f_k$ such that $V = \mathbb{V}(\langle f_1, f_2, \ldots, f_k \rangle)$. Then $f_1, f_2, \ldots, f_k \in \mathbb{I}(V)$, thus $\langle f_1, f_2, \ldots, f_k \rangle \subset \mathbb{I}(V)$. By using Theorem 3.2(2), $V = \mathbb{V}(\langle f_1, f_2, \ldots, f_k \rangle) \supset \mathbb{V}(\mathbb{I}(V))$.
(2) By Remark 3.3, $I \subset \sqrt{I}$. Let us prove $\sqrt{I} \subset \mathbb{I}(\mathbb{V}(I))$. If $f(x) \in \sqrt{I}$, then there exists m such that $f(x)^m \in I$. Therefore $f(x) = 0$ for any $x \in \mathbb{V}(I)$, which is equivalent to $f(x) \in \mathbb{I}(\mathbb{V}(I))$. □

Example 3.5 In $\mathbb{R}[x, y]$, ideals

$$I_1 = \langle x^2 y^4 \rangle, \quad I_2 = \langle (x+1)^2, y^3 z^2 \rangle, \quad I_3 = \langle x^4 + 2x^2 + 1 \rangle$$

are not radical ideals. Radicals of them are respectively

$$\sqrt{I_1} = \langle xy \rangle, \quad \sqrt{I_2} = \langle x + 1, yz \rangle, \quad \sqrt{I_3} = \langle x^2 + 1 \rangle.$$

The real algebraic sets are respectively

$$\begin{aligned}
\mathbb{V}(I_1) &= \mathbb{V}(xy) = \mathbb{V}(x) \cup \mathbb{V}(y), \\
\mathbb{V}(I_2) &= \mathbb{V}(x+1) \cap \{\mathbb{V}(y) \cup \mathbb{V}(z)\}, \\
\mathbb{V}(I_3) &= \emptyset,
\end{aligned}$$

whereas corresponding ideals are respectively given by

$$\begin{aligned}
\mathbb{I}(\mathbb{V}(I_1)) &= \langle xy \rangle, \\
\mathbb{I}(\mathbb{V}(I_2)) &= \langle x+1, yz \rangle, \\
\mathbb{I}(\mathbb{V}(I_3)) &= \langle 1 \rangle = \mathbb{R}[x, y].
\end{aligned}$$

In $\mathbb{R}[x, y, z, w]$

$$I = \langle (xy + zw)^2 + (xy^3 + zw^3)^2 \rangle$$

is a radical ideal.

$$\mathbb{I}(\mathbb{V}(I)) = \langle xy + zw, xy^3 + zw^3 \rangle \neq \sqrt{I} = I.$$

Remark 3.4 (Hilbert's Nullstellensatz)
(1) In this book, we study mainly the polynomial ring with real coefficients $\mathbb{R}[x_1, x_2, \ldots, x_d]$ and the real algebraic set contained in $\mathbb{R}^d$. If we have the polynomial ring with complex coefficients and the complex algebraic set contained in $\mathbb{C}^d$, Hilbert's Nullstellensatz

$$\mathbb{I}(\mathbb{V}(I)) = \sqrt{I}$$

holds. This equation is equivalent to the proposition that the map $V \mapsto \mathbb{I}(V)$ is surjective onto the set of radical ideals.
(2) Hilbert's Nullstellensatz holds for any polynomial ring with coefficients of an algebraic closed field. If Hilbert's Nullstellensatz holds, then the correspondence

$$\text{complex algebraic set} \mapsto \text{radical ideal}$$

is one-to-one and onto, giving the result that an algebraic set can be identified with a radical ideal.
(3) We study mainly real algebraic geometry, therefore Hilbert's Nullstellensatz does not hold. However, there are some properties between ideals and real algebraic sets.

Definition 3.5 (Prime ideal and irreducible set)
(1) An ideal I in $\mathcal{R} = \mathbb{R}[x_1, x_2, \ldots, x_d]$ is called a prime ideal if $I \neq \mathcal{R}$ and if $f(x)g(x) \in I \Longrightarrow f(x) \in I$ or $g(x) \in I$.

(2) A real algebraic set V in $\mathbb{R}^d$ is said to be irreducible if

$$V = V_1 \cup V_2 \Longrightarrow V = V_1 \text{ or } V = V_2.$$

Remark 3.5 (1) A real algebraic set V is irreducible if and only if $\mathbb{I}(V)$ is a prime ideal.
(2) For an arbitrary real algebraic set V, there exist irreducible real algebraic sets $V_1, \ldots, V_k$ such that

$$V = V_1 \cup V_2 \cup \cdots \cup V_k.$$

Example 3.6 (1) In $\mathbb{R}[x, y]$, $\langle x \rangle$ is a prime ideal. $\langle x^3 - y^2 \rangle$ is also a prime ideal. $\langle xy \rangle$ is not a prime ideal. $\langle x^2 \rangle$ is not a prime ideal. $\langle x^2 + y^2, xy \rangle$ is not a prime ideal, because $(x + y)^2 \in \langle x^2 + y^2, xy \rangle$.
(2) For a real algebraic set $V = \mathbb{V}(\langle (ab + c)^2 + c^4 \rangle)$,

$$V_1 = \mathbb{V}(\langle a, c \rangle),$$
$$V_2 = \mathbb{V}(\langle b, c \rangle).$$

it follows that

$$V = V_1 \cup V_2,$$

hence V is not irreducible. Note that $\langle (ab + c)^2 + c^4 \rangle$ is a prime ideal, whereas $\mathbb{I}(V) = \langle ab, c \rangle$ is not a prime ideal.

Definition 3.6 (Maximal ideal) An ideal I in $\mathbb{R}[x_1, x_2, \ldots, x_d]$ is said to be a maximal ideal if $I \neq \mathcal{R}$ and, for any ideal J,

$$I \subset J \Longrightarrow I = J \text{ or } J = \mathcal{R}.$$

Example 3.7 In $\mathbb{R}[x, y, z]$, the following results hold.
(1) $\langle x - a, y - b, z - c \rangle$ is a maximal ideal for arbitrary a, b, c.
(2) $\langle x^2 + y^2 + z^2 \rangle$ is not a maximal ideal, because it is contained in $\langle x, y, z \rangle$.
(3) $\langle x^2 + 1, y, z \rangle$ is a maximal ideal.
(4) In the polynomial ring $\mathbb{C}[x_1, x_2, \ldots, x_d]$ with complex coefficients, an ideal I is a maximal ideal if and only if

$$I = \langle x_1 - a_1, x_2 - a_2, \ldots, x_d - a_d \rangle$$

for some $(a_1, a_2, \ldots, a_d) \in \mathbb{C}^d$.

Remark 3.6 (1) A maximal ideal is a prime ideal.
(2) A prime ideal is a radical ideal.

Remark 3.7 A polynomial f is said to be irreducible if

$$f(x) = g(x)h(x)$$

implies $f(x) \propto g(x)$ or $f(x) \propto h(x)$. Any polynomial $f(x)$ can be represented by

$$f(x) = \prod_{i=1}^{k} f_i(x)^{a_i}$$

using irreducible polynomials $f_1, \ldots, f_k$, where $\{a_i\}$ is a set of natural numbers. For an arbitrary polynomial $f(x)$, $f_i(x)$ and a_i are determined uniquely without trivial multiplication. By using this representation,

$$\sqrt{\langle f \rangle} = \langle f_1 f_2 \cdots f_k \rangle$$

holds. If the coefficient of the ring is an algebraically closed field, then the condition that a polynomial f is irreducible is equivalent to the condition that the real algebraic set $\mathbb{V}(\langle f \rangle)$ is irreducible. However, if the coefficient is not an algebraic closed set, then its equivalence does not hold; see Example 3.6 (2). In order to check whether a real algebraic set is irreducible or not, $\mathbb{I}(V)$ should be studied. In real algebraic geometry, the geometry of $V = \mathbb{V}(I)$ corresponds not to $\sqrt{I}$ but to $\mathbb{I}(V)$.

Definition 3.7 (Coordinate ring) Let V be a real algebraic set in $\mathbb{R}^d$. The polynomial ring restricted on V is called a coordinate ring of V, which is denoted by $\mathbb{R}[V]$.

Example 3.8 Let $V \subset \mathbb{R}^2$ be

$$V = \mathbb{V}(x^2 - y^3).$$

Then, in $\mathbb{R}[V]$,

$$x^3 + y^3 = x^3 - x^2 = y^3(x - 1).$$

Remark 3.8 In the polynomial ring $\mathbb{R}[x_1, x_2, \ldots, x_d]$, we introduce an equivalence relation

$$f \sim g \Longleftrightarrow f(x) = g(x) \quad (x \in V).$$

Then the quotient set $\mathcal{R}/_\sim$ is equivalent to the coordinate ring. If $\mathbb{R}[V] \ni f(x)g(x) = 0 \Longrightarrow f(x) = 0$ or $g(x) = 0$ then $\mathbb{R}[V]$ is said to be an integral domain. The ideal $\mathbb{I}(V)$ is a prime ideal if and only if $\mathbb{R}[V]$ is an integral domain.

3.3 Singularities and dimension

In this section, we study a necessary and sufficient condition of a singularity in a real algebraic set.

Definition 3.8 (Dimension of a real algebraic set) Let V be a nonempty real algebraic set in $\mathbb{R}^d$. Assume that polynomials $f_1, f_2, \ldots, f_r$ satisfy

$$\mathbb{I}(V) = \langle f_1, f_2, \ldots, f_r \rangle.$$

The Jacobian matrix is defined as

$$J(x) = \begin{pmatrix} \frac{\partial f_1(x)}{\partial x_1} & \cdots & \frac{\partial f_1(x)}{\partial x_d} \\ \vdots & \ddots & \vdots \\ \frac{\partial f_r(x)}{\partial x_1} & \cdots & \frac{\partial f_r(x)}{\partial x_d} \end{pmatrix}$$

The maximum value of the rank of the Jacobian matrix,

$$d_0 = \max_{x \in V} \operatorname{rank} J(x),$$

is said to be the dimension of the real algebraic set V.

Theorem 3.4 *Let V be a nonempty real algebraic set in $\mathbb{R}^d$ whose dimension is equal to d_0. Then $x \in V$ is a nonsingular point of V if and only if $\operatorname{rank} J(x) = d_0$. In other words, $x \in V$ is a singularity if and only if $\operatorname{rank} J(x) < d_0$.*

(Explanation of Theorem 3.4) For the proof of this theorem, see [21]. In this book, the definition of a singularity is given in Definition 2.6. Therefore the above statement is a theorem. In some books, the above statement is used as a definition of singularity and Definition 2.6 is a theorem. The condition $\operatorname{rank} J(x) < d_0$ holds if and only if all minor determinants of $d_0 \times d_0$ are equal to zero. Since the elements of a Jacobian matrix are polynomials, all minor determinants are also polynomials. Therefore the set of singularities Sing(V) is a real algebraic subset of V, and Sing(V) is not equal to V. That is to say, $\mathrm{Sing}(V) \subset V$ and $\mathrm{Sing}(V) \neq V$.

Example 3.9 (1) In $\mathbb{R}^3$, let us study a real algebraic set

$$V = \mathbb{V}(f, g),$$

where $f(x, y, z) = x^2 - y$, $g(x, y, z) = x^3 - z^2$. The Jacobian matrix is given by

$$J(x) = \begin{pmatrix} 2x & -1 & 0 \\ 3x^2 & 0 & -2z \end{pmatrix}.$$

Therefore,

$$\mathrm{rank} J(x) = \begin{cases} 1 & (x = (0,0,0)) \\ 2 & (x \neq (0,0,0)). \end{cases}$$

Hence the set of singularities is $\mathrm{Sing}(V) = \{(0,0,0)\}$.
(2) In $\mathbb{R}^3$

$$V = \mathbb{V}(f, g),$$

where $f(x, y, z) = x^2 - 2xy + z$ and $g(x, y, z) = y^2 - z$. Then

$$\mathbb{I}(V) = \langle x - y, y^2 - z \rangle,$$

where we used $f(x, y, z) + g(x, y, z) = (x - y)^2$. Therefore

$$J(x) = \begin{pmatrix} 1 & -1 & 0 \\ 0 & 2y & -1 \end{pmatrix}.$$

Hence $\mathrm{rank} J(x) = 2$ for arbitrary x, which means that V does not contain a singularity.

3.4 Real projective space

Let $\mathbb{R}^{d+1}$ be the real Euclidean space and O be the origin. We introduce an equivalence relation $\sim$ to the set $\mathbb{R}^{d+1} \setminus O$.

$$(x_0, x_1, x_2, \ldots, x_d) \sim (y_0, y_1, y_2, \ldots, y_d)$$

if and only if

$$(\exists c \neq 0) \quad (x_0, x_1, x_2, \ldots, x_d) = c\,(y_0, y_1, y_2, \ldots, y_d).$$

The quotient set is said to be a d-dimensional projective space,

$$\mathbb{P}^d = \mathbb{R}^{d+1}/\sim.$$

The equivalence class that contains $(x_0, x_1, x_2, \ldots, x_d) \in \mathbb{R}^{d+1} \setminus O$ is denoted by

$$(x_0 : x_1 : x_2 : \cdots : x_d).$$

By the definition,

$$(x_0; x_1 : x_2 : \cdots : x_d) = (cx_0 : cx_1 : cx_2 : \cdots : cx_d)$$

holds for $c \neq 0$. The subsets U_k ($k = 0, 1, 2, \ldots, d$) of $\mathbb{P}^d$ are defined by

$$U_0 = \{(1 : x_1 : x_2 : \cdots : x_d); x_1, x_2, \ldots, x_d, \in \mathbb{R}\}, \tag{3.1}$$

$$U_1 = \{(x_0 : 1 : x_2 : \cdots : x_d); x_0, x_2, \ldots, x_d \in \mathbb{R}\}, \tag{3.2}$$

$$\vdots \tag{3.3}$$

$$U_d = \{(x_0 : x_1 : \cdots : 1); x_0, x_1, \ldots, x_{d-1} \in \mathbb{R}\}. \tag{3.4}$$

Then

$$\mathbb{P}^d = U_0 \cup U_1 \cup \cdots \cup U_d$$

holds. Since U_k is analytically isomorphic to Euclidean space $\mathbb{R}^d$, $\mathbb{P}^d$ is a d-dimensional manifold and U_k is a local coordinate.

Example 3.10 (1) One-dimensional projective space $\mathbb{P}^1$ is

$$\mathbb{P}^1 = \{(x : y) \; ; \; (x, y) \neq (0, 0)\},$$

which can be rewritten as

$$\begin{aligned} \mathbb{P}^1 &= \{(1 : a); a \in \mathbb{R}\} \cup \{(0 : 1)\} \\ &\cong \mathbb{R}^1 \cup \mathbb{R}^0, \end{aligned}$$

where $A \cong B$ means that there is a one-to-one and onto map between A and B.
(2) Two-dimensional projective space $\mathbb{P}^2$ is

$$\begin{aligned} \mathbb{P}^2 &= \{(1 : a : b); (a, b) \in \mathbb{R}^2\} \cup \{(0 : a : b); (a : b) \in \mathbb{P}^1\} \\ &\cong \mathbb{R}^2 \cup \mathbb{P}^1 \cong \mathbb{R}^2 \cup \mathbb{R}^1 \cup \mathbb{R}^0. \end{aligned}$$

(3) In the same way,

$$\begin{aligned} \mathbb{P}^d &= \{(1 : x_1 : \cdots : x_d); (x_1, \cdots, x_d) \in \mathbb{R}^d\} \\ &\quad \cup \{(0 : x_1 : \cdots : x_d); (x_1 : \cdots : x_d) \in \mathbb{P}^{d-1}\} \\ &\cong \mathbb{R}^d \cup \mathbb{R}^{d-1} \cup \cdots \cup \mathbb{R}^0. \end{aligned}$$

Remark 3.9 The real projective space $\mathbb{P}^d$ is orientable if and only if d is an odd number.

Definition 3.9 (Homogeneous ideal) For a multi-index $\alpha = (\alpha_0, \alpha_1, \ldots, \alpha_d)$, $|\alpha| = \alpha_0 + \alpha_1 + \cdots + \alpha_d$. If a polynomial $f(x) \in \mathbb{R}[x_0, x_1, \ldots, x_d]$ is given by

$$f(x) = \sum_{|\alpha|=n} a_\alpha x^\alpha,$$

then $f(x)$ is said to be a homogeneous polynomial of degree n. If an ideal $I \subset \mathbb{R}[x_0, x_1, \dots, x_d]$ is generated by homogeneous polynomials $f_1(x), f_2(x), \dots, f_k(x)$, so that

$$I = \langle f_1, f_2, \dots, f_k \rangle,$$

then I is called a homogeneous ideal.

Remark 3.10 (1) In $\mathbb{R}[x, y, z]$, both $x^2y + z^3$ and $x + y$ are homogeneous polynomials, hence $I = \langle x^2y + z^3, x + y \rangle$ is a homogeneous ideal. The ideal contains polynomials that are not homogeneous, for example, $x^2y + z^3 + x(x + y)$.
(2) For a given polynomial

$$f(x) = \sum_{\alpha} a_\alpha x^\alpha,$$

its homogeneous part of degree n is

$$\deg_n(f)(x) = \sum_{|\alpha|=n} a_\alpha x^\alpha.$$

An ideal I is a homogenous ideal if and only if, for an arbitrary $f \in I$, its homogeneous part of any degree is contained in I.
(3) For a homogenous ideal I, its radical $\sqrt{I}$ is homogeneous.

Definition 3.10 (Real projective variety) If a subset $V \subset \mathbb{P}^d$ is represented by homogeneous polynomials $f_1(x), f_2(x), \dots, f_k(x) \in \mathbb{R}[x_0, x_2, \dots, x_d]$,

$$V = \{(x_0 : x_1 : \cdots : x_d) \in \mathbb{P}^d; f_1(x) = f_2 = \cdots = f_k(x) = 0\},$$

then V is said to be a real projective variety. This set V is written

$$V = \mathbb{V}(f_1, f_2, \dots, f_k).$$

Since $f_1, \dots, f_k$ are homogeneous polynomials, V is a well-defined subset in the real projective set. Also, for a given homogeneous ideal I,

$$\mathbb{V}(I) = \{(x_0 : x_1 : \cdots : x_d) \in \mathbb{P}^d; \deg_n(f) = 0, \ (\forall f \in I, \ \forall n \geq 1)\}.$$

Example 3.11 Let $V = \mathbb{V}(x^3 + xyw + w^3) \subset \mathbb{P}^3$ be a real projective variety. By definition,

$$V = \{(x : y : z : w) \in \mathbb{P}^3; x^3 + xyw + w^3 = 0\}.$$

In the local coordinate U_0 in eq.(3.1),

$$V \cap U_0 = \{(1 : y : z : w); 1 + yw + w^3 = 0\}.$$

In the same way,

$$V \cap U_1 = \{(x : 1 : z : w); x^3 + xw + w^3 = 0\}.$$

In each local coordinate, a real projective variety is defined by non-homogeneous polynomials. On the other hand, a real projective variety is determined by one of its local coordinates. For example, if a real projective variety V satisfies

$$V \cap U_0 = \{(1 : y : z : w); y^3 + yz + w^3 + 2 = 0\},$$

then

$$\begin{aligned} V &= \left\{(x : y : z : w); \left(\frac{y}{x}\right)^3 + \frac{y}{x} \cdot \frac{z}{x} + \left(\frac{w}{x}\right)^3 + 2 = 0\right\} \\ &= \{(x : y : z : w); y^3 + xyz + w^3 + 2x^3 = 0\}, \end{aligned}$$

which is defined by a homogeneous polynomial.

Definition 3.11 For a real projective variety $V \subset \mathbb{P}^d$, its defining ideal is defined by

$$\mathbb{I}(V) = \{f(x) \in \mathbb{R}[x_0, x_1, \ldots, x_d]; f(x) = 0 \ (\forall x \in V)\}.$$

By definition, $\mathbb{I}(V)$ is a homogeneous ideal.

Remark 3.11 Between real projective varieties V and a homogenous ideal I, the following relations hold. Let V_1 and V_2 be real projective varieties and I_1 and I_2 be homogeneous ideals.
(1) $V_1 \subset V_2 \Longleftrightarrow \mathbb{I}(V_1) \supset \mathbb{I}(V_2)$.
(2) $I_1 \subset I_2 \Longrightarrow \mathbb{V}(I_1) \supset \mathbb{V}(I_2)$.
(3) $V = \mathbb{V}(\mathbb{I}(V)) = V$.
(4) $\sqrt{I} \subset \mathbb{I}(\mathbb{V}(I))$.

Remark 3.12 (Zariski topology) In real Euclidean space $\mathbb{R}^d$, the topology is usually determined by the Euclidean norm. Since the real projective space $\mathbb{P}^d$ is the manifold whose local coordinate is analytically isomorphic to $\mathbb{R}^d$, its topology is determined from $\mathbb{R}^d$. However, in algebraic geometry, the other topology is employed. The Zarisiki topology of $\mathbb{R}^d$ is determined by the condition that the family of all real algebraic sets is equal to the family of all closed sets. Also the Zariski topology of $\mathbb{P}^d$ is determined in the same way. Note that a closed set by Zariski topology is also a closed set by Euclidean topology, but that a closed set by Euclidean topology may not be a closed set by Zariski topology. The closure of a set A by Zariski topology is the smallest real algebraic set that contains A. An open set in $\mathbb{P}^d$ is called a quasiprojective variety. For example, local coordinates $U_0, U_1, \ldots, U_d$ of $\mathbb{P}^d$ are quasiprojective varieties.

3.5 Blow-up

In this section, we introduce a blow-up. Intuitively speaking, a blow-up reduces the complexity of a singularity. Any singularities can be desingularized by finite recursive blow-ups.

Definition 3.12 (Blow-up of Euclidean space) Let $\mathbb{R}^d$ be a real Euclidean space, and r be an integer which satisfies $2 \leq r \leq d$. Let V be a real algebraic set,

$$V = \{x = (x_1, x_2, \ldots, x_d) \in \mathbb{R}^d; x_1 = x_2 = \cdots = x_r = 0\}.$$

The blow-up of $\mathbb{R}^d$ with center V is defined by

$$B_V(\mathbb{R}^d) \equiv \overline{\{(x, (x_1 : x_2 : \cdots : x_r)) \in \mathbb{R}^d \times \mathbb{P}^{r-1} \; ; \; x \in \mathbb{R}^d \setminus V\}}$$

where $\overline{A}$ shows the closure of a set A in $\mathbb{R}^d \times \mathbb{P}^{r-1}$. Here the topology of $\mathbb{R}^d \times \mathbb{P}^{r-1}$ is the direct product of both sets. In this definition, closures by both Zariski and Euclidean topologies result in the same set.

Remark 3.13 This definition represents the procedure of the blow-up.
(1) Firstly, we prepare $\mathbb{R}^d \setminus V$, in other words, the center V is removed from $\mathbb{R}^d$.
(2) Secondly, a point in real projective space $(x_1 : x_2 : \cdots : x_r) \in \mathbb{P}^{r-1}$ can be defined for $x = (x_1, x_2, \ldots, x_d) \in \mathbb{R}^d \setminus V$.
(3) Thirdly, the pair $(x, (x_1 : x_2 : \ldots, x_d))$ is collected,

$$A = \{(x, (x_1 : x_2 : \cdots : x_r)) \; ; \; x \in \mathbb{R}^d \setminus V\}.$$

This set is a subset of $\mathbb{R}^d \times \mathbb{P}^{r-1}$.
(4) And lastly, the closure of A in $\mathbb{R}^d \times \mathbb{P}^{r-1}$ gives the blow-up. By taking the closure, a set

$$\{(x, y); x \in V, y \in \mathbb{P}^{r-1}\}$$

is added to A.

Remark 3.14 Let us define a map

$$\pi : B_V(\mathbb{R}^d) \to \mathbb{R}^d$$

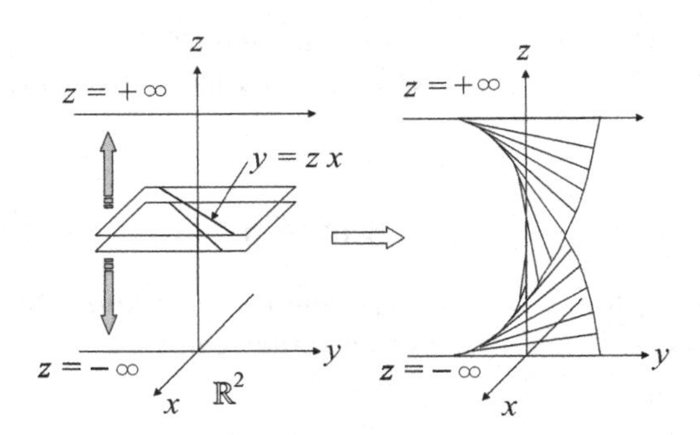

Fig. 3.1. Blow-up of $\mathbb{R}^2$

by

$$\pi((x, y)) = x.$$

Then

$$\pi : \{(x, (x_1 : x_2 : \cdots : x_r)); x \in \mathbb{R}^d \setminus V\} \to \mathbb{R}^d \setminus V$$

is a one-to-one and onto map. Moreover

$$\pi^{-1}(V) = \mathbb{P}^{r-1}$$

holds.

Example 3.12 To understand what a blow-up is, the following example is very helpful. The blow-up of $\mathbb{R}^2$ with center $V = \mathbb{V}(x, y) = \{(0, 0)\}$ is illustrated in Figure 3.1. Firstly, $(x : y) \in \mathbb{P}^1$ can be defined for $(x, y) \notin V = \{(0, 0)\}$. Secondly, the set

$$A = \{(x, y, (x : y)) \; ; \; (x, y) \notin V\} \subset \mathbb{R}^2 \times \mathbb{P}^1$$

is introduced. By using a notation $(x : y) = (1 : z) \in \mathbb{P}^1$

$$z = \begin{cases} y/x & (x \neq 0) \\ \infty & (x = 0), \end{cases} \tag{3.5}$$

where $\infty = (0 : 1)$, the set A is rewritten as

$$A = \{(x, y, z) \; ; \; y = zx \;\; (x, y) \neq (0, 0)\} \subset \mathbb{R}^2 \times \mathbb{P}^1,$$

where, if $z = \infty$, then $y = zx$ is replaced by $x = 0$. Lastly the blow-up $B_V(\mathbb{R}^d)$ is obtained by its closure,

$$B_V(\mathbb{R}^2) = \{(x, y, z) \; ; \; y = zx\} \subset \mathbb{R}^2 \times \mathbb{P}^1,$$

which is a Möbius' strip. In this case, taking closure is equivalent to removing the condition $(x, y) \neq (0, 0)$. The projection map is given by

$$\pi : (x, y, z) \mapsto (x, y),$$

therefore

$$\pi^{-1}(\{(0, 0)\}) = \{(0, 0)\} \times \mathbb{P}^1.$$

Definition 3.13 (Blow-up of a real algebraic set) Let $\mathbb{R}^d$ be a d-dimensional real Euclidean space. Let r be an integer which satisfies $2 \leq r \leq d$.

$$V = \{x \in \mathbb{R}^d; x_1 = x_2 = \cdots = x_r = 0\}.$$

Let W be a real algebraic set such that $V \subset W$. The blow-up of W with center V is defined by

$$B_V(W) \equiv \overline{\{(x, (x_1 : x_2 : \cdots : x_r)); x \in W \setminus V\}}.$$

Remark 3.15 (Strict and total transform, exceptional set) Let π be a map defined in Remark 3.14. Then

$$B_V(W) \subset \pi^{-1}(W)$$

holds, but $B_V(W) \neq \pi^{-1}(W)$. The set $B_V(W)$ is said to be a strict transform of W, whereas $\pi^{-1}(W)$ is a total transform.

$$\overline{\pi^{-1}(W) \setminus B_V(W)}$$

is called an exceptional set.

Example 3.13 Let a real algebraic set in $\mathbb{R}^2$ be

$$W = \mathbb{V}(x^3 - y^2).$$

In Figure 3.2, we illustrate the blow-up of W with center

$$V = \mathbb{V}(x, y) = \{(0, 0)\}.$$

Firstly, $V = \{(0, 0)\}$ is removed from W. Secondly, the subset A in $\mathbb{R}^2 \times \mathbb{P}^1$ is defined by

$$\begin{aligned} A &= \{(x, y, (x : y)) \; ; \; (x, y) \in W \setminus V\} \\ &= \{(x, y, (x : y)) \; ; \; x^3 - y^2 = 0, \; (x, y) \neq (0, 0)\}. \end{aligned}$$

By using the same notation $(1 : z) \in \mathbb{P}^1$ as in eq.(3.5), the set A is rewritten as

$$A = \{(x, y, z) \; ; \; x^3 - y^2 = 0, \; y = zx, \; (x, y) \neq (0, 0)\}.$$

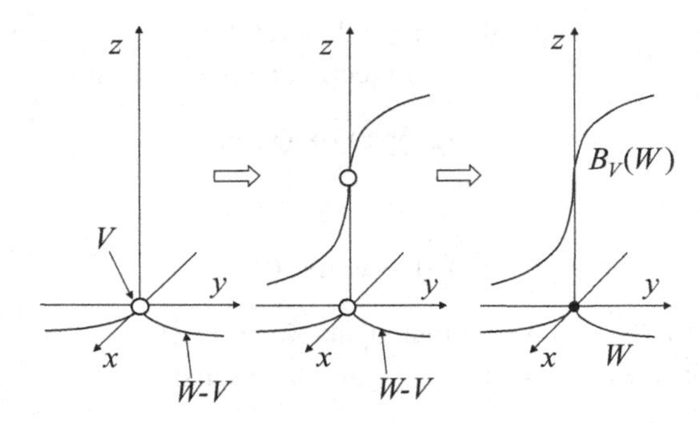

Fig. 3.2. Blow-up of a real algebraic set

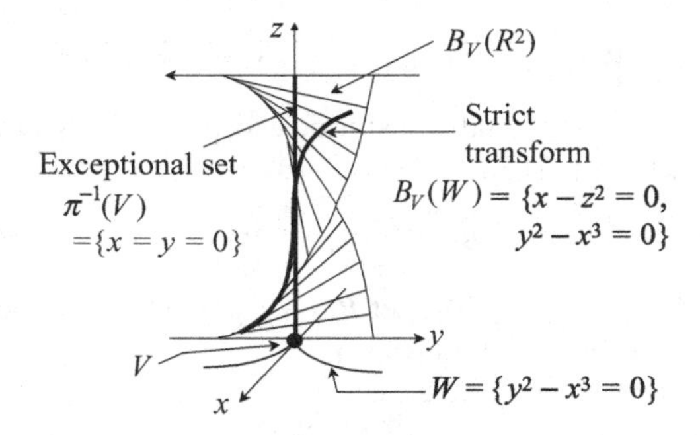

Fig. 3.3. Blow-up of $\mathbb{V}(x^3 - y^2)$

On the set A,

$$x^3 - y^2 = x^2(x - z^2).$$

By using $x \neq 0$, the set A is equal to

$$A = \{(x, y, z) \; ; \; x = z^2, \;\; y = zx, \;\; (x, y) \neq (0, 0)\}.$$

Lastly, its closure is $B_V(W)$:

$$B_V(W) = \overline{A} = \{(x, y, z) \; ; \; x = z^2, \;\; y = zx\}.$$

This set is a strict transform. In this example, W contains singularities, whereas $B_V(W)$ does not contain a singularity. The real algebraic set W which contains a singularity is the image of a nonsingular real algebraic set $B_V(W)$. Such $B_V(W)$ is called a resolution of singularity of W. Figure 3.3 shows the strict transform, the exceptional set, and the total transform. The projection is defined by

$$\pi : B_V(W) \mapsto W.$$

The exceptional set is

$$\pi^{-1}(\{(0,0)\}) = \{(0,0)\} \times \mathbb{P}^1,$$

and the total transform is

$$\pi^{-1}(W) = B_V(W) \cup \pi^{-1}(\{(0,0)\}).$$

Remark 3.16 Let us study the blow-up from an algebraic point of view.
(1) The blow-up of $\mathbb{R}^2$ with center $V = \mathbb{V}(x, y)$ is equivalent to the substitution

$$x = u = st,$$
$$y = uv = s.$$

The blow-up $B_V(\mathbb{R}^2)$ is made by gluing two coordinates (u, v) and (s, t).
(2) In the same way, the blow-up of $\mathbb{R}^3$ with center $V = \mathbb{V}(x, y, z)$ is equivalent to substitution

$$x = x_1 = x_2 y_2 = x_3 z_3,$$
$$y = y_1 x_1 = y_2 = y_3 z_3,$$
$$z = z_1 x_1 = z_2 y_2 = z_3.$$

The blow-up $B_V(\mathbb{R}^3)$ is made by gluing three coordinates.
(3) The blow-up of $\mathbb{R}^3$ with center $V = \mathbb{V}(x, y)$ is equivalent to the substitution

$$x = x_1 = x_2 y_2,$$
$$y = y_1 x_1 = y_2,$$
$$z = z_1 = z_2.$$

The blow-up $B_V(\mathbb{R}^3)$ is made by gluing two coordinates.

Example 3.14 Figure 3.4 shows a blow-up of a real algebraic set V

$$V = \mathbb{V}(x^4 - x^2 y + y^3)$$

in $\mathbb{R}^2$. The singularity of V is the origin O. The blow-up $g : P^2 \to \mathbb{R}^2$ with center O can be represented by using local coordinates

$$x = u = st,$$
$$y = uw = s.$$

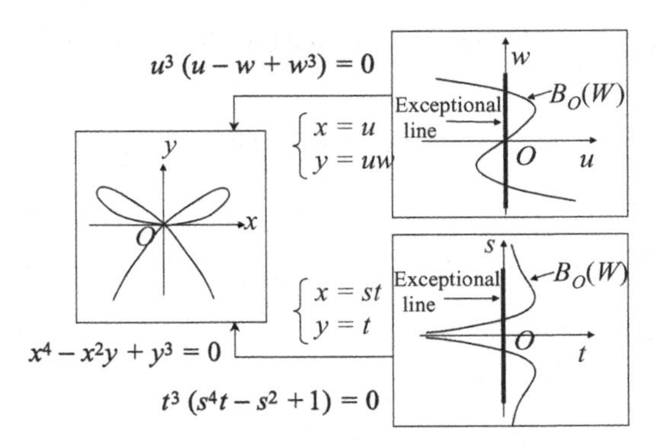

Fig. 3.4. Blow-up using local coordinates

The projective variety $g^{-1}(V)$ is represented on each local coordinate,

$$u^3(u - w + w^3) = 0,$$
$$s^3(st^4 - t^2 + 1) = 0,$$

whose singularities are all normal crossing. The blow-up $B_O(V)$ is given by

$$u - w + w^3 = 0,$$
$$st^4 - t^2 + 1 = 0,$$

which is a nonsingular real algebraic set.

Definition 3.14 (General blow-up in Euclidean space) Let $\mathbb{R}^d$ be a real Euclidean space and r be an integer which satisfies $2 \leq r \leq d$. Assume that both V and W $(V \subset W)$ are real algebraic sets in $\mathbb{R}^d$. Let $f_1, f_2, \ldots, f_r$ be a set of polynomials which satisfy

$$\mathbb{I}(V) = \langle f_1, f_2, \ldots, f_r \rangle.$$

The blow-up of W with center V, $B_V(W)$, is defined by

$$B_V(W) \equiv \overline{\{(x, (f_1 : f_2 : \cdots : f_r)); x \in W \setminus V\}}.$$

If V does not contain a singularity, then $B_V(\mathbb{R}^d)$ is an analytic manifold.

Remark 3.17 (1) Let V and W be real projective varieties which satisfy $V \subset W \subset \mathbb{P}^d$. In each local coordinate $U_0, U_1, \ldots, U_d$ of $\mathbb{P}^d$, real algebraic sets are given by

$$V_i = U_i \cap V,$$
$$W_i = U_i \cap W.$$

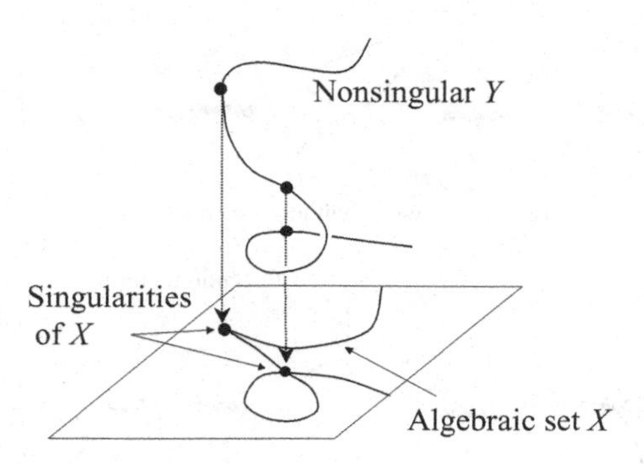

Fig. 3.5. Resolution of singularities

Then for each $0 \leq i \leq d$, the blow-up of W_i with center V_i is defined by $B_{V_i}(W_i)$, which is a real projective variety. The blow-up of projective variety W with center V is defined by gluing all real projective varieties $\{B_{V_i}(W_i)\}$. Such a set made by gluing real projective varieties is called a real algebraic variety. A real algebraic set and a real projective variety are special examples of a real algebraic variety. The blow-up of a real algebraic variety can be defined in the same way, and is also a real algebraic variety. Hence a real algebraic variety is a closed concept by blow-ups.
(2) There is a more abstract definition of algebraic variety. In modern mathematics, a property of a set is determined by the family of all functions with the desired property on the set. By using this relation, a set is sometimes defined by the family of all functions on the set.

Now, let us introduce two processes of desingularization. The first is the following theorem.

Theorem 3.5 (Hironaka's theorem, I) *For an arbitrary real algebraic set V, there is a sequence of real algebraic varieties $V_0, V_1, V_2, \ldots, V_n$ which satisfies the following conditions.*
(1) $V = V_0$.
(2) V_n is nonsingular.
(3) For $i = 1, 2, \ldots, n$, $V_i = B_{C_{i-1}}(V_{i-1})$, where V_i is a blow-up of V_{i-1} with center C_i.
(4) C_i is a nonsingular real algebraic variety which is contained in $\mathrm{Sing}(V_{i-1})$.

(Explanation of Theorem 3.5) This theorem is proved by Hironaka [40]. By this theorem, any real algebraic set can be understood as an image of a nonsingular real algebraic variety. Figure 3.5 shows the result of this theorem.

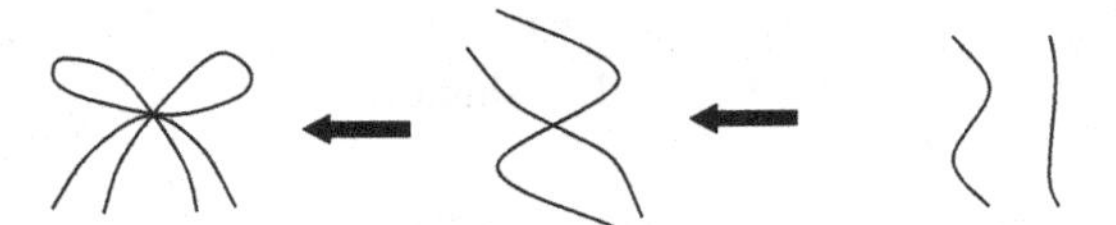

Recursive blow-ups without exceptional sets

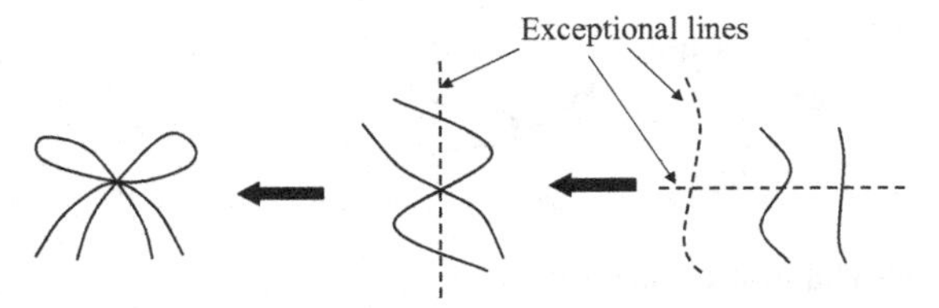

Recursive blow-ups with exceptional sets

Fig. 3.6. Two processes of desingularization

Theorem 3.6 (Hironaka's theorem, II) *Let $f(x)$ be an arbitrary polynomial in $\mathbb{R}[x_1, x_2, \ldots, x_d]$. There is a sequence of pairs of real algebraic varieties $(V_0, W_0), (V_1, W_1), \ldots, (V_n, W_n)$ which satisfies the following conditions.*

(1) $V_i \subset W_i$ $(i = 1, 2, \ldots, n)$.

(2) $V_0 = \mathbb{V}(f)$, $W_0 = \mathbb{R}^d$.

(3) $\{W_i; i = 0, 1, 2, \ldots, n\}$ are nonsingular algebraic varieties.

(4) V_n is defined by a normal crossing polynomial on each local coordinate of W_n.

(5) For $i = 1, 2, \ldots, n$, $W_i = B_{C_{i-1}}(W_{i-1})$, where W_i is a blow-up of W_{i-1} with center C_i.

(6) Let $\pi_i : W_i \to W_{i-1}$ be a projection map defined in the blow-up $B_{C_{i-1}}(W_{i-1})$. Then V_i is the total transform of π_i, $V_i = \pi_i^{-1}(V_{i-1})$.

(7) The center C_i of each blow-up is a nonsingular real algebraic variety which is contained in the set of critical points of $f \circ \pi_1 \circ \pi_2 \circ \cdots \circ \pi_i$.

(Explanation of Theorem 3.6) This theorem is also proved by Hironaka [40]. This theorem shows that there exists a recursive procedure by which the resolution of singularities in Theorem 2.3 is attained. Note that Theorem 3.5 does not contain exceptional sets in the blow-ups, whereas this theorem does. In Theorem 3.5, any real algebraic set can be made to be an image of a nonsingular real algebraic variety because exceptional sets are removed. In Theorem 3.6, since exceptional sets are contained, any singularities of a real algebraic set are images of normal crossing singularities. Figure 3.6 shows the difference between two resolutions of singularities. See also [19, 37].

Remark 3.18 In statistical learning theory, Theorem 2.3 is needed. The map g in Theorem 2.3 can be algorithmically found by Theorem 3.6. An arbitrary polynomial can be made normal crossing by recursive blow-ups with the center of nonsingular sets in a singular locus of a previous algebraic variety. It is not easy to find such a process; however, in some practical statistical models, a concrete resolution map was found, which is introduced in Chapter 7.

3.6 Examples

3.6.1 Simple cases

Example 3.15 The blow-up of $\mathbb{R}^2$ with center $O = \{(0,0)\}$ is

$$\mathcal{M} = B_O(\mathbb{R}^2) = U_1 \cup U_2,$$

where each local coordinate is given by

$$U_1 = \{(x_1, y_1)\},$$
$$U_2 = \{(x_2, y_2)\},$$

which satisfy the relations

$$x = x_1 = x_2 y_2,$$
$$y = x_1 y_1 = y_2.$$

Therefore the resolution map has the form,

$$\mathbb{R}^2 \leftarrow \mathcal{M} = \begin{cases} U_1 \\ U_2 \end{cases}$$

The function $f(x,y) = x^2 + y^2$ is represented on each coordinate,

$$\begin{aligned} f &= f(x_1, x_1 y_1) = x_1^2(1 + y_1^2), \\ &= f(x_2 y_2, y_2) = y_2^2(x_2^2 + 1). \end{aligned}$$

The function f is not normal crossing on $\mathbb{R}^2$, but it is on $\mathcal{M}$.

Example 3.16 Using the same coordinates U_1 and U_2 as above, the function $f(x,y) = x^3 - y^2$ is represented on $B_O(\mathbb{R}^2)$,

$$f = f(x_1, x_1 y_1) = x_1^2(x_1 - y_1^2),$$

hence f is not normal crossing on U_1. On the other hand, on U_2,

$$f = f(x_2 y_2, y_2) = y_2^2(x_2^3 y_2 - 1)$$

is normal crossing because $x_2^2 y_2^3 - 1 = 0$ does not have singularities. The recursive blow-ups are as follows. For U_1,

$$x_1 = x_3 = x_4 y_4,$$
$$y_1 = x_3 y_3 = y_4.$$

Then, on U_3

$$f(x, y) = x_3^3 (1 - x_3 y_3^2)$$

is normal crossing. But on U_4,

$$f = x_4^2 y_4^3 (x_4 - y_4)$$

is not normal crossing. One more blow-up is needed:

$$x_4 = x_5 = x_6 y_6,$$
$$y_4 = x_5 y_5 = y_6.$$

Then

$$\begin{aligned} f &= x_5^6 y_5^4 (1 - y_5^2) \\ &= x_6^2 y_6^6 (x_6 - 1), \end{aligned}$$

which shows f is normal crossing on both U_5 and U_6. The obtained sequence of blow-ups is

$$\mathbb{R}^2 \leftarrow \begin{cases} U_1 \leftarrow \begin{cases} U_3 \\ U_4 \leftarrow \begin{cases} U_5 \\ U_6 \end{cases} \end{cases} \\ U_2 \end{cases}$$

Note that the manifold $\mathcal{M}$ made of local coordinates

$$\mathcal{M} = U_2 \cup U_3 \cup U_5 \cup U_6$$

is a real analytic manifold; in other words, it does not have singularities. The map

$$g : \mathcal{M} \to \mathbb{R}^2$$

is defined on each local coordinate,

$$x = x_2 y_2 = x_3 = x_5^2 y_5^2 = x_6 y_6^2,$$
$$y = y_2 = x_3^2 y_3 = x_5^3 y_5^2 = x_6 y_6^3,$$

which makes f normal crossing,

$$f = y_2^2 (x_2^3 y_2 - 1) = x_3^3 (1 - x_3 y_3^2) = x_5^6 y_5^4 (1 - y_5^2) = x_6^2 y_6^6 (x_6 - 1),$$

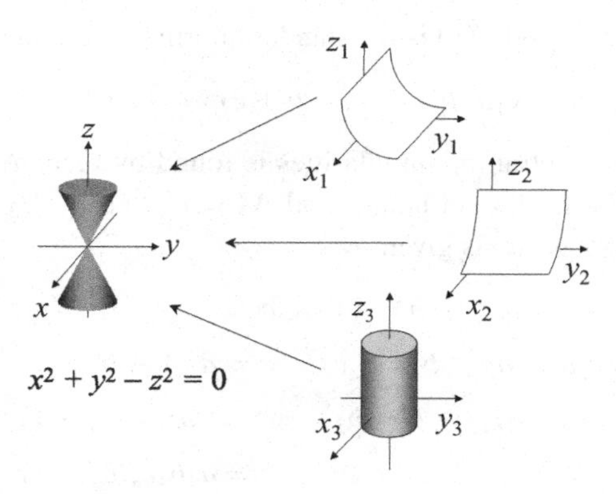

Fig. 3.7. Example of blow-up

and the Jacobian determinant is

$$|g'| = |y_2| = |x_3^2| = 2|x_4^4 y_4^3| = |x_6 y_6^4|.$$

Example 3.17 In $\mathbb{R}^3$, let us study the function in Figure 3.7:

$$f = x^2 + y^2 - z^2.$$

The blow-up with center $O = \{(0, 0, 0)\}$ is given by

$$\begin{aligned} x &= x_1 = x_2 y_2 = x_3 z_3, \\ y &= x_1 y_1 = y_2 = y_3 z_3, \\ z &= x_1 z_1 = z_2 y_2 = z_3. \end{aligned}$$

Then the function f on each coordinate is

$$\begin{aligned} f &= x_1^2(1 + y_1^2 - z_1^2) \\ &= y_2^2(x_2^2 + 1 - z_2^2) \\ &= z_3^1(x_3^2 + y_3^2 - 1), \end{aligned}$$

which is normal crossing on each coordinate.

3.6.2 A sample of a statistical model

Let us study the resolution process of statistical models.

Example 3.18 For a Kullback–Leibler distance,

$$K(a, b, c) = \frac{1}{2} \int (as(bx) + cx)^2 q(x) dx,$$

where $s(t) = t + t^2$ and $q(x)$ is the standard normal distribution. Then

$$K(a, b, c) = \frac{1}{2}(ab + c)^2 + \frac{3}{2}a^2b^4.$$

The following resolution of singularities is found by recursive blow-ups. Let U_1, U_2, U_3, U_4 be local coordinates and $\mathcal{M} = U_1 \cup U_2 \cup U_3 \cup U_4$. The resolution map $g : \mathcal{M} \to \mathbb{R}^3$ is given by

$$\begin{array}{lll} a = a_1c_1 & b = b_1 & c = c_1, \\ a = a_2 & b = b_2c_2 & c = a_2(1 - b_2)c_2, \\ a = a_3 & b = b_3 & c = a_3b_3(b_3c_3 - 1), \\ a = a_4 & b = b_4c_4 & c = a_4b_4c_4(c_4 - 1). \end{array}$$

Then $K(g(u))$ is made to be normal crossing,

$$\begin{aligned} 2K(g(u)) &= z_1^2\{(a_1b_1 + 1)^2 + 3a_1^2b_1^4\} \\ &= a_2^2c_2^2(1 + 3b_2^2c_2^2) \\ &= a_3^2b_3^4(c_3^2 + 3) \\ &= a_4^2b_4^2c_4^4(1 + 3b_4^2). \end{aligned}$$

The Jacobian determinant $g'(u)$ is also made to be normal crossing.

$$\begin{aligned} g'(u) &= c_1 \\ &= a_2c_2 \\ &= a_3b_3^2 \\ &= a_4b_4c_4^2. \end{aligned}$$

Therefore, the real log canonical threshold is $3/4$.

Example 3.19 Let us study a function which is the Kullback–Leibler distance of a layered neural network, Example 7.1,

$$f = f(a, b, c, d) = (ab + cd)^2 + (ab^2 + cd^2)^2.$$

(1) Firstly a blow-up with center $\mathbb{V}(b, d) \subset \mathbb{V}(f)$ is tried. By $b = b_1d$,

$$f = d^2\{(ab_1 + c)^2 + d^2(ab_1^2 + c)^2\}.$$

Since f is symmetric for (b, d), we need not try $d = bd_1$.
(2) The transform $c_1 = ab_1 + c$ is an analytic isomorphim and its Jacobian determinant is equal to 1.

$$f = d^2\{c_1^2 + d^2(ab_1^2 + c_1 - ab_1)^2\}.$$

(3) The second step is to try the blow-up with center $\mathbb{V}(c_1, d) \subset \mathbb{V}(f)$. In the first local coordinate, by $d = c_1 d_1$, it follows that

$$f = c_1^4 d_1^2 \{1 + d_1^2 (ab_1^2 + c_1 - ab_1)^2\},$$

which is normal crossing. In the second local coordinate, by $c_1 = c_2 d$, it follows that

$$f = d^4 \{c_2^2 + (ab_1^2 + c_2 d - ab_1)^2\},$$

which is not normal crossing.
(4) The third step is the blow-up with center $\mathbb{V}(a, c_2) \subset \mathbb{V}(f)$. By $a = c_2 a_1$, f is made normal crossing. By $c_2 = ac_3$, it follows that

$$f = d^4 a^2 \{c_3^2 + (b_1^2 + c_3 d - b_1)^2\},$$

which is not yet normal crossing.
(5) The fourth step is the blow-up with center $\mathbb{V}(b_1, c_3) \subset \mathbb{V}(f)$. By $b_1 = c_3 b_2$, f is made normal crossing. By $c_3 = b_1 c_4$, it follows that

$$f = a^2 b_1^2 d^4 \{c_4^2 + (b_1 + c_4 d - 1)^2\}, \tag{3.6}$$

which is not yet normal crossing.
(6) The last step is the blow-up with center $\mathbb{V}(b_1 - 1, c_4) \subset \mathbb{V}(f)$. The relation $b_1 - 1 = c_4 b_2$ makes f normal crossing. Also $c_4 = c_5(b_1 - 1)$ results in

$$f = d^4 a^2 b_1^2 (b_1 - 1)^2 \{c_5^2 + (1 + c_5 d)^2\},$$

which is normal crossing. The last coordinate is given by

$$\begin{aligned} a &= a, \\ b &= b_1 d, \\ c &= a(b_1 - 1) b_1 c_5 d - ab_1, \\ d &= d, \end{aligned}$$

whose Jacobian determinant is given by

$$|g'| = |ab_1(b_1 - 1)d^2|.$$

The real log canonical threshold is $3/4$.

Remark 3.19 (1) In this example, the resolution process started from the blow-up with center $b = d = 0$. If one starts from the blow-up with center $a = c = 0$ or $a = d = 0$, the other desingularization is found.

(2) The concept of blow-up is generalized. A transformation defined by

$$
\begin{aligned}
x_1 &= y_1^{a_{11}} y_2^{a_{12}} \cdots y_d^{a_{1d}}, \\
x_2 &= y_1^{a_{21}} y_2^{a_{22}} \cdots y_d^{a_{2d}}, \\
&\vdots \\
x_d &= y_1^{a_{d1}} y_2^{a_{d2}} \cdots y_d^{a_{dd}},
\end{aligned}
$$

is called a toric modification if $\det(A) = 1$ where $A = (a_{ij})$. In this method, a toric variety is introduced, which is the generalized concept from the real algebraic variety. If the Newton diagram of the target function is not degenerate, then the resolution of singularities can be found by the toric modification.
(3) In complex algebraic geometry, a log canonical threshold can be defined in the same way, that is to say, it is defined by using resolution of singularities. Even for the same polynomial, the complex resolution is different from the real resolution. For example, $f(x, y) = x(y^2 + 1)$ is normal crossing in real resolution, whereas $f(x, y) = x(y + \sqrt{-1})(y - \sqrt{-1})$ is the complex resolution. Hence the real log canonical threshold is different from the complex one. As is shown in Chapter 6, the real log canonical threshold is equal to the learning coefficient, if the *a priori* distribution is positive at singularity.

4

Zeta function and singular integral

In singular learning theory, we need an asymptotic expansion of an integral for $n \to \infty$,

$$Z(n) = \int \exp(-nK(w))\varphi(w)dw, \tag{4.1}$$

where $K(w)$ is a function of $w \in \mathbb{R}^d$ and $\varphi(w)$ is a probability density function. If the minimum point of $K(w)$ is unique and the Hessian matrix at the minimum point is positive definite, then the saddle point approximation or the Gaussian approximation can be applied. However, to study the case when $K(w) = 0$ contains singularities, we need a more precise mathematical foundation. An integral such as eq.(4.1) is called a singular integral. To analyze a singular integral, we need the zeta function, Schwartz distribution, Mellin transform, and resolution of singularities. For example, if

$$K(a, b) = (a^3 - b^2)^2,$$

then the function $\exp(-nK(a, b))$ has the form shown in Figure 4.1. Note that the neighborhood of singularities of $K(a, b) = 0$ occupy almost all parts of the integral when $n \to \infty$.

4.1 Schwartz distribution

Definition 4.1 (Function space $\mathcal{D}$) A function $\varphi : \mathbb{R}^d \to \mathbb{C}$ is said to be of class C_0^∞ if
(1) $\varphi(w)$ is a function of class C^∞.
(2) The support of $\varphi(w)$,

$$\text{supp}\,\varphi \equiv \overline{\{w \in \mathbb{R}^d\ ;\ \varphi(w) \neq 0\}},$$

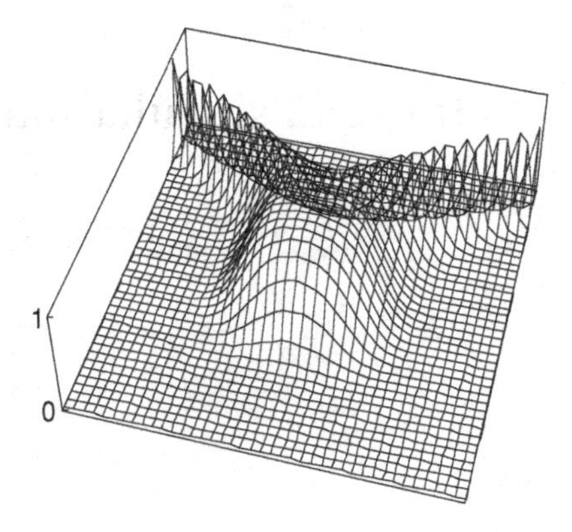

Fig. 4.1. Singular integral

is compact, where $\overline{A}$ is the closure of a set A with Euclidean topology. The set of all functions of class C_0^∞ is denoted by $\mathcal{D}$, which is a vector space over the field $\mathbb{C}$.

Definition 4.2 (Topology of $\mathcal{D}$) Let $\{\varphi_i; i = 1, 2, \ldots\}$ be a sequence of functions in $\mathcal{D}$ and φ be a function in $\mathcal{D}$. The convergence $\varphi_k \to \varphi$ is defined by the following conditions.
(1) There exists a compact set $K \subset \mathbb{R}^d$ such that

$$\begin{aligned} &\text{supp } \varphi \subset K, \\ &\text{supp } \varphi_k \subset K \quad (k = 1, 2, 3, \ldots). \end{aligned}$$

(2) For each multi-index α,

$$\lim_{k\to\infty} \max_{w\in K} |\partial_\alpha \varphi_k(w) - \partial_\alpha \varphi(w)| = 0.$$

Example 4.1 (1) A nonzero analytic function on $\mathbb{R}^d$ does not have a compact support, hence such a function is not contained in $\mathcal{D}$.
(2) Let $a > 0$. A function $\rho_a(w)$ defined on $\mathbb{R}^d$,

$$\rho_a(w) = \begin{cases} \exp\left(-\dfrac{1}{a^2 - \|w\|^2}\right) & (\|w\| < a) \\ 0 & (\|w\| \geq a), \end{cases}$$

is contained in $\mathcal{D}$.
(3) If $f(w)$ is an analytic function, then $f(w)\rho_a(w)$ is contained in $\mathcal{D}$. If $a \to \infty$, then $f(w)\rho_a(w) \to f(w)$ for each w.
(4) Assume $\varphi(w) \in \mathcal{D}$. If $a_k \in \mathbb{R}^d$ satisfies $|a_k| \to \infty$ $(k \to \infty)$, then for each $w \in \mathbb{R}^d$

$$\lim_{k\to\infty} \varphi(w - a_k) = 0.$$

However, since the support of $\varphi(w - a_k)$ is not contained in any compact set, it does not converge to $\varphi(w) \equiv 0$ by the topology of $\mathcal{D}$.

Definition 4.3 (Schwartz distribution) A function $T : \mathcal{D} \to \mathbb{C}$ is said to be a Schwartz distribution on $\mathbb{R}^d$ if it satisfies the following conditions.
(1) T is a linear function from $\mathcal{D}$ to $\mathbb{C}$. In other words, for each $\varphi \in \mathcal{D}$, $T(\varphi)$ is contained in $\mathbb{C}$ and for arbitrary $a, b \in \mathbb{C}$ and arbitrary $\varphi(w), \psi(w) \in \mathcal{D}$

$$T(a\varphi + b\psi) = aT(\varphi) + bT(\psi).$$

(2) T is a continuous function from $\mathcal{D}$ to $\mathbb{C}$. In other words, if a sequence φ_k is such that $\varphi_k \to \varphi$ by the topology of $\mathcal{D}$, then the complex sequence $T(\varphi_k)$ converges to $T(\varphi)$.

The set of all Schwartz distributions is denoted by $\mathcal{D}'$.

Example 4.2 (1) A function $f : \mathbb{R}^d \to \mathbb{C}$ is said to be locally integrable if, for an arbitrary compact set K, the integral

$$\int_K f(w)dw$$

is well defined and finite, where dw is Lebesgue measure on $\mathbb{R}^d$. If a function $f(w)$ is locally integrable, then

$$T(\varphi) = \int f(w)\varphi(w)dw \quad (\varphi \in \mathcal{D})$$

satisfies the conditions of Definition 4.3, hence it determines a Schwartz distribution. If a Schwartz distribution T is represented by a locally integrable function in the same way, then T is called a Schwartz distribution with a regular integral.
(2) A Schwartz distribution

$$T(\varphi) = \varphi(0)$$

satisfies the conditions of Definition 4.3, but it cannot be represented by any locally integrable function. Usually this Schwartz distribution is denoted by a formal integration,

$$T(\varphi) = \int \delta(w)\varphi(w)dw. \tag{4.2}$$

Note that, in this equation, $\delta(w)$ on the right-hand side is defined by the left-hand side. Here $\delta(w)$ is not an ordinary function of w. Therefore, T is a Schwartz distribution but not a Schwartz distribution with a regular integral. This Schwartz distribution T and $\delta(w)$ is called a delta function or Dirac's delta function.

(3) For $\varphi(x, y)$, a Schwartz distribution defined by

$$T(\varphi) = \int \frac{\partial}{\partial x}\varphi(x, y)\Big|_{x=0} dy$$

is not a Schwartz distribution with a regular integral. A Schwartz distribution

$$S(\varphi) = \int \varphi(0, y)dy$$
$$= \int\int \delta(x)\varphi(x, y)dxdy$$

is not a Schwartz distribution with a regular integral.

Definition 4.4 (Topology of Schwartz distribution) Let $T_1, T_2, \ldots$ and T be Schwartz distributions in $\mathcal{D}'$. The convergence in $\mathcal{D}'$, $T_k \to T$ is defined by the condition that, for each $\varphi \in \mathcal{D}$, the complex sequence $T_k(\varphi) \to T(\varphi)$.

Note that, in order to prove the convergence of Schwartz distribution, it is sufficient to prove $T_k(\varphi) \to T(\varphi)$ for each φ. No uniform convergence for φ is needed.

Theorem 4.1 (Completeness of $\mathcal{D}$) *Let $T_1, T_2, \ldots,$ be a sequence of Schwartz distributions. If the sequence $T_k(\varphi)$ converges in $\mathbb{C}$ for each $\varphi \in \mathcal{D}$, then there exists a Schwartz distribution $T \in \mathcal{D}'$ such that $T_k \to T$.*

(Explanation of Theorem 4.1) This is a fundamental theorem of Schwartz distribution theory. For the proof, see [32]. If $\{T_k(\varphi)\}$ is a Cauchy sequence for an arbitrary $\varphi \in \mathcal{D}$, then there exists a Schwarz distribution T such that $T_k \to T$.

Example 4.3 For $a > 0$, a function S_a on $\mathbb{R}^1$

$$S_a(w) = \begin{cases} 1/(2a) & |w| < a \\ 0 & |w| \geq a \end{cases}$$

is locally integrable, hence

$$T_a(\varphi) = \int S_a(w)\varphi(w)dw$$

is a Schwartz distribution with a regular integral. When $a \to 0$, a function $S_a(w)$ does not converge to any ordinary function, whereas T_a converges to a delta function as a Schwartz distribution.

Theorem 4.2 *For any Schwartz distribution T, there exists a sequence of Schwartz distributions with regular integral $\{T_k\}$ such that $T_k \to T$.*

(Explanation of Theorem 4.2) This theorem shows that the set of Schwartz distributions with regular integral is a dense subset in $\mathcal{D}'$. For the proof, see [32].

Definition 4.5 (Derivative and integral on $\mathcal{D}'$) Let T_t be a function from the real numbers to a Schwartz distribution,

$$\mathbb{R} \ni t \mapsto T_t \in \mathcal{D}'.$$

Based on the topology of $\mathcal{D}'$, we can define a derivative and integral of T_t. For each $\varphi \in \mathcal{D}$,

$$\Big(\frac{d}{dt}T_t\Big)(\varphi) = \frac{d}{dt}\Big(T_t(\varphi)\Big),$$

and

$$\Big(\int T_t\, dt\Big)(\varphi) = \int T_t(\varphi)\, dt.$$

Example 4.4 By the identity,

$$\int_0^\infty dt \int dx \delta(t-x)\varphi(x) = \int_0^\infty dx\varphi(x) \quad (\varphi \in \mathcal{D}),$$

it follows that

$$\int_0^\infty dt\delta(t-x) = \theta(x),$$

where $\theta(x)$ is a locally integrable function by which a Schwartz distribution is defined with a regular integral,

$$\theta(x) = \begin{cases} 1 & (x > 0) \\ 0 & \text{otherwise.} \end{cases} \tag{4.3}$$

This function $\theta(x)$ is called a Heaviside function or a step function.

On the other hand, from the identity

$$\frac{d}{dt}\int_{-\infty}^{\infty} \theta(t-x)\varphi(x)dx = \varphi(t),$$

it follows that

$$\frac{d}{dt}\theta(t-x) = \delta(t-x).$$

Example 4.5 (Definition of $\delta(t - f(x))$) Let $f : \mathbb{R}^1 \to \mathbb{R}^1$ be a differentiable function which satisfies

$$\frac{df}{dx} > 0.$$

We assume that, for arbitrary t, there exists x such that

$$f(x) = t.$$

Let us study how to define a Schwartz distribution $\delta(t - f(x))$ on $x \in \mathbb{R}^1$ so that it satisfies

$$\delta(t - f(x)) = \frac{d}{dt}\theta(t - f(x)), \tag{4.4}$$

for any parameter $t \in \mathbb{R}^1$. By the definition, eq.(4.4) is equivalent to

$$\int \delta(t - f(x))\varphi(x)dx = \frac{d}{dt}\int \theta(t - f(x))\varphi(x)dx.$$

By the definition of eq.(4.3),

$$\begin{aligned}\frac{d}{dt}\int \theta(t - f(x))\varphi(x)dx. &= \frac{d}{dt}\int_{-\infty}^{f^{-1}(t)} \varphi(x)dx \\ &= \varphi(f^{-1}(t))f^{-1}(t)' \\ &= \int \delta(x - f^{-1}(t))\frac{\varphi(x)}{f'(x)}dx.\end{aligned}$$

Therefore, we adopt the definition,

$$\delta(t - f(x)) = \frac{\delta(x - f^{-1}(t))}{|f'(x)|}.$$

Then it satisfies eq.(4.4) as a theorem. As a special case, we obtain the definition,

$$\delta(f(x)) = \frac{\delta(x - x_0)}{|f'(x)|}$$

where $x_0 = f^{-1}(0)$. Note that, if $f'(x) = 0$, then $\delta(f(x))$ cannot be defined.

Remark 4.1 A function of class C_0^∞ and a Schwartz distribution can be locally defined.
(1) The following theorem shows that a C_0^∞ class function can be localized. By using the partition of unity, Theorem 2.10, a function $\varphi \in \mathcal{D}$ can be decomposed as

$$\varphi(x) = \sum_{j=1}^{J} \varphi(x)\ \rho_j(x).$$

Therefore, a Schwartz distribution can also be represented as a sum of localized ones,

$$T(\varphi) = \sum_{j=1}^{J} T(\varphi \cdot \rho_j). \tag{4.5}$$

(2) By this decomposition, we can define a Schwartz distribution by its local behavior in any open set. Assume that a compact set K and its open covering $K \subset \cup_j U_j$ are given. Let $C_0^\infty(U_j)$ be the set of all C_0^∞ class functions whose supports are contained in U_j. If a set of Schwartz distributions $\{T_j\}$ is given where

$$T_j : C_0^\infty(U_j) \to \mathbb{R}$$

and if, for arbitrary $\varphi \in C_0^\infty(U_i \cap U_j)$, the consistent condition

$$T_i(\varphi) = T_j(\varphi)$$

is satisfied, then there exists a unique Schwartz distribution T such that

$$T(\varphi) = \sum_{j=1}^{J} T_j(\varphi \cdot \sigma_j). \tag{4.6}$$

By using this property, a Schwartz distribution can be defined on each local open set. Also a Schwartz distribution on a manifold is defined in the same way on each local coordinate.

4.2 State density function

As is shown in Example 4.5, if a real function of class C^1, $f : \mathbb{R}^1 \to \mathbb{R}^1$ satisfies

$$f(x) = 0 \iff x = x_0,$$
$$f'(x_0) \neq 0,$$

then the Schwartz distribution $\delta(f(x))$ is defined by

$$\delta(f(x)) = \frac{\delta(x - x_0)}{|f'(x_0)|}.$$

In this section, let us generalize this definition to the case of several variables.

Let U be an open set of $\mathbb{R}^d$ and $f : U \to \mathbb{R}^1$ be a real analytic function. Let us define a Schwartz distribution $\delta(t - f(x))$. Assume that $\nabla f(x) \neq 0$ for $f(x) = t$.

For the definition of the Schwartz distribution we can assume U is sufficiently small and local such that at least one of $\frac{\partial f}{\partial x_1}(x)$, $\frac{\partial f}{\partial x_2}(x), \ldots, \frac{\partial f}{\partial x_d}(x)$ is not equal to zero. Therefore, we can assume

$$\frac{\partial f}{\partial x_1}(x) \neq 0 \quad (x \in U)$$

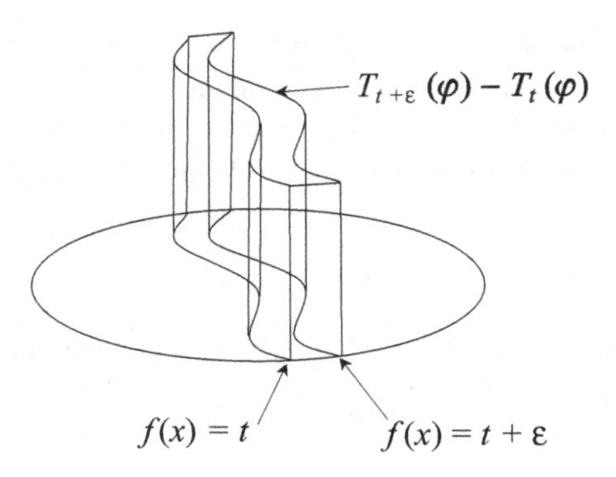

Fig. 4.2. Definition of $\delta(t - f(x))$

without loss of generality. Let us choose a coordinate $(f, u_2, u_3, \ldots, u_d)$ instead of $(x_1, x_2, x_3, \ldots, x_d)$, where $x_2 = u_2, x_3 = u_3, \ldots, x_d = u_d$. Then

$$dx_1 dx_2 \cdots dx_d = \frac{\partial(x_1, x_2, x_3, \ldots, x_d)}{\partial(f, u_2, u_3, \ldots, u_d)} df du_2 du_3 \cdots du_d$$
$$= \frac{df du_2 du_3 \cdots du_d}{|df/dx_1|}.$$

Let $\varphi(f, x_2, x_3, \ldots, x_d)$ be a function of class C_0^∞ whose support is contained in U. A Schwartz distribution D_t on $\mathbb{R}^d$ is defined by

$$D_t(\varphi) = \int \frac{\varphi(t, u_2, u_3, \ldots, u_d)}{\left|\frac{\partial f}{\partial x_1}(t, u_2, u_3, \ldots, u_d)\right|} du_2 du_3 \cdots du_d. \tag{4.7}$$

Also we define another Schwartz distribution T_t in Figure 4.2 by

$$T_t(\varphi) = \int_{f(x)<t} \varphi(x) dx$$
$$= \int \theta(t - f(x)) \varphi(x) dx,$$

where

$$\theta(a) = \begin{cases} 1 & (a \geq 0) \\ 0 & (a < 0). \end{cases}$$

The following theorem shows that the derivative of T_t is equal to D_t.

Theorem 4.3 *Assume that $f : \mathbb{R}^d \to \mathbb{R}^1$ is a real function of class C^1 which satisfies $\nabla f(x) \neq 0$ for $f(x) = t$. Then, for arbitrary $\varphi \in \mathcal{D}$,*

$$\lim_{\epsilon \to 0} \frac{1}{\epsilon}(T_{t+\epsilon}(\varphi) - T_t(\varphi)) = D_t(\varphi).$$

Therefore,

$$\frac{d}{dt}T_t = D_t$$

as a Schwartz distribution holds.

Proof of Theorem 4.3 By the definition of T_t,

$$\begin{aligned}\frac{1}{\epsilon}(T_{t+\epsilon}(\varphi) - T_t(\varphi)) &= \frac{1}{\epsilon}\int_{t\le f(x)<t+\epsilon}\varphi(x)dx \\ &= \frac{1}{\epsilon}\int_{t\le f(x)<t+\epsilon}\varphi(f, u_2, u_3, \ldots, u_d)\frac{df du_2 du_3 \cdots du_d}{|df/dx_1|}.\end{aligned}$$

In general, for a continuous function $F(f)$ of f, there exists $t \le t^* \le t + \epsilon$ such that

$$\int_t^{t+\epsilon} F(f)df = F(t^*)\epsilon,$$

therefore,

$$\frac{1}{\epsilon}(T_{t+\epsilon}(\varphi) - T_t(\varphi)) = \frac{1}{\epsilon}\int \varphi(t^*, u_2, u_3, \ldots, u_d)\frac{du_2 du_3 \cdots du_d}{|df/dx_1|} = D_{t^*}(\varphi).$$

Then by $\epsilon \to 0$, we obtain the theorem. □

Definition 4.6 (Definition of $\delta(t - f(x))$) Let D_t be a Schwartz distribution defined by eq.(4.7). The expression $\delta(t - f(x))$ is defined by

$$D_t(\varphi) = \int \delta(t - f(x))\varphi(x)dx,$$

where the right-hand side is defined by the left-hand side.

Corollary 4.1 *For real numbers* $n, t > 0$,

$$\delta(t - nf(x)) = \frac{1}{n}\delta\Big(\frac{t}{n} - f(x)\Big).$$

Proof of Corollary 4.1 For $\varphi(x)$, by using the definition in eq.(4.7),

$$\int \delta(t - nf(x))\varphi(x)dx = \int \frac{\varphi(t/n, u_2, u_3, \ldots, u_d)}{\left|\frac{\partial f}{\partial x_1}(t/n, u_2, u_3, \ldots, u_d)\right| \times n} du_2 du_3 \cdots du_d.$$

□

From Theorem 4.3 and definitions

$$v(t) = \int \delta(t - f(x))\varphi(x)dx, \tag{4.8}$$

$$V(t) = \int \Theta(t - f(x))\varphi(x)dx, \tag{4.9}$$

it follows that

$$\frac{dV}{dt}(t) = v(t).$$

In particular, if

$$\{x \in U; f(x) = t\} \cap \text{supp } \varphi(x) = \emptyset,$$

then $v(t) = 0$. The function $v(t)$ is called a state density function.

Theorem 4.4 *Let $F : \mathbb{R}^1 \to \mathbb{R}^1$ be a locally integrable function, and $v(t)$ be the state density function defined by eq.(4.8). Then*

$$\int_{\mathbb{R}^1} F(t)v(t)dt = \int_{\mathbb{R}^d} F(f(x))\varphi(x)dx.$$

Proof of Theorem 4.4 Since the support of $v(t)$ is compact, we can assume that $F(t)$ has a compact support. Since $F(t)$ can be decomposed as

$$F(t) = F_1(t) - F_2(t),$$

where

$$F_1(t) = \max\{0, F(t)\},$$
$$F_2(t) = -\max\{0, -F(t)\},$$

we can assume $F(t) \geq 0$ without loss of generality. The integrable function $F(t)$ can be approximated by

$$F(t) = \lim_{n\to\infty} F_n(t),$$

where $F_n(t)$ is a simple function, given by

$$F_n(t) = \sum_{i=1}^{n} c_i \chi_{a_i,b_i}(t).$$

We can assume $F_n(t)$ is a nondecreasing sequence for each x, and

$$\chi_{a_i,b_i}(t) = \begin{cases} 1 & (a_i < t < b_i) \\ 0 & \text{otherwise} \end{cases}.$$

Since $F(t)$ is a locally integrable function,

$$\int F(t)v(t)dt = \lim_{n\to\infty} \sum_{i=1}^{n} c_i \int_{a_i}^{b_i} v(t)dt.$$

Then since

$$\int_{a_i}^{b_i} v(t)dt = \int_{a_i<f(x)<b_i} \varphi(x)dx,$$

we have

$$\begin{aligned}\int F(t)v(t)dt &= \lim_{n\to\infty} \int \sum_{i=1}^{n} c_i \chi_{a_i,b_i}(f(x))\varphi(x)dx\\ &= \int F(f(x))\varphi(x)dx,\end{aligned}$$

which completes the theorem. □

Corollary 4.2 *Let $f(x) : \mathbb{R}^d \to \mathbb{R}^1$ be a real analytic function and $\varphi(x)$ be a function of the class C_0^∞. The state density function*

$$v(t) = \int \delta(t - f(x))\varphi(x)dx$$

has a compact support and satisfies

$$\int t^z v(t)dt = \int f(x)^z \varphi(x)dx,$$

$$\int \exp(-nt)v(t)dt = \int \exp(-nf(x))\varphi(x)dx.$$

Proof of Corollary 4.2 By applying $F(t) = t^z$ or $F(t) = \exp(-nt)$, in Theorem 4.4, Corollary 4.2 is obtained. □

Definition 4.7 (Zeta function of $f(x)$ and $\varphi(x)$) For a given real analytic function $f(x) \geq 0$ and a function $\varphi(x)$ of C_0^∞, the zeta function is defined by

$$\zeta(z) = \int f(x)^z \varphi(x)dx.$$

The inverse Mellin transform of $\zeta(z)$ is called the state density function,

$$v(t) = \int \delta(t - f(x))\varphi(x)dx.$$

The Laplace transform of $v(t)$ is called the partition function,

$$Z(n) = \int \exp(-nf(x))\varphi(x)dx.$$

4.3 Mellin transform

In this section, let us summarize the properties of the Mellin transform. Let $i = \sqrt{-1}$.

Definition 4.8 (Mellin transform) For a measurable function $f : (0, \infty) \to \mathbb{C}$, if the integral

$$F(z) = \int_0^{\infty} f(t)t^z dt \tag{4.10}$$

satisfies $|F(z)| < \infty$, then $F(z)$ is said to be the Mellin transform of $f(t)$.

Definition 4.9 (Function of bounded variation) A function $f : \mathbb{R}^1 \to \mathbb{R}^1$ is said to have bounded variation in a neighborhood of $t = s$, if there exists $\epsilon > 0$ such that, in any partition of $[s - \epsilon, s + \epsilon]$, where

$$s - \epsilon = t_1 < t_2 < \cdots < t_k = s + \epsilon,$$

the total variation is finite,

$$TV(f) \equiv \sum_{i=1}^{k-1} |f(t_{i+1}) - f(t_i)| < \infty.$$

Remark 4.2 The following are well-known properties of a function of bounded variation.
(1) A continuous function does not have bounded variation in general.
(2) A function of class C^1 has bounded variation.
(3) If a function $f(t)$ has bounded variation in a neighborhood of $t = s$, then there exists $\delta > 0$ such that $f(s + i\delta)$ is well defined, and limit values

$$f(s + i0) \equiv \lim_{\delta \to +0} f(s + i\delta),$$

$$f(s - i0) \equiv \lim_{\delta \to +0} f(s - i\delta),$$

exist.

Theorem 4.5 (Inverse Mellin transform) *Assume that eq.(4.10) absolutely converges in a region $a < Re(z) < b$, in other words, assume that*

$$\int_0^{\infty} |f(t)|t^{Re(z)} dt < \infty.$$

If $f(t)$ has bounded variation in a neighborhood $t = s$, then for arbitrary c $(a < c < b)$, the equation

$$\frac{1}{2}\{f(s + i0) + f(s - i0)\} = \frac{1}{2\pi i} \int_{c-i\infty}^{c+i\infty} F(z)s^{-z-1} dz \tag{4.11}$$

holds, where $F(z)$ is defined by eq. (4.10).

(Explanation of Theorem 4.5) This theorem is proved by using the inverse Laplace transform. For example, see [25].

Remark 4.3 (1) Equation (4.11) means that, using $z = c + iy$,

$$\lim_{M\to\infty} \frac{1}{2\pi i} \int_{-M}^{M} F(c+iy)s^{-c-iy-1}\, i\, dy$$

converges and coincides with $\{f(s+i0) + f(s-i0)\}/2$.
(2) In a lot of books, the ordinary Mellin transform and inverse Mellin transform are respectively defined by

$$F(z) = \int_0^\infty f(t)t^{z-1}dt,$$

$$f(t) = \frac{1}{2\pi i} \int_{c-i\infty}^{c+i\infty} F(z)t^{-z}dz.$$

However, in statistical learning theory, eq.(4.10) and eq.(4.11) are more appropriate, hence we use eq.(4.10) and eq.(4.11) in this book.

Example 4.6 Let $\lambda > 0$ and $a > 0$ be positive real numbers. A function

$$f(t) = \begin{cases} t^{\lambda-1} & (0 < t < a) \\ 0 & \text{otherwise} \end{cases}$$

has bounded variation at an arbitrary point.

$$F(z) = \int_0^a t^{\lambda-1}\, t^z\, dt$$

converges absolutely in $-\lambda < \text{Re}(z) < \infty$, and is equal to

$$F(z) = \frac{a^{z+\lambda}}{z+\lambda}.$$

Hence $F(z)$ can be analytically continued to the meromorphic function on the entire complex plane, which is equal to $a^{z+\lambda}/(z+\lambda)$. The inverse transform of $F(z)$ is given by

$$f(s) = \frac{1}{2\pi i} \int_{c-i\infty}^{c+i\infty} \frac{a^{z+\lambda}}{z+\lambda} s^{-z-1} dz,$$

which is equal to

$$f(s) = a^{z+\lambda}s^{-z-1}\Big|_{z=-\lambda} = s^{\lambda-1}$$

by the residue theorem.

Example 4.7 This example is very important in this book. Let $\lambda > 0$ and $a > 0$ be positive real numbers, and m be a natural number. The Mellin transform of

$$f(t) = \begin{cases} \dfrac{a^{-\lambda}}{(m-1)!}\, t^{\lambda-1}\left(\log\dfrac{a}{t}\right)^{m-1} & (0 < t < a) \\ 0 & \text{otherwise} \end{cases}$$

is given by

$$F(z) = \frac{a^{-\lambda}}{(m-1)!}\int_0^a t^{\lambda-1}\left(\log\frac{a}{t}\right)^{m-1} t^z\, dt.$$

By using partial integration,

$$\begin{aligned} F(z) &= \left[\frac{a^{-\lambda}}{(m-1)!}\frac{t^{z+\lambda}}{(z+\lambda)}\left(\log\frac{a}{t}\right)^{m-1}\right]_0^a \\ &\quad + \frac{a^{-\lambda}}{(m-2)!}\int_0^a \frac{t^{z+\lambda-1}}{(z+\lambda)}\left(\log\frac{a}{t}\right)^{m-2} dt \\ &\ \ \vdots \\ &= \frac{a^{-\lambda}}{(z+\lambda)^{m-1}}\int_0^a t^{z+\lambda-1}\, dt. \end{aligned}$$

This integral converges absolutely in $-\lambda < \mathrm{Re}(z) < \infty$, and is equal to

$$F(z) = \frac{a^z}{(z+\lambda)^m}.$$

The inverse transform is given by

$$f(s) = \left(\frac{d}{dz}\right)^{m-1}\left\{\left(\frac{a}{s}\right)^z\frac{1}{s}\right\}\Big|_{z=-\lambda} = \frac{a^{-\lambda}}{(m-1)!}s^{\lambda-1}\left(\log\frac{a}{s}\right)^{m-1}.$$

4.4 Evaluation of singular integral

Based on the resolution theorem, any singularities are images of normal crossing singularities. Any singular integral is equal to the sum of integrals of normal crossing singularities. However, evaluation of a singular integral contains a rather complicated calculation, hence we introduce an example before the general theory.

Example 4.8 Let us consider a normal crossing function

$$K(a, b) = a^2 b^2.$$

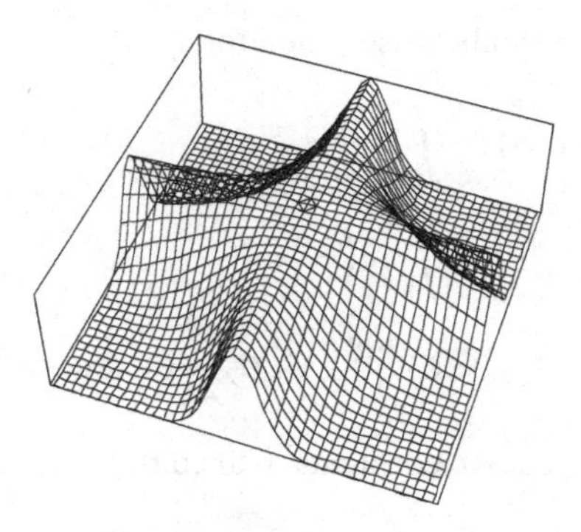

Fig. 4.3. Normal crossing singular integral

Then the function $\exp(-nK(a, b))$ in a singular integral is illustrated in Figure 4.3,

$$Z(n) = \int_{[0,1]^2} \exp(-nK(a, b))\, da\, db.$$

The zeta function

$$\zeta(z) = \int_{[0,1]^2} K(a, b)^z\, da\, db \tag{4.12}$$

is the Mellin transform of the state density function,

$$v(t) = \int_{[0,1]^2} \delta(t - K(a, b))\, da\, db. \tag{4.13}$$

Moreover, the singular integral is the Laplace transform of the state density function.

$$Z(n) = \int_0^1 \exp(-nt)\, v(t)\, dt. \tag{4.14}$$

Among the three integrals, eqs.(4.12), (4.13), (4.14), the zeta function can be calculated explicitly for $\mathrm{Re}(z) > -1/2$,

$$\zeta(z) = \frac{1}{4(z + 1/2)^2}.$$

For $\mathrm{Re}(z) > -1/2$, this function is holomorphic. It can be analytically continued to the meromorphic function on the entire complex plane. By using the inverse Mellin transform, the state density function is given by

$$v(t) = \tfrac{1}{4} t^{-1/2}(-\log t),$$

if $0 < t < 1$, or $v(t) = 0$ otherwise. Therefore

$$\begin{aligned} Z(n) &= \int_0^1 v(t)e^{-nt}dt \\ &= \int_0^n v\left(\frac{t}{n}\right)e^{-t}\frac{dt}{n} \\ &= C_1\frac{\log n}{4\sqrt{n}} - C_2\frac{1}{4n} + C_3(n), \end{aligned}$$

where C_1 and C_2 are positive constants defined by

$$\begin{aligned} C_1 &= \int_0^\infty t^{-1/2}e^{-t}dt, \\ C_2 &= \int_0^\infty t^{-1/2}e^{-t}\log t\, dt, \end{aligned}$$

and $C_3(n)/n$ converges to zero when $n \to \infty$. Therefore, the largest order of the singular integral $Z(n)$ is equal to $\log n/\sqrt{n}$. Any singular integral can be asymptotically expanded in the same way as this example.

We need to generalize the previous example. The following equations give constructive evaluation of the singular integral. Let r be a natural number. For a vector

$$x = (x_1, x_2, \ldots, x_r) \in \mathbf{R^r}$$

and multi-indices,

$$\begin{aligned} k &= (k_1, k_2, \ldots, k_r), \\ h &= (h_1, h_2, \ldots, h_r), \end{aligned}$$

x^{2k} and x^h are monomials defined by

$$\begin{aligned} x^{2k} &= x_1^{2k_1}\, x_2^{2k_2} \cdots x_r^{2k_r}, \\ x^h &= x_1^{h_1}\, x_2^{h_2} \cdots x_r^{h_r}. \end{aligned}$$

Also we define

$$|k| = k_1 + k_2 + \cdots + k_r.$$

For a given function $f : \mathbb{R}^r \to \mathbb{C}$, the integral in $[0, b]^r$ is denoted by

$$\int_{[0,b]^r} f(x)dx = \int_0^b dx_1 \cdots \int_0^b dx_r\, f(x_1, \ldots, x_r).$$

Theorem 4.6 *Let $k_1, \ldots, k_r$ be natural numbers and $h_1, \ldots, h_r$ be nonnegative integers. Assume that there exists a rational number $\lambda > 0$ such that*

$$\frac{h_1+1}{2k_1} = \frac{h_2+1}{2k_2} = \cdots = \frac{h_r+1}{2k_r} = \lambda.$$

Then for arbitrary real number $a, b > 0$, the state density function

$$v(t) = \int_{[0,b]^r} \delta(t - a\, x^{2k})\, x^h\, dx$$

is equal to

$$v(t) = \begin{cases} \gamma_b \, \dfrac{t^{\lambda-1}}{a^\lambda} \left(\log \dfrac{ab^{2|k|}}{t}\right)^{r-1} & (0 < t < ab^{2|k|}) \\ 0 & \textit{otherwise}, \end{cases} \tag{4.15}$$

where $\gamma_b > 0$ is a constant

$$\gamma_b = \frac{b^{|h|+r-2|k|\lambda}}{2^r\,(r-1)!\, k_1 k_2 \cdots k_r}. \tag{4.16}$$

Proof of Theorem 4.6 The Mellin transform of $v(t)$ for the case $b = 1$ is given by

$$\begin{aligned} \zeta(z) &= \int_{[0,1]^r} (a\, x^{2k})^z\, x^h\, dx \\ &= \frac{1}{(2k_1)(2k_2)\cdots(2k_r)} \, \frac{a^z}{(z+\lambda)^r}. \end{aligned}$$

By using the result of Example 4.7, if $0 < t < a$,

$$\int_{[0,1]^r} \delta(t - a\, x^{2k})\, x^h\, dx = \gamma_1 \, \frac{t^{\lambda-1}}{a^\lambda} \left(\log \frac{a}{t}\right)^{r-1};$$

otherwise $v(t) = 0$. Then by putting $x' = bx$ and $a' = ab^{-2|k|}$, we obtain the theorem. □

Corollary 4.3 *Let $\theta(t)$ be a step function defined by $\theta(t) = 1$ $(t \geq 0)$ or $\theta(t) = 0$ $(t < 0)$. Assume the same condition as Theorem 4.6. Then*

$$\int_{[0,b]^r} \theta(t - x^{2k})\, x^h dx = \sum_{m=1}^{r} \frac{\gamma_b\,(r-1)!}{\lambda^m\,(r-m)!} \, t^\lambda \left(\log \frac{b^{2|k|}}{t}\right)^{r-m}. \tag{4.17}$$

Proof of Corollary 4.3 Let $V(t)$ be the left-hand side of eq.(4.17), then $V(t)'$ is equal to $v(t)$ in Theorem 4.6 with $a = 1$,

$$V(t)' = \gamma_b t^{\lambda-1} \left(\log \frac{b^{2|k|}}{t}\right)^{r-1}.$$

Since $V(0) = 0$,

$$V(T) = \int_0^T \gamma_b t^{\lambda-1} \left(\log \frac{b^{2|k|}}{t}\right)^{r-1} dt.$$

By using recursive partial integration such as

$$V(T) = \left[\gamma_b \frac{t^\lambda}{\lambda} \left(\log \frac{b^{2|k|}}{t}\right)^{r-1}\right]_0^T + \frac{\gamma_b(r-1)}{\lambda} \int_0^T t^{\lambda-1} \left(\log \frac{b^{2|k|}}{t}\right)^{r-2} dt,$$

we complete the corollary. □

Theorem 4.7 *Let r and s be natural numbers. Assume that four multi-indices k, k', h, h' satisfy*

$$\frac{h_1+1}{2k_1} = \frac{h_2+1}{2k_2} = \cdots = \frac{h_r+1}{2k_r} = \lambda$$

and that

$$\frac{h'_j+1}{2k'_j} > \lambda \quad (j = 1, 2, \ldots, s).$$

A singular integral $Z^p(n)$ is defined by

$$Z^p(n) = \int_{[0,b]^r} dx \int_{[0,b]^s} dy\ K(x, y)^p \exp(-n\beta K(x, y)^2)\, x^h y^{h'}, \tag{4.18}$$

where $p \geq 0$, $\beta > 0$ and

$$K(x, y) = x^k y^{k'}.$$

Then there exist constants $a_1, a_2 > 0$ such that, for arbitrary natural number $n > 1$,

$$a_1 \frac{(\log n)^{r-1}}{n^{\lambda+p}} \leq Z^p(n) \leq a_2 \frac{(\log n)^{r-1}}{n^{\lambda+p}}.$$

Proof of Theorem 4.7. Firstly we prove the case $b = 1$. Since $y^{2k'} \leq 1$ in $[0, 1]^s$,

$$\begin{aligned}
Z^p(n) &\geq \int_{[0,1]^r} dx \int_{[0,1]^s} dy \exp(-n\beta x^{2k})\, x^{h+kp} y^{h'+k'p} \\
&= a_1' \int_{[0,1]^r} dx \exp(-n\beta x^{2k})\, x^{h+kp} \\
&= a_1' \int_0^\infty dt \int_{[0,1]^r} dx\, \delta(t - nx^{2k})\, t^p\, e^{-\beta t}\, x^h \\
&= a_1'' \int_0^n dt\, \frac{t^{\lambda-1}}{n^{\lambda+p}} \left(\log \frac{n}{t}\right)^{r-1} t^p\, e^{-\beta t} \\
&\geq a_1'' \int_0^1 dt\, \frac{t^{\lambda-1}}{n^{\lambda+p}} (\log n)^{r-1}\, t^p\, e^{-\beta t} \\
&= a_1 \frac{(\log n)^{r-1}}{n^{\lambda+p}},
\end{aligned}$$

where we used Theorem 4.6, $-\log t > 0$ for $0 < t < 1$, and a_1' and a_1'' are constants. On the other hand, by using Theorem 4.6,

$$\begin{aligned}
Z^p(n) &= \int_0^\infty dt\, e^{-\beta t}\, t^p \int_{[0,1]^r} dx \int_{[0,1]^s} dy\, \delta(t - nx^{2k} y^{2k'})\, x^h y^{h'} \\
&\leq \int_0^\infty dt \int_{[0,1]^s} dy\, \frac{c_0\, t^{\lambda-1}}{n^{\lambda+p}(y^{2k'})^\lambda} \left|\log \frac{ny^{2k'}}{t}\right|^{r-1} y^{h'}\, t^p\, e^{-\beta t} \\
&\leq \sum_{m=0}^{r-1} \binom{r-1}{m} \frac{(\log n)^m}{n^{\lambda+p}} \int_0^\infty dt \int_{[0,1]^s} dy \\
&\qquad \times c_0\, t^{\lambda-1}\, e^{-\beta t} \left|\log \frac{y^{2k'}}{t}\right|^{r-1-m} y^{h'-2\lambda k'}.
\end{aligned}$$

The last integration is finite because $\lambda > 0$ and $h' - 2\lambda k' > -1$. Therefore we have proved the theorem for the case $b = 1$. Secondly, we study the case for general $b > 0$. By putting $bx' = x$, $by' = y$, in eq.(4.18), we have

$$Z^p(n) = B_1 \int_{[0,1]^r} dx' \int_{[0,1]^s} dy'\, K^p \exp(-n\beta B_2 K^2)\, (x')^h (y')^{h'},$$

where $B_1 = b^{r+s+|h|+|h'|+|k|p+|k'|p} > 0$, $B_2 = b^{2|k|+2|k'|} > 0$, and $K' = K(x', y')$. Therefore we obtain the theorem. □

Definition 4.10 (Partition function) Let ξ and φ be functions of C^1 class from $[0, b]^{r+s}$ to $\mathbb{R}$. Assume that $\varphi(x, y) > 0$, $(x, y) \in [0, b)^{r+s}$. The partition function of ξ, φ, $n > 1$, and $p \geq 0$, is defined by

$$Z^p(n,\xi,\varphi) = \int_{[0,b]^r} dx \int_{[0,b]^s} dy\; K(x,y)^p\, x^h y^{h'}\, \varphi(x,y) \\ \times \exp(-n\beta\; K(x,y)^2 + \sqrt{n}\beta\; K(x,y)\,\xi(x,y)),$$

where

$$K(x,y) = x^k y^{k'}.$$

We use the definitions

$$\|\xi\| = \max_{(x,y)\in[0,b]^{r+s}} |\xi(x,y)|$$

and

$$\|\nabla\xi\| = \max_{1\leq j\leq r}\ \max_{(x,y)\in[0,b]^{r+s}} \left|\frac{\partial\xi}{\partial x_j}\right|.$$

Theorem 4.8 *Assume the same condition as Theorem 4.7 for k, k', h, h'. Then there exist constants $a_1, a_2 > 0$ such that, for arbitrary ξ and φ ($\varphi(x) > 0 \in [0,b]^d$) and an arbitrary natural number $n > 1$,*

$$a_1 \frac{(\log n)^{r-1}}{n^{\lambda+p}}\, e^{-\beta\|\xi\|^2/2} \min\varphi \leq Z^p(n,\xi,\varphi) \leq a_2 \frac{(\log n)^{r-1}}{n^{\lambda+p}}\, e^{\beta\|\xi\|^2/2}\, \|\varphi\|$$

holds, where

$$\min\varphi = \min_{x\in[0,b]^d} \varphi(x).$$

Proof of Theorem 4.8. By using the Cauchy–Schwarz inequality,

$$|\sqrt{n}\; K(x,y)\,\xi(x,y)| \leq \frac{1}{2}\{n\; K(x,y)^2 + \|\xi\|^2\},$$

hence

$$Z^p\left(\frac{3n}{2}\right) \exp\left(-\frac{\beta\|\xi\|^2}{2}\right) \min\varphi \leq Z^p\,(n,\xi,\varphi) \\ \leq Z^p\left(\frac{n}{2}\right) \exp\left(\frac{\beta\|\xi\|^2}{2}\right) \|\varphi\|,$$

where $Z^p(n)$ is given by eq.(4.18). By appying Theorem 4.7, we obtain the theorem. □

Theorem 4.9 *Assume the same condition as Theorem 4.7 for k, k', h, h'. Let ξ and φ be functions of class C^1. The central part of the partition function is* *defined by*

$$Y^p(n,\xi,\varphi) \equiv \frac{\gamma_b(\log n)^{r-1}}{n^{\lambda+p}} \int_0^\infty dt \int_{[0,b]^s} dy\ t^{\lambda+p-1} y^\mu e^{-\beta t+\beta\sqrt{t}\xi_0(y)} \varphi_0(y), \tag{4.19}$$

where we use the notation,

$$\xi_0(y) = \xi(0,y),$$
$$\varphi_0(y) = \varphi(0,y),$$
$$\mu = h' - 2\lambda k'.$$

Then there exist a constant $c_1 > 0$ such that, for arbitrary $n > 1$, ξ, φ, and $p \geq 0$,

$$|Z^p(n,\xi,\varphi) - Y^p(n,\xi,\varphi)|$$
$$\leq c_1 \frac{(\log n)^{r-2}}{n^{\lambda+p}} e^{\beta\|\xi\|^2/2} \{\beta\|\nabla\xi\|\|\varphi\| + \|\nabla\varphi\| + \|\varphi\|\}.$$

Proof of Theorem 4.9 Firstly, we prove that

$$|Z^p(n,\xi_0,\varphi_0) - Y^p(n,\xi,\varphi)| \leq c_1' \frac{(\log n)^{r-2}}{n^{\lambda+p}} e^{\beta\|\xi_0\|^2/2} \|\varphi_0\|. \tag{4.20}$$

By the definition of $Z^p(n,\xi_0,\varphi_0)$,

$$Z^p(n,\xi_0,\varphi_0) = \int_0^\infty dt \int_{[0,b]^r} dx \int_{[0,b]^s} dy\ \delta(t - nK(x,y))$$
$$\times t^p\ e^{-\beta t+\beta\sqrt{t}\ \xi_0(y)}\ x^h y^{h'}\ \varphi_0(y).$$

By using Theorem 4.6,

$$Z^p(n,\xi_0,\varphi_0) = \frac{\gamma_b}{n^{\lambda+p}} \int_{[0,b]^*} dt\ dy\ t^{\lambda+p-1}\ e^{-\beta t+\beta\sqrt{t}\ \xi_0(y)}$$
$$\times y^\mu \left(\log \frac{ny^{2k'} b^{2|k|}}{t}\right)^{r-1} \varphi_0(y),$$

where the integrated region is

$$[0,b]^* = \{(t,y)\ ;\ 0 < t < ny^{2k'} b^{2|k|},\ y \in [0,b]^s\}.$$

By expanding $\{\log n + \log(y^{2'} b^{2|k|}/t)\}^{r-1}$, $Z^p(n,\xi_0,\varphi_0)$ is decomposed as

$$Z^p(n,\xi_0,\varphi_0) = Y^p(n,\xi,\varphi) + Z_1 + Z_2$$

where

$$Z_1 = -\gamma_b \frac{(\log n)^{r-1}}{n^{\lambda+p}} \int_{[0,b]^{**}} dtdy\ t^{\lambda+p-1} y^{\mu} e^{-\beta t+\beta\sqrt{t}\xi_0(y)} \varphi_0(y)$$

with

$$[0, b]^{**} = \{(t, y)\ ;\ ny^{2k'} b^{2|k|} \le t, y \in [0, b]^s\},$$

and

$$Z_2 = \sum_{m=0}^{r-2} \binom{r-1}{m} \frac{c_0(\log n)^m}{n^{\lambda+p}} \int_{[0,b]^*} dt\ dy\ t^{\lambda+p-1} \times e^{-\beta t+\beta\sqrt{t}\ \xi_0(y)}\ y^{\mu} \left(\log \frac{y^{2k'}}{t}\right)^{r-1-m} \varphi_0(y).$$

Therefore Z_1 is evaluated as follows:

$$|Z_1| \le \gamma_b \frac{(\log n)^{r-1}}{n^{\lambda+p}} e^{\beta\|\xi_0\|^2/2} \|\varphi_0\| \times \int_0^{\infty} t^{\lambda+p-1} e^{-\beta t} dt \int_{[0,b]^s} dy\ y^{\mu}\ \theta\Big(\frac{t}{n} - y^{2k'} b^{2|k|}\Big)$$

By using eq.(4.17), there exist $c_1' > 0$, $\delta > 0$ such that

$$\int_{[0,b]^s} dy\ y^{\mu}\ \theta\Big(\frac{t}{n} - y^{2k'} b^{2|k|}\Big) \le c_1' \Big(\frac{t}{n}\Big)^{\delta} \Big(\log \frac{nb^{2|k|}}{t}\Big)^{s-1},$$

it follows that

$$|Z_1| \le c_1'' \frac{(\log n)^{r-1}}{n^{\lambda+p+\delta}} e^{\beta\|\xi_0\|^2/2} \|\varphi_0\|.$$

Also the term Z_2 can be evaluated

$$|Z_2| \le c_1''' \frac{(\log n)^{r-2}}{n^{\lambda}} e^{\beta\|\xi_0\|^2/2} \|\varphi_0\|.$$

Hence we obtain eq.(4.20). Secondly, let us prove

$$|Z^p(n, \xi, \varphi) - Z^p(n, \xi_0, \varphi_0)| \le a_2 \frac{(\log n)^{r-2}}{n^{\lambda+p}} e^{\beta\|\xi\|^2/2} (\beta\|\nabla\xi\|\|\varphi\| + \|\nabla\varphi\|). \tag{4.21}$$

A function $H = H(z, y)$ is defined by

$$H(z, y) = e^{\beta\sqrt{n}\ K(x,y)\ \xi(z,y)} \varphi(z, y).$$

Let ΔZ be the left-hand side of eq.(4.21) and $K = K(x, y)$. Then

$$\Delta Z \leq \iint K^p \, e^{-n\beta K^2} x^h y^{h'} |H(x, y) - H(0, y)| \, dxdy.$$

There exists $z^* \in [0, b]^r$ such that

$$|H(x, y) - H(0, y)| \leq \sum_{j=1}^{r} x_j \left| \frac{\partial}{\partial z_j} H(z^*, y) \right|$$

$$\leq \sum_{j=1}^{r} x_j \{\beta \sqrt{n} K \|\nabla \xi\| \|\varphi\| + \|\nabla \varphi\|\}.$$

Therefore, by using Theorem 4.7, eq.(4.21) is obtained. Finally, by combining eq.(4.20) with eq.(4.21), the proof of Theorem 4.9 is completed. □

Theorem 4.10 *Assume the same condition as Theorem 4.9. Let* $Y^p(n, \xi, \varphi)$ *be the central part of the partition function in eq.(4.19) in Theorem 4.9. Then there exist constants* $a_3, a_4 > 0$ *such that, for arbitrary* ξ, φ, $n > 1$, *and* $p \geq 0$,

$$a_3 \frac{(\log n)^{r-1}}{n^{\lambda+p}} \, e^{-\beta \|\xi\|^2/2} \min |\varphi| \leq Y^p(n, \xi, \varphi) \leq a_4 \frac{(\log n)^{r-1}}{n^{\lambda+p}} \, e^{\beta \|\xi\|^2/2} \, \|\varphi\|$$

holds.

Proof of Theorem 4.10. This theorem is proved in the same way as Theorem 4.8. □

Remark 4.4 In statistical learning theory, we need to evaluate the asymptotic behavior of a general singular integral

$$Z = \int \exp(-n\beta K(w) + \beta \sqrt{n K(w)} \xi(w)) \varphi(w) dw,$$

when $n \to \infty$. By applying resolution of singularities to $K(w)$, which makes $K(g_\alpha(u))$ normal crossing on each local coordinate U_α,

$$K(g_\alpha) = x^{2k} y^{2k'},$$

the term Z can be written as a finite sum of the partition functions,

$$Z = \sum_\alpha Z(n, \ \xi \circ g_\alpha, \ \varphi \circ g_\alpha |g'_\alpha|),$$

to which theorems in this section can be applied.

4.5 Asymptotic expansion and b-function

In the previous section, the singular integral was studied from the viewpoint of resolution of singularities. In this section, its asymptotic expansion is analyzed without resolution of singularities.

Let $K(w) \geq 0$ be a real analytic function on an open set $U \subset \mathbb{R}^d$ and $\varphi(w)$ be a function of class C_0^∞ whose support is contained in U. The zeta function of $K(w)$ and $\varphi(w)$ is defined by

$$\zeta(z) = \int K(w)^z \varphi(w) dw. \tag{4.22}$$

For $\mathrm{Re}(z) > 0$, $\zeta(z)$ is a holomorphic function. In fact, for $z_0 = a + bi$ $(a > 0)$, $\zeta(z)$ is differentiable as a complex function because, if $z \to z_0$, there exists z^* such that

$$\left| \frac{\zeta(z) - \zeta(z_0)}{z - z_0} \right| \leq \int |K(w)^{z^*} \log K(w)| \varphi(w) dw.$$

The analytic continuation of $\zeta(z)$ can be obtained from the following theorem.

Theorem 4.11 (b-function, Bernstein–Sato polynomial) *Let $K(w)$ be a real analytic function on an open set U in $\mathbb{R}^d$. Then there exist a differential operator $D(z, w, \partial_w)$ and a polynomial $b(z)$ of z such that*

$$D(z, w, \partial_w) K(w)^{z+1} = b(z) K(w)^z \quad (w \in U, z \in \mathbb{C}), \tag{4.23}$$

where $D(z, w, \partial_w)$ and $b(z)$ satisfies the following conditions.
(1) $D(z, w, \partial_w)$ is a finite sum

$$D(z, w, \partial_w) = \sum_\alpha a_\alpha(w) b_\alpha(z) \frac{\partial^\alpha}{\partial w^\alpha}$$

using real analytic functions $a_\alpha(w)$ and polynomials $b_\alpha(z)$.
(2) The smallest-order function among all functions $b(z)$ that satisfy eq.(4.23) is uniquely determined if the coefficient of the largest order is 1. This polynomial is called the b-function or Bernstein-Sato polynomial.
(3) The roots of $b(z) = 0$ are real, negative, and rational numbers.

(Explanation of Theorem 4.11) This theorem was proved by Bernstein [16], Sato [80], Kashiwara [46], and Björk [17]. In the proof that all roots of the b-function are rational numbers, resolution of singularities is employed. The algorithm and software for calculating the b-function for a given polynomial were realized by Oaku [67, 68] and Takayama [89]. In the analysis of the b-function, D-module theory plays the central role. From a historical point of view, in 1954, Gel'fand conjectured that the Schwartz distribution $K(w)^z$ can

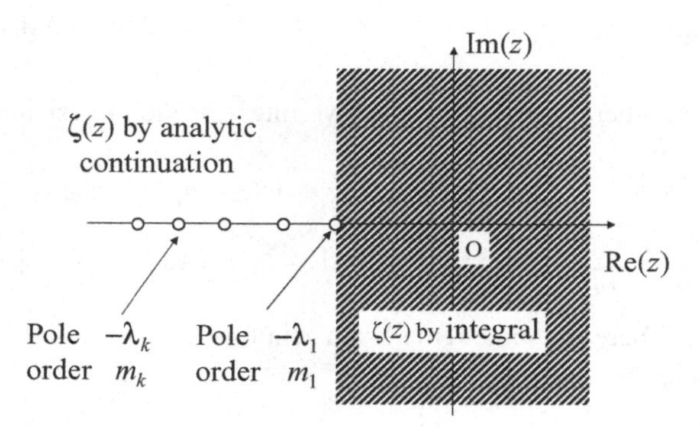

Fig. 4.4. Analytic continuation of zeta function

be analytically continued to the entire complex plane. The first answer to the conjecture was given by Atiyah using resolution of singularities [14]. Without resolution theorem, Bernstein and Sato created the b-function.

Let $D^*(z, w, \partial_w)$ be the adjoint operator of $D(z, w, \partial_w)$,

$$D^*(z, w, \partial_w)\varphi(w) = \sum_{\alpha}(-1)^{|\alpha|} b_\alpha(z) \frac{\partial^\alpha}{\partial w^\alpha}\{a_\alpha(w)\varphi(w)\}.$$

Then

$$\int \psi(w)(D\varphi(w))dw = \int (D^*\psi(w))\varphi(w)dw$$

for arbitrary $\psi(w), \varphi(w) \in C_0^\infty$. By using the existence of the b-function,

$$\begin{aligned} \zeta(z) &= \int K(w)^z \varphi(w)dw \\ &= \frac{1}{b(z)} \int D(z, w, \partial_w) K(w)^{z+1} \varphi(w)dw \\ &= \frac{1}{b(z)} \int K(w)^{z+1} D^*(z, w, \partial_w)\varphi(w)dw. \end{aligned} \tag{4.24}$$

Therefore the zeta function $\zeta(z)$ can be analytically continued to $\mathrm{Re}(z) > -1$ by eq.(4.24). By recursively using this procedure illustrated in Figure 4.4, the zeta function can be analytically continued to the meromorphic function on the entire complex plane. All poles of the zeta function are real, negative, and rational numbers.

Example 4.9 (1) For x^2,

$$\frac{d^2}{dx^2}(x^2)^{z+1} = (2z+2)(2z+1)(x^2)^z$$

shows that its b-function is $(z+1)(z+1/2)$. The largest root of the b-function is $-1/2$.
(2) For $x^j y^k$, where j and k are positive integers, the largest root of the b-function is $-\min(1/j, 1/k)$.
(3) For $x^2+y^2+z^2$, the b-function is $(z+1)(z+3/2)$, since

$$\left(\frac{\partial^2}{\partial x^2}+\frac{\partial^2}{\partial y^2}+\frac{\partial^2}{\partial w^2}\right)(x^2+y^2+w^2)^{z+1} = (z+1)(4z+6)(x^2+y^2+w^2)^z.$$

Remark 4.5 There are some remarks on zeta functions.
(1) A holomorphic function

$$\zeta(z) = \int_0^1 x^z dx \quad (\mathrm{Re}(z) > -1)$$

is analytically continued to the meromorphic function $1/(z+1)$ on the entire complex plane; however, the integration

$$\int_0^1 x^z dx$$

is not a finite value if $\mathrm{Re}(z) \le -1$. In other words,

$$\zeta(z) \neq \int_0^1 x^z dx \quad (\mathrm{Re}(z) \le -1).$$

(2) Assume that a function $\varphi(x)$ is of class C^∞. The analytic continuation of a Schwartz distribution T_z defined by

$$T_z(\varphi) = \int_0^1 x^z \varphi(x) dx$$

can be found:

$$\begin{aligned}\zeta(z) &= \int_0^1 x^z(\varphi(0) + \varphi'(0)x + \frac{\varphi''(0)x^2}{2} + \cdots)dx \\ &= \frac{\varphi(0)}{z+1} + \frac{\varphi'(0)}{z+2} + \frac{\varphi''(0)}{2(z+3)} + \cdots .\end{aligned}$$

This relation shows that the asymptotic expansion of a Schwartz distribution is

$$T_z \cong \sum_{k=0}^{\infty} \frac{\delta^{(k)}(x)}{k!(z+k+1)}.$$

By the existence of the b-function, the zeta function defined by eq.(4.22) can be understood as the meromorphic function on the entire complex plane. Let its poles be

$$0 > -\lambda_1 > -\lambda_2 > \cdots$$

and the order of $(-\lambda_k)$ be m_k. Then

$$\zeta(z) = \int K(w)^z \varphi(w) dw$$

has Laurent series expansion

$$\zeta(z) = \sum_{k=1}^{\infty} \frac{D_k(\varphi)}{(z+\lambda_k)^{m_k}},$$

where $\{D_k\}$ is a set of Schwartz distributions. By using an inverse Mellin transform, the state density function

$$v(t) = \int \delta(t - K(w)) \varphi(w) dw$$

has an asymptotic expansion,

$$v(t) = \sum_{k=1}^{\infty} \sum_{m=1}^{m_k} D_{km}(\varphi) t^{\lambda_k - 1} (-\log t)^{m-1},$$

where $\{D_{km}\}$ is a set of Schwartz distributions. Hence

$$\begin{aligned} Z(n) &= \int \exp(-nK(w)) \varphi(w) dw \\ &= \int_0^M \exp(-nt) v(t) dt, \end{aligned}$$

where $M = \max_w K(w)$, has also an asymptotic expansion

$$Z(n) = \sum_{k=1}^{\infty} \sum_{m=1}^{m_k} \frac{1}{n^{\lambda_k}} D_{km}(\varphi) \int_0^{Mn} t^{\lambda_k - 1} e^{-t} \left(\log \frac{n}{t}\right)^{m-1} dt.$$

Remark 4.6 In this chapter we have shown the asymptotic expansion of

$$Z(n) = \int \exp(-nK(w)) \varphi(w) dw.$$

The largest term of $Z(n)$ is $(\log n)^{m_1 - 1}/n^{\lambda - 1}$. In statistical learning theory, we need the probability distribution of

$$Z_n = \int \exp(-nK_n(w)) \varphi(w) dw.$$

In the following chapter, we prove that

$$Z_n = \frac{(\log n)^{m_1 - 1}}{n^{\lambda}} Z_n^*$$

where Z_n^* is a random variable which converges in law to a random variable.

Remark 4.7 (Riemann zeta function) In modern mathematics, a lot of zeta functions play important roles in many mathematical research fields. The original concept of many zeta functions is the Riemann zeta function

$$\zeta(z) = \sum_{n=1}^{\infty} \frac{1}{n^z},$$

which is equal to

$$\zeta(z) = \frac{1}{\Gamma(z)} \int_0^{\infty} \frac{x^{z-1}}{e^x - 1} dx.$$

This function is holomorphic in $\mathrm{Re}(z) > 1$ and analytically continued to the meromorphic function of the entire complex plane by the relation,

$$\frac{1}{\pi^{z/2}} \Gamma\left(\frac{z}{2}\right) \zeta(z) = \frac{1}{\pi^{(1-z)/2}} \Gamma\left(\frac{1-z}{2}\right) \zeta(1-z),$$

where $\Gamma(z)$ is the gamma function that satisfies $\Gamma(z+1) = z\Gamma(z)$. The Riemann hypothesis states that all roots of $\zeta(z)$ in $0 < \mathrm{Re}(z) < 1$ are on the line $\mathrm{Re}(z) = 1/2$. The Riemann hypothesis is one of the most important conjectures in mathematics. In this book we show that the zeta function of a statistical model determines the asymptotic learning efficiency. The Riemann zeta function determines the asymptotic distribution of the prime numbers.

5
Empirical processes

In singular statistical models, the set of true parameters is not one point but a real analytic set with singularities. In conventional statistical learning theory, asymptotic normality is proved by applying central limit theorem to the log likelihood function in the neighborhood of the true parameter, whereas in singular learning theory, the main formulas are proved by applying empirical process theory to that in the neighborhood of the true analytic set with singularities.

5.1 Convergence in law

Definition 5.1 Let $(\Omega, \mathcal{B})$ be a measurable space.
(1) Let $\{P_n\}$ and P respectively be a sequence of probability distributions and a probability distribution on $(\Omega, \mathcal{B})$. It is said that $\{P_n\}$ converges to P in law, or $\{P_n\}$ weakly converges to P, if, for any bounded and continuous function $f : \Omega \to \mathbb{R}^1$,

$$\int f(x)P(dx) = \lim_{n\to\infty} \int f(x)P_n(dx). \tag{5.1}$$

(2) Let $\{X_n\}$ be a sequence of random variables and X be a random variable which take values on the measurable space $(\Omega, \mathcal{B})$. It is said that $\{X_n\}$ converges to X in law, or $\{X_n\}$ weakly converges to X, if P_{X_n} converges to P_X in law, or equivalently, if, for any bounded and continuous function f,

$$E_X[f(X)] = \lim_{n\to\infty} E_{X_n}[f(X_n)].$$

Remark 5.1 (1) It is well known that if, for any bounded and uniformly continuous function f,

$$E_{X_n}[f(X_n)] = E_X[f(X)],$$

then X_n converges to X in law. In other words, to prove the convergence of $\{X_n\}$ in law, we can restrict f as a uniformly bounded function. Here a function f is said to be uniformly continuous if, for a given $\epsilon > 0$, there exists $\delta > 0$ such that

$$|x - y| < \delta \Longrightarrow |f(x) - f(y)| < \epsilon. \tag{5.2}$$

Note that δ does not depend on x, y.
(2) Even if X_n and Y_n converge in law to X and Y respectively, $X_n + Y_n$ may not converge in law in general. For example, let X be an $\mathbb{R}^1$-valued random variable which is subject to the normal distribution with mean 0 and variance 1. Then $Y = -X$ is subject to the same probability distribution as X and both $X_n = X$ and $Y_n = (-1)^n X$ converge to X in law. However, $X_n + Y_n$ does not converge in law.
(3) In the definition of convergence in law, random variables $X_1, X_2, \ldots, X_n, \ldots$ and X have the same image space $(\Omega, \mathcal{B})$. They may be defined on different probability spaces. In other words, in the definition,

$$X_n : \Omega_n \to \Omega,$$

the space Ω_n may depend on n. Even if $\{X_n\}$ are defined on different spaces, the probability distributions of $\{X_n\}$ are defined on the common measurable set Ω.
(4) If all random variables $X_1, X_2, \ldots, X_n, \ldots$, and X are defined on the same probability space, and if $X_n \to X$ in probability, then $X_n \to X$ in law. In fact, for any bounded and uniformly continuous function f which satisfies eq.(5.2)

$$\begin{aligned}
|E_X[f(X)] - E_{X_n}[f(X_n)]| &\leq |E[f(X)] - E[f(X_n)]|_{\{|X - X_n| < \delta\}} \\
&\quad + E[|f(X_n) - f(X_n)|]_{\{|X - X_n| \geq \delta\}} \\
&\leq \epsilon + 2\Big(\sup_x |f(x)|\Big)\, E[1]_{\{|X - X_n| \geq \delta\}} \\
&\leq \epsilon + 2\Big(\sup_x |f(x)|\Big)\, P(|X - X_n| \geq \delta).
\end{aligned}$$

Hence by preparing arbitrary small $\epsilon > 0$ and then taking n sufficiently large, we obtain convergence in law.

Example 5.1 Let $X_n : \mathbb{R}^1 \to \mathbb{R}^1$ be a random variable which is subject to a probability distribution

$$p_n(x)dx = \frac{1}{C_n}\Big(\sum_{i=1}^{n} \frac{1}{2^i}\delta(x - i)\Big)dx,$$

where C_n is a normalizing constant. Then X_n converges in law because

$$\int f(x)p_n(x)dx = \frac{1}{C_n}\sum_{i=1}^{n}\frac{f(i)}{2^i} \to \sum_{i=1}^{\infty}\frac{f(i)}{2^i}.$$

On the other hand, a random variable which is subject to

$$p_n(x)dx = \delta(x-n)dx$$

does not converge in law, because

$$\int \sin(x)p_n(x)dx = \sin(n)$$

does not converge for a bounded and continuous function $\sin(x)$. Note that convergence in law is a mathematically different concept from the topology of a Schwartz distribution. In fact, $\sin(x)$ is not contained in C_0^∞. As a Schwartz distribution, $p_n(x) \to 0$ holds.

There are several elemental theorems for convergence in law. We use Thereoms 5.1 and 5.2 in Chapter 6.

Theorem 5.1 *Let $(\Omega_1, \mathcal{F}_1, P)$ be a probability space. Also let $(\Omega_2, \mathcal{F}_2)$ and $(\Omega_3, \mathcal{F}_3)$ be measurable spaces. Assume that $\{X_n : \Omega_1 \to \Omega_2\}$ is a sequence of random variables which converges to X in law. If $g : \Omega_2 \to \Omega_3$ is a continuous function, then $\{g(X_n) : \Omega_1 \to \Omega_3\}$ converges to $g(X)$ in law.*

Proof of Theorem 5.1 If $f : \Omega_2 \to \Omega_3$ is a bounded and continuous function, then $f(g(\ ))$ is also a bounded and continuous function from Ω_1 to Ω_3. Therefore

$$\lim_{n\to\infty} E_{X_n}[f(g(X_n))] = E_X[f(g(X))],$$

which shows that $g(X_n)$ converges in law to $g(X)$. □

Theorem 5.2 *Let $\{X_n\}$ and $\{Y_n\}$ be sequences of random variables which take values on the Euclidean space $\mathbb{R}^N$.*
(1) If X_n converges to 0 in law, then X_n converges to 0 in probability, where the probability distribution of 0 is defined by $\delta(x)$.
(2) If X_n and Y_n respectively converge to X and 0 in law, then $X_n + Y_n$ converges to X in law.

Proof of Theorem 5.2 Let $p_n(dx)$ be a probability distribution of X_n. For an arbitrary $\epsilon > 0$, there exists a bounded and continuous function $\rho(x)$ $(0 \leq$

$\rho(x) \le 1$) which satisfies

$$\rho(x) = \begin{cases} 1 & (\|x\| \le \epsilon/2) \\ 0 & (\|x\| \ge \epsilon). \end{cases}$$

By the convergence in law of $X_n \to 0$,

$$\begin{aligned} 1 - P(\|X_n\| > \epsilon) &= \int_{\|x\| \le \epsilon} p_n(dx) \\ &\ge \int \rho(x) p_n(dx) \\ &\to \int \rho(x)\delta(x)dx = 1, \end{aligned}$$

which shows $P(\|X_n\| > \epsilon) \to 0$.
(2) Let $f : \mathbb{R}^N \to \mathbb{R}$ be a bounded and uniformly continuous function. For an arbitrary $\epsilon > 0$, there exists $\delta > 0$ such that

$$|x - y| < \delta \Longrightarrow |f(x) - f(y)| < \epsilon,$$

where δ does not depend on x, y. Then

$$\begin{aligned} A(n) &\equiv |E[f(X)] - E[f(X_n + Y_n)]| \\ &\le |E[f(X)] - E[f(X_n)]| + |E[f(X_n)] - E[f(X_n + Y_n)]| \\ &= |E[f(X)] - E[f(X_n)]| + B(n), \end{aligned}$$

where $B(n)$ is defined by

$$B(n) = E[|f(X_n) - f(X_n + Y_n)|].$$

Also

$$\begin{aligned} B(n) &= E[|f(X_n) - f(X_n + Y_n)|]_{\{|Y_n| < \delta\}} \\ &\quad + E[|f(X_n) - f(X_n + Y_n)|]_{\{|Y_n| \ge \delta\}} \\ &\le \epsilon + 2\Big(\sup_x |f(x)|\Big)\, E[1]_{\{|Y_n| \ge \delta\}}. \end{aligned}$$

Since f is a bounded function, $C \equiv 2 \sup |f(x)|$ is finite and

$$A(n) \le |E[f(X)] - E[f(X_n)]| + \epsilon + C\ P(|Y_n| \ge \delta)$$

By the convergence of $X_n \to X$ in law and the convergence of $Y_n \to 0$ in probability, we obtain the theorem. □

Assume that a sequence of random variables $\{X_n\}$ converges to X in law. If f is bounded and continuous, then $E_{X_n}[f(X_n)] \to E_X[f(X)]$ holds by definition. However, if f is not bounded, $E_{X_n}[f(X_n)]$ does not converge in general.

Theorem 5.3 *Assume that a sequence of random variables $\{X_n\}$ converges to X in law. If f is a continuous function and if*

$$\sup_n E[|f(X_n)|] < C,$$

then the following hold.
(1) The integration $E[|f(X)|]$ takes a finite value.
(2) $E[|f(X)|] \leq \limsup_{n\to\infty} E[|f(X_n)|]$.

Proof of Theorem 5.3 For a given continuous function $f(x)$, we define $f_M(x)$ by

$$f_M(x) = \begin{cases} f(x) & (|f(x)| \leq M) \\ M & (f(x) > M) \\ -M & (f(x) < -M). \end{cases} \tag{5.3}$$

Then $|f_M(X)| \leq |f(X)|$ and $f_M(x)$ is a continuous and bounded function. For each x, $|f_M(x)|$ is a non-decreasing function of M and

$$\lim_{M\to\infty} |f_M(x)| = |f(x)|. \tag{5.4}$$

Then $E[|f_M(X)|] \to E[|f(X)|]$ by the monotone lemma of Lebesgue measure theory. By the convergence $X_n \to X$ in law,

$$\begin{aligned} E[|f_M(X)|] &= \lim_{n\to\infty} E[|f_M(X_n)|] \\ &\leq \lim_{n\to\infty} \sup E[|f(X_n)|] < C \end{aligned}$$

holds for each M. □

Example 5.2 If a function f is unbounded, then three conditions, (1) $X_n \to X$ in law, (2) $E[|f(X_n)|] < \infty$, and (3) $E[|f(X)|] < \infty$, do not ensure $E[f(X_n)] \to E[f(X)]$ in general. For example, a sequence of probability distributions on $\mathbb{R}^1$

$$P_n(dx) = \frac{1}{n+1}\{\, n\,\delta(x) + \delta(x-n)\,\}\, dx$$

and a probability distribution

$$P(dx) = \delta(x)\, dx$$

satisfies the convergence in law, $P_n \to P$. Let X_n and X be random variables which are subject to P_n and P respectively. For a continuous and unbounded function $f(x) = x$,

$$E[f(X_n)] = \frac{n}{n+1} \to 1,$$

whereas

$$E[f(X)] = 0.$$

When we need convergence of the expectation value of a sequence of weak convergent random variables, the following definition is important.

Definition 5.2 (Asymptotically uniformly integrable, AUI) A sequence of real-valued random variables $\{X_n\}$ is said to be asymptotically uniformly integrable if it satisfies

$$\lim_{M\to\infty} \lim_{n\to\infty} \sup_{N\geq n} E[|X_N|]_{\{|X_N|\geq M\}} = 0. \tag{5.5}$$

Theorem 5.4 *(1) If two sequences of real-valued random variables $\{X_n\}$ and $\{Y_n\}$ satisfy $|X_n| \leq Y_n$ and $\{Y_n\}$ is asymptotically uniformly integrable, then $\{X_n\}$ is asymptotically uniformly integrable.*
(2) Let $\{X_n\}$ be a sequence of real-valued random variables. If there exists a random variable Y such that

$$|X_n| \leq Y, \quad E[Y] < \infty,$$

then $\{X_n\}$ is asymptotically uniformly integrable.
(3) Let $0 < \delta < s$ be positive constants. If a sequence of real-valued random variables $\{X_n\}$ satisfies $E[|X_n|^s] < C$ where C does not depend on n, then $X_n^{s-\delta}$ is asymptotically uniformly integrable.

Proof of Theorem 5.4 (1) Since $|X_n(w)| \leq Y_n(w)$ for any $w \in \Omega$, for any M,

$$\{w \in \Omega\ ;\ |X_n(w)| \geq M\} \subset \{w \in \Omega\ ;\ Y_n(w) \geq M\}.$$

It follows that

$$\begin{aligned} E[|X_n|]_{\{|X_n|\geq M\}} &\leq E[|X_n|]_{\{Y_n\geq M\}} \\ &\leq E[Y_n]_{\{Y_n\geq M\}} \end{aligned}$$

which shows (1) of the theorem.
(2) This is a special case of (1) by putting $Y_n = Y$.
(3) By the assumption, it follows that

$$E[|X_n|^{s-\delta}]_{\{|X_n|\geq M\}} \leq E[|X_n|^s/M^\delta]_{\{|X_n|\geq M\}} \leq \frac{C}{M^\delta}$$

which completes (3) of the theorem. □

Theorem 5.5 (Convergence of expectation value) *Assume that a sequence of Ω-valued random variables $\{X_n\}$ converges to X in law, and that $f : \Omega \to \mathbb{R}^1$ is a continuous function satisfying $E[|f(X_n)|] < C$. If $f(X_n)$ is asymptotically uniformly integrable, then*

$$\lim_{n\to\infty} E[f(X_n)] = E[f(X)]$$

holds.

Proof of Theorem 5.5 By Theorem 5.3, $E[f(X)]$ is well defined and finite. Let $f_M(x)$ be the function defined in eq.(5.3). Then

$$\begin{aligned}|E[f(X)] - E[f(X_n)]| \le\; & |E[f(X)] - E[f_M(X)]| \\ & + |E[f_M(X)] - E[f_M(X_n)]| \\ & + |E[f_M(X_n)] - E[f(X_n)]|.\end{aligned}$$

Let the last term be $A(n)$.

$$\begin{aligned}A(n) &\le E[|f_M(X_n) - f(X_n)|] \\ &\le E[|f(X_n)|]_{\{|f(X_n)|\ge M\}}.\end{aligned}$$

Therefore,

$$\begin{aligned}|E[f(X)] - E[f(X_n)]| \le\; & |E[f(X)] - E[f_M(X)]| \\ & + |E[f_M(X)] - E[f_M(X_n)]| \\ & + E[|f(X_n)|]_{\{|f(X_n)|\ge M\}}.\end{aligned}$$

By using Theorem 5.3 and the definition of AUI, eq.(5.5), for arbitrary $\epsilon > 0$, there exists M which satisfies both

$$|E[f(X)] - E[f_M(X)]| < \epsilon$$

and

$$\lim_{n\to\infty} \sup_{N\ge n} E[|f(X_N)|]_{\{|f(X_N)|\ge M\}} < \epsilon.$$

For such an M, we can choose n which satisfies both

$$|E[f_M(X)] - E[f_M(X_n)]| < \epsilon,$$

and

$$\sup_{N\ge n} E[|f(X_N)|]_{\{|f(X_N)|\ge M\}} < \epsilon,$$

which completes the theorem. □

Remark 5.2 In statistical learning theory, we often use this theorem in the case

$$Z_n = f(\xi_n) + a_n X_n.$$

We prove the following conditions.
(1) The sequence of random variables $\{Z_n\}$ is asymptotically uniformly integrable.
(2) The sequence of random variables $\{\xi_n\}$ converges to ξ in law.
(3) The function f is continuous.
(4) The real sequence a_n converges to zero.
(5) The sequence of random variables X_n converges in law.
Then, by Theorem 5.1, we have the convergence in law, $Z_n \to f(\xi)$, and by Theorem 5.5,

$$E[Z_n] \to E[f(\xi)].$$

5.2 Function-valued analytic functions

In statistical learning theory, a statistical model or a learning machine is a probability density function-valued analytic function. In this section, we introduce a function-valued analytic function.

Let $\mathbb{R}^N$ be an N-dimensional Euclidean space and $q(x) \geq 0$ be a probability density function on $\mathbb{R}^N$,

$$\int_{\mathbb{R}^N} q(x)dx = 1.$$

Hereafter, a real number $s \geq 1$ is fixed. The set of all measurable functions f from $\mathbb{R}^N$ to $\mathbb{C}^1$ which satisfy

$$\int |f(x)|^s q(x)dx < \infty$$

is denoted by $L^s(q)$. The set $L^s(q)$ is a vector space over $\mathbb{C}$. For an element $f \in L^s(q)$, we define the norm $\| \cdot \|_s$ by

$$\|f\|_s \equiv \left\{ \int |f(x)|^s q(x)dx \right\}^{1/s},$$

which satisfies the following conditions.
(1) For any $f \in L^s(q)$, $\|f\|_s \geq 0$.
(2) $\|f\|_s = 0 \Longleftrightarrow f = 0$.
(3) For $a \in \mathbb{C}$, $\|af\|_s = |a|\|f\|_s$.
(4) For any $f, g \in L^s(q)$, $\|f + g\|_s \leq \|f\|_s + \|g\|_s$.
By this norm, $L^s(q)$ is a Banach space.

Remark 5.3 In $L^s(q)$, the following hold.
(1) Let $\alpha > 0$ and $\beta > 0$ be constants which satisfy $1/\alpha + 1/\beta = 1$. For arbitrary $f, g \in L^s(q)$,

$$\left|\int f(x)g(x)q(x)dx\right| \leq \|f\|_\alpha \ \|g\|_\beta.$$

This is called Hölder's inequality. The special case $\alpha = \beta = 1/2$ is called the Cauchy–Schwarz inequality.
(2) From Hölder's inequality, for arbitrary $1 < s < s'$,

$$\|f\|_s \leq \|f\|_{s'}$$

holds. Therefore,

$$L^{s'}(q) \subset L^s(q).$$

(3) From a mathematical point of view, $L^s(q)$ is not a set of functions but the quotient set of the equivalent relation,

$$f \sim g \iff f(x) = g(x) \quad \text{(a.s. } q(x)dx).$$

Although f represents the equivalence class, the value $f(x)$ is well defined almost surely for x with $q(x)dx$.

Definition 5.3 (Function-valued analytic function) Let $s \geq 1$ be a real constant. A function

$$f : \mathbb{R}^N \times \mathbb{R}^d \ni (x, w) \mapsto f(x, w) \in \mathbb{R}^1$$

is said to be $L^s(q)$-valued real analytic if there exists an open set $W \subset \mathbb{R}^d$ such that

$$W \ni w \mapsto f(\ , w) \in L^s(q)$$

is analytic. Here 'analytic' means that, for arbitrary $w^* \in W$, there exists $\{a_\alpha(x) \in L^s(q)\}$ such that

$$f(x, w) = \sum_{\alpha \in \mathbf{N}^d} a_\alpha(x)(w - w^*)^\alpha \tag{5.6}$$

is a convergent power series in a Banach space; in other words,

$$\sum_{\alpha \in \mathbf{N}^d} \|a_\alpha\|_s \prod_{j=1}^{d} |w_j - w_j^*|^{\alpha_j} < \infty \tag{5.7}$$

in some open set

$$|w_j - w_j^*| < \delta_j \quad (j = 1, 2, \ldots, N).$$

By the completeness of $L^s(q)$, if eq.(5.7) holds, then the sum in eq.(5.6) absolutely converges.

Remark 5.4 (Analytic function of several complex variables) Let $f(z_1, z_2, \ldots, z_d)$ be a function from $\mathbb{C}^d$ to $\mathbb{C}^1$ which is given by the Taylor expansion,

$$f(z) = \sum_{\alpha \in \mathbf{N}^d} a_\alpha (z - b)^\alpha, \tag{5.8}$$

where $(b_1, b_2, \ldots, b_d) \in \mathbb{C}^d$. If the power series in eq.(5.8) absolutely converges in the complex region

$$\{(z_1, z_2, \ldots, z_d) \in \mathbb{C}^d \ ; \ |z_i - b_i| < r_i; \, i = 1, 2, \ldots, d\}, \tag{5.9}$$

and it does not in

$$\{(z_1, z_2, \ldots, z_d) \in \mathbb{C}^d \ ; \ |z_i - b_i| > r_i; \, i = 1, 2, \ldots, d\},$$

then $(r_1, \ldots, r_d)$ are said to be associated convergence radii of f. If $d = 1$, then r_1 is called the convergence radius that is uniquely determined. However, if $d \geq 2$, then there are another $(r_1', r_2', \ldots, r_d')$ which are also associated convergence radii. If real numbers $r = (r_1, r_2, \ldots, r_d)$ are associated convergence radii, then

$$\limsup_{|\alpha| \to \infty} |a_\alpha| r^\alpha = 1,$$

where, for a multi-index α, $|\alpha|$ is defined by $|\alpha| = \alpha_1 + \alpha_2 + \cdots + \alpha_d$. If $C_1, C_2, \ldots, C_d$ are closed and continuous lines in the region defined by eq.(5.9), then Cauchy's integral formula

$$f(b) = \frac{1}{(2\pi i)^d} \int_{C_1} \cdots \int_{C_d} \frac{f(z_1, \ldots, z_d)}{(z_1 - b_1) \cdots (z_d - b_d)} dz_1 \cdots dz_d$$

holds, where $i = \sqrt{-1}$.

Remark 5.5 (Function-valued analytic function of several complex variables) Let $f(x, z)$ be a function from $\mathbb{C}^d$ to $L^s(q)$ defined by

$$f(x, z) = \sum_{\alpha} a_\alpha(x)(z - b)^\alpha,$$

where $a_\alpha \in L^s(q)$. Then, by using the completeness of $L^s(q)$, the same associated convergence radii of a function-valued analytic function can be defined as above. For the associated convergence radii,

$$\limsup_{|\alpha| \to \infty} \|a_\alpha\|_s r^\alpha = 1.$$

Cauchy's integral formula also holds,

$$f(x,b) = \frac{1}{(2\pi i)^d} \int_{C_1} \cdots \int_{C_d} \frac{f(x,z)}{(z_1 - b_1) \cdots (z_d - b_d)} dz_1 \cdots dz_d.$$

Moreover, by

$$a_\alpha(x) = \frac{1}{(2\pi i)^d} \int_{C_1} \cdots \int_{C_d} \frac{f(x,z)}{(z_1 - b_1)^{\alpha_1+1} \cdots (z_d - b_d)^{\alpha_d+1}} dz_1 \cdots dz_d,$$

it follows that

$$|a_\alpha(x)| \leq \frac{1}{(2\pi)^d} \int_{C_1} \cdots \int_{C_d} \frac{\sup_z |f(x,z)|}{|z_1 - b_1|^{\alpha_1+1} \cdots |z_d - b_d|^{\alpha_d+1}} d|z_1| \cdots d|z_d|.$$

Then, if the radius of integrated path $\mathcal{C} = (C_1, C_2, \ldots, C_d)$ is given as $R^\alpha < r^\alpha$, we obtain an inequality,

$$|a_\alpha(x)| \leq \frac{\sup_{z \in \mathcal{C}} |f(x,z)|}{R^\alpha}. \tag{5.10}$$

Since the Taylor expansion of $f(x,z)$ at $z = b$ converges absolutely, completeness of $L^s(q)$ ensures that

$$\sup_{z \in \mathcal{C}} |f(x,z)|$$

is contained in $L^s(q)$. Therefore,

$$\|a_\alpha\|_s \leq \frac{\| \sup_{z \in \mathcal{C}} |f(\cdot,z)| \|_s}{r^\alpha} < \infty \tag{5.11}$$

holds. Let W_R^* be an open set in $\mathbb{R}^d$ and $f(\cdot, w)$ be an $L^s(q)$-valued real analytic function on W_R^*. For a given compact set $W \subset W_R^*$, W can be covered by an open set $W^* \subset \mathbb{C}^d$ on which $f(\cdot, w)$ can be defined as a complex analytic function. Then

$$M(x) \equiv \sup_{w \in W^*} |f(x,w)|$$

is contained in $L^s(q)$. The coefficient $a_\alpha(x)$ in the Taylor expansion of $f(x,w)$ for $w_0 \in W$

$$f(x,w) = \sum_\alpha a_\alpha(x)(w - w_0)^\alpha$$

satisfies

$$|a_\alpha(x)| \leq \frac{M(x)}{R^\alpha},$$

where $R = (R_1, \ldots, R_d)$ shows an multi-indexed radii of $C_1, C_2, \ldots, C_d$.

5.3 Empirical process

The following is the well-known central limit theorem of $\mathbb{R}^d$-valued random variables.

Theorem 5.6 (Central limit theorem) *Let $\{X_n\}$ be a sequence of $\mathbb{R}^d$-valued random variables which are independently subject to the same probability distribution of a random variable X. The expectation $m = (m_i) \in \mathbb{R}^d$ and the covariance matrix $\{\sigma_{ij}\}$ of*

$$X = (X_{(1)}, X_{(2)}, \ldots, X_{(d)}) \in \mathbb{R}^d$$

are respectively defined by

$$m_i = E[X_{(i)}],$$

$$\sigma_{ij} = E[(X_{(i)} - m_{(i)})(X_{(j)} - m_{(j)})] \quad (1 \le i, j \le d).$$

If X has finite expectation and finite covariance matrix, then the $\mathbb{R}^d$-valued random variable

$$Y_n = \frac{1}{\sqrt{n}} \sum_{i=1}^{n} (X_i - m)$$

converges in law to the normal distribution with expectation 0 and covariance matrix $\{\sigma_{ij}\}$.

Proof of Theorem 5.6 This is the well-known central limit theorem. □

In this section, we study the extension of the central limit theorem to $L^s(q)$-valued random variables.

Definition 5.4 (Pre-empirical process) Let X be an $\mathbb{R}^N$-valued random variable and W be a subset of $\mathbb{R}^d$. Let $f(x, w)$ be a function

$$\mathbb{R}^N \times W \ni (x, w) \mapsto f(x, w) \in \mathbb{R}^1,$$

which satisfies

$$m(w) = E_X[f(X, w)] < \infty$$

and

$$\rho(w, w') \equiv E_X[(f(X, w) - m(w))(f(X, w') - m(w'))] < \infty$$

for each $w, w' \in W$. Let $(\mathbb{R}^W, \mathcal{B}^W)$ be a measurable space where $\mathbb{R}^W$ is the set of all real functions on W and $\mathcal{B}$ is the minimal σ-algebra that contains all cylindrical sets in the function space

$$\{g \in \mathbb{R}^W \; ; \; a_i < g(w_i) < b_i, \; w_i \in W, \text{(for arbitrary finite } i)\}.$$

Assume that $\{X_n\}$ is a sequence of $\mathbb{R}^N$-valued random variables which are independently subject to the same probability distribution as X. Then an $\mathbb{R}^W$-valued random variable

$$\psi_n(w) = \frac{1}{\sqrt{n}} \sum_{i=1}^{n} (f(X_i, w) - E_X[f(X, w)]) \tag{5.12}$$

is said to be a pre-empirical process. For arbitrary $w, w' \in W$

$$E[\psi_n(w)] = 0, \tag{5.13}$$

$$E[\psi_n(w)\psi_n(w')] = \rho(w, w'). \tag{5.14}$$

Remark 5.6 By the central limit theorem (Theorem 5.6), for arbitrary finite points $w_1, w_2, \ldots, w_T \in W$,

$$(\psi_n(w_1), \psi_n(w_2), \ldots, \psi_n(w_T))$$

converges in law to the T-dimensional normal distribution with expectation 0 and covariance matrix $\rho(w_i, w_j)$ $(1 \leq i, j \leq T)$. The limiting normal distribution on $(w_1, w_2, \ldots, w_T)$ is equal to the marginalized distribution of any limiting distribution that contains $(w_1, w_2, \ldots, w_T)$. Such a property is called the consistency of the distribution. Kolmogorv's extension theorem ensures that, given this consistency condition, there exists a unique probability distribution ψ on $(\mathbb{R}^W, \mathcal{B}^W)$. The pre-empirical process $\psi_n(w)$ converges to $\psi(w)$ in law on every set of finite points. Note that this weak convergence holds only on every set of finite points. To prove the uniform version of the convergence in law $\psi_n(w) \to \psi(w)$, in other words, to prove the central limit theorem in a Banach space, we need some assumptions about $f(x, w)$.

Theorem 5.7 *Let s be a positive and even integer, and X be an $\mathbb{R}^d$-valued random variable which is subject to the probability distribution $q(x)dx$. Assume that $f(x) \in L^s(q)$ and $E_X[f(X)] = 0$. Also assume that $X_1, X_2, \ldots, X_n$ are subject to the same probability distribution as X. Then the following hold.*
(1) If $s = 2$, for any natural number n,

$$E\Big[\,\Big|\frac{1}{\sqrt{n}}\sum_{i=1}^{n} f(X_i)\Big|^2\,\Big]^{1/2} = \|f\|_2. \tag{5.15}$$

(2) If $s \geq 4$, for any natural number n,

$$E\Big[\,\Big|\frac{1}{\sqrt{n}}\sum_{i=1}^{n} f(X_i)\Big|^s\,\Big]^{1/s} \leq (s-1)\|f\|_s. \tag{5.16}$$

Proof of Theorem 5.7 (1) Let A be the left-hand side of eq.(5.15). Since $E[f(X_i)f(X_j)] = 0$ for $i \neq j$,

$$A^2 = \frac{1}{n}\sum_{i=1}^{n} E[f(X_i)^2]$$

$$= \int f(x)^2 q(x)dx = \|f\|_2^2.$$

(2) Let us prove the case $s \geq 4$. Let Y_n be a random variable,

$$Y_n = \sum_{i=1}^{n} f(X_i).$$

By $E[f(X)] = 0$ and Hölder's inequality,

$$E[Y_{n+1}^s] = E[(Y_n + f(X_{n+1}))^s]$$

$$= \sum_{k\neq 1} \binom{s}{k} E[Y_n^{s-k} f(X_{n+1})^k]$$

$$\leq \sum_{k\neq 1} \binom{s}{k} E[Y_n^s]^{(s-k)/s} E[f(X_{n+1})^s]^{k/s}$$

$$= \sum_{k\neq 1} \binom{s}{k} E[Y_n^s]^{(s-k)/s} \|f\|_s^k$$

$$= (E[Y_n^s]^{1/s} + \|f\|_s)^s - sE[Y_n^s]^{(s-1)/s}\|f\|_s.$$

By the definition of Y_n,

$$y_n = E\Big[\Big(\frac{Y_n}{\sqrt{n}\,\|f\|_s}\Big)^s\Big]$$

satisfies $y_n \geq 0$ and $y_1 = 1$. Moreover,

$$y_{n+1} = E\Big[\Big(\frac{Y_n + f(X_{n+1})}{\sqrt{n+1}\,\|f\|_s}\Big)^s\Big] \tag{5.17}$$

$$\leq \frac{1}{(n+1)^{s/2}}\big[(\sqrt{n}\,y_n^{1/s} + 1)^s - s(\sqrt{n}\,y_n^{1/s})^{s-1}\big]$$

$$= \frac{y_n}{(1+1/n)^{s/2}}\Big[(1 + \frac{1}{\sqrt{n}\,y_n^{1/s}})^s - \frac{s}{\sqrt{n}\,y_n^{1/s}}\Big]. \tag{5.18}$$

For a function $f(x) = (1+x)^s$ with $x \geq 0$, there exists $0 \leq x^* \leq x$ such that

$$f(x) = 1 + f'(0)x + \frac{f''(x^*)x^2}{2},$$

giving

$$f(x) \leq 1 + sx + \frac{s(s-1)(1+x)^{s-2}x^2}{2}. \tag{5.19}$$

Also we have

$$\left(1+\frac{1}{n}\right)^{s/2} \geq 1 + \frac{s}{2n}. \tag{5.20}$$

Applying eq.(5.19) with $x = 1/(\sqrt{n}y_n^{1/s})$ and eq.(5.20) to eq.(5.18) and by defining

$$F_n(y) = \frac{y}{(1+s/2n)}\left[1 + \frac{s(s-1)}{2n\ y^{2/s}}\left(1+\frac{1}{\sqrt{n}\ y^{1/s}}\right)^{s-2}\right],$$

we obtain

$$y_{n+1} \leq F_n(y_n).$$

The function $F_n(y)$ is monotone-increasing for $y > 0$, and $F_n(y) > 0$. Moreover,

$$\begin{aligned} F_n((s-1)^s) &= \frac{(s-1)^s}{1+s/2n}\left[1+\frac{s}{2n(s-1)}\left(1+\frac{1}{\sqrt{n}\ (s-1)}\right)^{s-2}\right] \\ &\leq \frac{(s-1)^s}{1+s/2n}\left[1+\frac{se}{2n(s-1)}\right] \leq (s-1)^s, \end{aligned}$$

where we used $(1+1/t)^t < e < s-1$ $(t \geq 0)$. Therefore, for arbitrary n,

$$0 < y_n \leq (s-1)^s,$$

which completes the theorem. □

Theorem 5.8 *Let $s \geq 2$ be a positive and even integer. Assume that the function $f(x,w)$ defines an $L^s(q)$-valued real analytic function on the open set $W \subset \mathbb{R}^d$. For the pre-empirical process $\psi_n(w)$ in eq.(5.12), there exists a constant $C > 0$ such that*

$$E[\sup_{w\in K} |\psi_n(w)|^s] \leq C,$$

where $K \subset W$ is a compact set.

Proof of Theorem 5.8 Since $\psi_n(w)$ is defined so that $E[\psi_n(w)] = 0$, we can assume $E_X[f(X,w)] = 0$ without loss of generality. The Taylor expansion of $f(x,w)$ for $w^* \in K$ absolutely converges in a region,

$$B(w^*, r(w^*)) \equiv \{w \in W; |w_j - w_j^*| < r_j\},$$

where $\{r_j\}$ are associated convergence radii. The set K can be covered by the union of all open sets,

$$K \subset \cup_{w^* \in K} B(w^*, r(w^*)).$$

Since K is compact, K is covered by a finite union of $B(w^*, r(w^*))$. Therefore, in order to prove the theorem, it is sufficient to prove that, in each $B = B(w^*, r(w^*))$,

$$E[\sup_{w \in B} |\psi_n(w)|^s] \leq C$$

holds. By the Taylor expansion of $f \in L^s(q)$ for w^*,

$$\begin{aligned} \psi_n(w) &\equiv \frac{1}{\sqrt{n}} \sum_{i=1}^{n} f(X_i, w) \\ &= \frac{1}{\sqrt{n}} \sum_{i=1}^{n} \sum_{\alpha} a_\alpha(X_i)(w - w^*)^\alpha \\ &= \sum_{\alpha} \Big\{ \frac{1}{\sqrt{n}} \sum_{i=1}^{n} a_\alpha(X_i) \Big\} (w - w^*)^\alpha, \end{aligned}$$

where $a_\alpha \in L^s(q)$ is a coefficient of Taylor expansion. Since

$$\sup_{w \in B} |\psi_n(w)| \leq \sum_{\alpha} \Big| \frac{1}{\sqrt{n}} \sum_{i=1}^{n} a_\alpha(X_i) \Big| \prod_j r_j^{\alpha_j},$$

we can apply Theorem 5.7

$$\begin{aligned} E\Big[\sup_{w \in B} |\psi_n(w)|^s \Big]^{1/s} &\leq \sum_{\alpha} E\Big[\Big\{ \Big| \frac{1}{\sqrt{n}} \sum_{i=1}^{n} a_\alpha(X_i) \Big| \prod_j r_j^{\alpha_j} \Big\}^s \Big]^{1/s} \\ &\leq \sum_{\alpha} E\Big[\Big| \frac{1}{\sqrt{n}} \sum_{i=1}^{n} a_\alpha(X_i) \Big|^s \Big]^{1/s} \prod_j r_j^{\alpha_j} \\ &\leq (s-1) \sum_{\alpha} \|a_\alpha\|_s \prod_j r_j^{\alpha_j} \end{aligned}$$

Here the $\{r_j\}$ were taken to be smaller than the associated convergence radii, and the last term is a finite constant. □

Definition 5.5 (Separable and complete metric space of continuous functions) Let $K \subset \mathbb{R}^d$ be a compact set and $C(K)$ be a set of all the continuous functions from K to $\mathbb{C}$. A metric of $f_1, f_2 \in C(K)$ is defined by

$$d(f_1, f_2) = \max_{x \in K} |f_1(x) - f_2(x)|.$$

Then $C(K)$ is a complete and separable metric space. In fact, it is well known in the Weierstrass approximation theorem that, for any function $f_1 \in C(K)$ and any $\epsilon > 0$, there exists a polynomial f_2 which satisfies $d(f_1, f_2) < \epsilon$. Let $\mathcal{B}$ be the Borel set of $C(K)$, or equivalently, the minimal sigma algebra that contains all open subsets in $C(K)$. Then $(C(K), \mathcal{B})$ is a measurable set.

Definition 5.6 (Empirical process) Let $(\Omega, \mathcal{B}, P)$ be a probability space and X be an $\mathbb{R}^N$-valued random variable. Assume that $\{X_n\}$ is a set of $\mathbb{R}^N$-valued random variables which are independently subject to the same probability distribution as X. The probability distribution of X is denoted by $q(x)dx$. Let $f : \mathbb{R}^N \times W \to \mathbb{R}^1$ define an $L^s(q)$-valued analytic function $w \mapsto f(\cdot, w)$ where $W \subset \mathbb{R}^d$ is an open set. Assume that $E[f(X, w)^2] < \infty$ for any $w \in W$ and that the compact set K is a subset of W. The empirical process ψ_n is a $C(K)$-valued random variable defined by

$$\psi_n(w) = \frac{1}{\sqrt{n}} \sum_{i=1}^{n} \{f(X_i, w) - E_X[f(X, w)]\}. \tag{5.21}$$

Remark 5.7 In the above definition, ψ_n is contained in the functional space $C(K)$ and it is measurable, therefore ψ_n is a $C(K)$-valued random variable. The empirical process can be defined even if $f(\cdot, w)$ is not an $L^s(q)$-valued analytic function; however, in this book, we adopt this restricted definition. For the more general definition, see [33, 92].

Definition 5.7 (Tight random variable) Let $(C(K), \mathcal{B})$ be a measurable space as defined in Definition 5.5.
(1) A probability measure μ on $(C(K), \mathcal{B})$ is said to be tight if, for any $\epsilon > 0$, there exists a compact set $C \subset C(K)$ such that $\mu(C) > 1 - \epsilon$.
(2) A sequence of measures $\{\mu_n\}$ is said to be uniformly tight if, for any $\epsilon > 0$, there exists a compact set $C \subset C(K)$ such that $\inf_n \mu_n(C) > 1 - \epsilon$.
(3) A $C(K)$-valued random variable X is said to be tight if the probability measure to which X is subject is tight. A sequence of $C(K)$-valued random variables $\{X_n\}$ is said to be uniformly tight if the sequence of probability distributions to which $\{X_n\}$ are subject is uniformly tight.

Example 5.3 The empirical process $\{\psi_n(w)\}$ in Definition 5.6 is uniformly tight if, for arbitrary $\epsilon > 0$, there exists a compact set $C \subset C(K)$ such that

$$P(\psi_n \in C) > 1 - \epsilon$$

for any n. If

$$E[\sup_{w\in K} |\psi_n(w)|^s] = C' < \infty,$$

then

$$\begin{aligned} C' &\geq E[\sup_{w\in K} |\psi_n(w)|^s]_{\{\sup_w |\psi_n(w)|\geq M\}} \\ &\geq M^s P(\sup_{w\in K} |\psi_n(w)| \geq M). \end{aligned}$$

Therefore

$$P(\sup_{w\in K} |\psi_n(w)| < M) \geq 1 - \frac{C'}{M^s},$$

which shows $\{\psi_n\}$ is uniformly tight.

Theorem 5.9 (Convergence in law of empirical process) *Let $s \geq 2$ be a positive and even constant. Assume that $f(x, w)$ is an $L^s(q)$-valued analytic function defined on an open set $W \subset \mathbb{R}^N$. If K is a compact subset in W, then the empirical process $\psi_n \in C(K)$ defined in eq.(5.21) satisfies the following conditions.*
(1) The set of $C(K)$-valued random variables $\{\psi_n\}$ is uniformly tight.
(2) There exists a $C(K)$-valued random variable ψ such that $\{\psi_n\}$ converges to ψ in law. In other words, for an arbitrary bounded and continuous function g from $C(K)$ to $\mathbb{R}^1$,

$$\lim_{n\to\infty} E_{\psi_n}[g(\psi_n)] = E_{\psi}[g(\psi)].$$

(3) The random process ψ is the normal distribution that has expectation eq.(5.13) and covariance eq.(5.14).

Proof of Theorem 5.9 Firstly, by Example 5.3 and Theorem 5.8, the sequence $\{\psi_n(w)\}$ is uniformly tight. Prohorov's theorem shows that, in a uniformly tight sequence of measures on a complete and separable metric space, there exists a subsequence which converges in law to a tight measure. Since $C(K) \subset \mathbb{R}^K$, $\mathcal{B} \subset \mathcal{B}^K$ and the limiting process is unique in $(\mathbb{R}^K, \mathcal{B}^K)$ by Remark 5.6, we obtain the theorem. □

Remark 5.8 Even if $f(x, w)$ is not an $L^s(q)$-valued analytic function, the same conclusion as Theorem 5.9 can be derived on the weaker condition [33, 92].

Theorem 5.10 *Let s be a positive even integer. Assume that a function $f(x, w)$ is an $L^{s+2}(q)$-valued analytic function of $w \in W$ and $K \subset W$ is a compact set. The empirical process ψ_n defined by eq.(5.21) satisfies*

$$\lim_{n\to\infty} E_{\psi_n}[\sup_{w\in K} |\psi_n(w)|^s] = E_{\psi}[\sup_{w\in K} |\psi(w)|^s].$$

Proof of Theorem 5.10 Let $h : C(K) \to \mathbb{R}^1$ be a function defined by

$$C(K) \ni \varphi \mapsto h(\varphi) \equiv \sup_{w \in K} |\varphi(w)| \in \mathbb{R}^1.$$

Then h is a continuous function on $C(K)$ because

$$\begin{aligned} |h(\varphi_1) - h(\varphi_2)| &= |\sup_{w \in K} |\varphi_1(w)| - \sup_{w \in K} |\varphi_2(w)|| \\ &\leq \sup_{w \in K} |\varphi_1(w) - \varphi_2(w)|. \end{aligned}$$

By the convergence in law of $\psi_n \to \psi$ in Theorem 5.9 and continuity of h, the convergence in law

$$\sup_{w \in K} |\psi_n(w)|^s \to \sup_{w \in K} |\psi(w)|^s$$

holds. On the other hand, by Theorem 5.8,

$$E_{\psi_n}[\sup_{w \in K} |\psi_n(w)|^{s+2}] \leq C$$

which implies that $h(\psi_n)^s$ is asymptotically uniformly integrable. By applying Theorem 5.5, we obtain the theorem. □

Example 5.4 Let X and Y be real-valued independent random variables which are respectively subject to the uniform distribution on $[-1, 1]$ and the standard normal distribution. Let $q(x)$ and $q(y)$ denote the probability density functions of X and Y respectively. When we use a statistical model,

$$p(y|x, a, b) = \frac{1}{\sqrt{2\pi}} \exp(-\tfrac{1}{2}(y - a\tanh(bx))^2),$$

then $q(y) = p(y|x, 0, 0)$. If $\{X_i, Y_i\}$ are independently taken from the distribution $q(x)q(y)$, then the Kullback–Leibler distance and the log likelihood ratio function are respectively given by

$$K(a, b) = \frac{a^2}{2} \int \tanh(bx)^2 q(x) dx,$$

$$K_n(a, b) = \frac{1}{2n} \sum_{i=1}^{n} \{a^2 \tanh(bX_i)^2 - 2aY_i \tanh(bX_i)\}.$$

If we define

$$[K(a, b)]^{1/2} = \begin{cases} +\sqrt{K(a, b)} & (ab \geq 0) \\ -\sqrt{K(a, b)} & (ab < 0), \end{cases}$$

then the origin $a = b = 0$ is a removable singularity of

$$\psi_n(a,b) = \sqrt{n}\,\frac{K(a,b) - K_n(a,b)}{[K(a,b)]^{1/2}},$$

which is an empirical process, and the log likelihood ratio function can be written as the standard form,

$$K_n(a,b) = K(a,b) + \frac{[K(a,b)]^{1/2}}{\sqrt{n}}\psi_n(a,b).$$

In Chapter 6, we show that any log likelihood ratio function can be written in the same standard form.

Example 5.5 Let X and Y be random variables which are independently subject to the standard normal distribution. We study a function

$$f(a,b,x,y) = ax + by.$$

Then the expectation and variance are respectively given by

$$E[f(a,b,X,Y)] = 0,$$
$$E[f(a,b,X,Y)^2] = a^2 + b^2.$$

If $\{X_i\}$ and $\{Y_i\}$ are independently subject to the same probability distribution as X and Y respectively,

$$\begin{aligned} f_n(a,b) &= \frac{1}{\sqrt{n}}\sum_{i=1}^{n}\frac{f(X_i,Y_i,a,b)}{\sqrt{a^2+b^2}} \\ &= \frac{a}{\sqrt{a^2+b^2}}\Big(\frac{1}{\sqrt{n}}\sum_{i=1}^{n}X_i\Big) + \frac{b}{\sqrt{a^2+b^2}}\Big(\frac{1}{\sqrt{n}}\sum_{i=1}^{n}Y_i\Big) \end{aligned}$$

defines an empirical process on a set

$$W = \{(a,b); |a|, |b| \le 1,\ a^2 + b^2 \neq 0\}.$$

Although $E[f_n(a,b)^2] = 1$ for an arbitrary $(a,b) \in W$,

$$\lim_{(a,b)\to(0,0)} f_n(a,b)$$

does not exist. The origin is a singularity of this empirical process. Let us apply the blow-up,

$$a = a_1 = a_2 b_2,$$
$$b = a_1 b_1 = b_2.$$

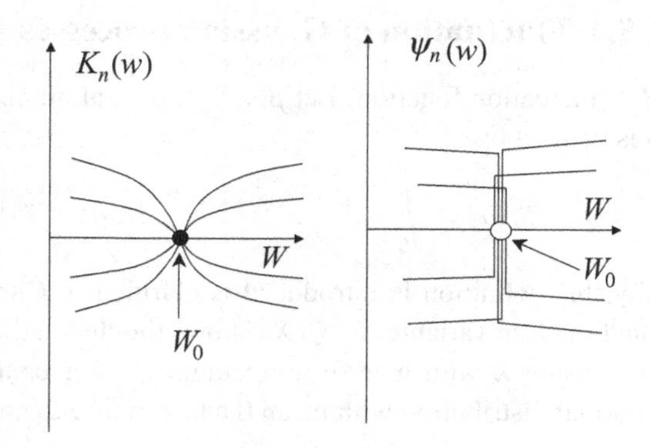

Fig. 5.1. Empirical process

Then

$$\begin{aligned}(a^2+b^2)^{1/2} &= \left(a_1^2\right)^{1/2}\left(1+b_1^2\right)^{1/2} \\ &= \left(b_2^2\right)^{1/2}\left(a_2^2+1\right)^{1/2}.\end{aligned}$$

Therefore, instead of $f_n(a,b)$, we can consider $\hat{f}_n$ on the real projective space represented on each coordinate by

$$\begin{aligned}\hat{f}_n &= \frac{1}{\sqrt{1+b_1^2}}\Big(\frac{1}{\sqrt{n}}\sum_{i=1}^{n}X_i\Big)+\frac{b_1}{\sqrt{1+b_1^2}}\Big(\frac{1}{\sqrt{n}}\sum_{i=1}^{n}Y_i\Big) \\ &= \frac{a_2}{\sqrt{a_2^2+1}}\Big(\frac{1}{\sqrt{n}}\sum_{i=1}^{n}X_i\Big)+\frac{1}{\sqrt{a_2^2+1}}\Big(\frac{1}{\sqrt{n}}\sum_{i=1}^{n}Y_i\Big),\end{aligned}$$

which is an empirical process without singularities.

Example 5.6 In singular learning theory, the log likelihood ratio function $K_n(w)$ is illustrated in the left of Figure 5.1. If $w \in W_0$, in other words, $K(w)=0$, then $K_n(w)=0$. To analyze $K_n(w)$, it is represented as

$$K_n(w) = K(w) - \frac{1}{\sqrt{n}}\sqrt{K(w)}\,\psi_n(w),$$

but $\psi_n(w)$ is ill-defined at a singularity of $K(w)=0$ as in the right of Figure 5.1. In Chapter 6, we show that the same procedure as in Example 5.4 and 5.5, where $\psi_n(w)$ becomes well-defined, is realized by resolution of singularities.

5.4 Fluctuation of Gaussian processes

Definition 5.8 (Fluctuation function) Let $\beta, \lambda > 0$ be real numbers. A fluctuation function is defined by

$$S_\lambda(a) = \int_0^\infty t^{\lambda-1}\, e^{-\beta t + \beta\, a\, \sqrt{t}}\, dt.$$

The reason why this definition is introduced is clarified in Chapter 6. For a given real-valued random variable X, $S_\lambda(X)$ shows the fluctuation of X under the singular dimension λ with inverse temperature β. For example, if X is subject to the normal distribution with mean 0 and variance 2, and $0 < \beta < 1$, then

$$E[S_\lambda(X)] = \frac{\Gamma(\lambda)}{(\beta - \beta^2)^\lambda},$$

where

$$\Gamma(\lambda) = \int_0^\infty t^{\lambda-1}\, e^{-t}\, dt$$

is the gamma function. It is easy to show that

$$\begin{aligned} S_\lambda(0) &= \Gamma(\lambda), \\ S'_\lambda(a) &= \beta\ S_{\lambda+1/2}(a), \\ S''_\lambda(a) &= \beta^2 S_{\lambda+1}(a). \end{aligned}$$

By using partial integration, it follows that

$$\begin{aligned} S_{\lambda+1}(a) &= \int_0^\infty \Big(-\frac{e^{-\beta t}}{\beta}\Big)' (t^\lambda\, e^{\beta\, a\, \sqrt{t}})\, dt \\ &= \int_0^\infty \frac{e^{-\beta t}}{\beta}\, (t^\lambda\, e^{\beta\, a\, \sqrt{t}})'\, dt \\ &= \frac{a}{2} S_{\lambda+1/2}(a) + \frac{\lambda}{\beta} S_\lambda(a). \end{aligned}$$

Therefore

$$S''_\lambda(a) = \frac{a\beta}{2} S'_\lambda(a) + \lambda\beta S_\lambda(a). \tag{5.22}$$

Let $\mathcal{M}$ be a compact subset of some manifold, and $a(x, u)$ be a function such that

$$\mathbb{R}^N \times \mathcal{M} \ni (x, u) \mapsto a(x, u) \in \mathbb{R}^1.$$

Assume that $\xi(u)$ is a Gaussian process on $\mathcal{M}$ which satisfies

$$E[\xi(u)] = 0,$$
$$E[\xi(u)\xi(v)] = E_X[a(X,u)a(X,v)] \equiv \rho(u,v).$$

Definition 5.9 (Singular fluctuation) Let $\mu(du)$ be a measure on $\mathcal{M}$. Assume $\rho(u,u) = 2$ for an arbitrary $u \in \mathcal{M}$. The singular fluctuation of the Gaussian process $\xi(u)$ on $\mathcal{M}$ is defined by

$$\nu(\beta) = \frac{1}{2\beta} E\Big[\frac{1}{Z(\xi)} \int_{\mathcal{M}} \mu(du)\, \xi(u)\, S'_\lambda(\xi(u))\Big],$$

where

$$Z(\xi) = \int_{\mathcal{M}} \mu(du)\, S_\lambda(\xi(u)).$$

From eq.(5.22) we obtain the following theorem.

Theorem 5.11 *The following equations hold.*

$$E\Big[\frac{\int \mu(du)\, S''_\lambda(\xi(u))}{Z(\xi)}\Big] = \lambda\beta + \beta^2\nu(\beta), \tag{5.23}$$

$$\tfrac{1}{2} E E_X\Big[\Big(\frac{\int \mu(du)\, a(X,u)\, S'_\lambda(\xi(u))}{Z(\xi)}\Big)^2\Big] = \lambda\beta + (\beta^2 - \beta)\nu(\beta). \tag{5.24}$$

Proof of Theorem 5.11 The first relation eq.(5.23) is derived directly from eq.(5.22). Let us prove eq.(5.24). Let $\{g_i\}_{i=1}^{\infty}$ be independent Gaussian random variables on $\mathbb{R}$ which satisfy $E[g_i] = 0$, $E[g_i g_j] = \delta_{ij}$. For such random variables,

$$E[g_i F(g_i)] = E\Big[\frac{\partial}{\partial g_i} F(g_i)\Big]$$

holds for a differentiable function of $F(\cdot)$. Since $L^2(q)$ is a separable Hilbert space, there exists a complete orthonormal system $\{e_k(x)\}_{k=1}^{\infty}$. By defining

$$b_k(u) = \int a(x,u)\, e_k(u)\, q(x)\, dx,$$

it follows that

$$a(x,u) = \sum_{k=1}^{\infty} b_k(u) e_k(x),$$

$$E_X[a(X,u)a(X,v)] = \sum_{k=1}^{\infty} b_k(u) b_k(v).$$

A Gaussian process defined by

$$\xi^*(u) = \sum_{k=1}^{\infty} b_k(u)\, g_k$$

has the same expectations and covariance matrices as $\xi(u)$, therefore it is subject to the same probability distribution as $\xi(u)$. Thus we can identify $\xi(u) = \xi^*(u)$ in the calculation of expectation values.

$$E[\xi(u)\xi(v)] = \sum_{i=1}^{\infty} b_i(u)b_i(v). \tag{5.25}$$

Then

$$\begin{aligned} A &\equiv E\Big[\frac{\int \mu(du)\, \xi(u)S'_\lambda(\xi(u))}{Z(\xi)}\Big] \\ &= E\Big[\int \mu(du)\, \Big\{\sum_{i=1}^{\infty} b_i(u)g_i\Big\} \frac{S'_\lambda(\xi(u))}{Z(\xi)}\Big] \\ &= E\Big[\int \mu(du)\Big\{\sum_{i=1}^{\infty} b_i(u)\frac{\partial}{\partial g_i}\Big\}\frac{S'_\lambda(\xi(u))}{Z(\xi)}\Big]. \end{aligned}$$

By using

$$\begin{aligned} \frac{\partial}{\partial g_i}\Big(\frac{S'_\lambda(\xi(u))}{Z(\xi)}\Big) &= \frac{S''_\lambda(\xi(u))b_i(u)}{Z(\xi)} \\ &\quad - \frac{S'_\lambda(\xi(u))}{Z(\xi)^2}\int \mu(dv)\, S'_\lambda(\xi(v))b_i(v), \end{aligned}$$

we obtain

$$\begin{aligned} A &= E\Big[\int \mu(du)\, \frac{S''_\lambda(\xi(u))\sum_{i=1}^{\infty} b_i(u)^2}{Z(\xi)}\Big] \\ &\quad - E\Big[\int \mu(du)\int \mu(dv)\frac{S'_\lambda(\xi(u))S'_\lambda(\xi(v))\sum_i b_i(u)b_i(v)}{Z(\xi)^2}\Big] \\ &= E\Big[\int \mu(du)\frac{S''_\lambda(\xi(u))\rho(u,u)}{Z(\xi)}\Big] \\ &\quad - E\Big[\int \mu(du)\,\mu(dv)\, \frac{S'_\lambda(\xi(u))S'_\lambda(\xi(v))\,\rho(u,v)}{Z(\xi)^2}\Big]. \end{aligned}$$

Therefore, by $\rho(u,u)=2$,

$$E\Big[\frac{\int \mu(du)\,\xi(u)\,S'_\lambda(\xi(u))}{Z(\xi)}\Big] = 2E\Big[\frac{\int \mu(du)\,S''_\lambda(\xi(u))}{Z(\xi)}\Big]$$
$$-EE_X\Big[\Big(\frac{\int \mu(du)\,a(X,u)\,S'_\lambda(\xi(u))}{Z(\xi)}\Big)^2\Big], \tag{5.26}$$

which completes the proof. □

Remark 5.9 (1) A singular fluctuation expresses the fluctuation on $\mathcal{M}$. It is represented by

$$\nu(\beta) = -\frac{1}{2\beta}\frac{\partial}{\partial a}\Big(E\Big[\log\int_{\mathcal{M}} \mu(du)\,S_\lambda(a\xi(u))\Big]\Big)_{a=1}.$$

Moreover, by using

$$\frac{\partial}{\partial\beta}S_\lambda(a) = -S_{\lambda+1}(a) + aS_{\lambda+1/2}(a)$$
$$= -(\lambda/\beta)S_\lambda(a) + (a/(2\beta))S'_\lambda(a),$$

the singular fluctuation and $F = -E[\log Z(\xi)]$ satisfy

$$\frac{\partial F}{\partial \beta} = \frac{\lambda}{\beta} - \nu(\beta).$$

(2) Define an expectation value,

$$\langle f(t,u)\rangle \equiv \frac{\int \mu(du)\int_0^\infty f(t,u)\,t^{\lambda-1}e^{-\beta t+\beta\xi(u)\sqrt{t}}dt}{\int \mu(du)\int_0^\infty t^{\lambda-1}e^{-\beta t+\beta\xi(u)\sqrt{t}}dt}.$$

Then, by Theorem 5.11 and $E_X[a(X,u)^2]=2$, $\nu(\beta)$ can be rewritten as

$$\nu(\beta) = \frac{\beta}{2}E_\xi E_X\Big[\big\langle a(X,u)^2 t\big\rangle - \big\langle a(X,u)\sqrt{t}\big\rangle^2\Big].$$

This is the variance of $\sqrt{t}a(X,u)$. For the meaning of this term, see Remark 6.13.

6
Singular learning theory

In this chapter we prove the four main formulas in singular learning theory. The formulas which clarify the singular learning process are not only mathematically beautiful but also statistically useful.

Firstly, we introduce the standard form of the log likelihood ratio function. A new foundation is established on which singular learning theory is constructed without the positive definite Fisher information matrix. By using resolution of singularities, there exists a map $w = g(u)$ such that the log likelihood ratio function of any statistical model is represented by

$$K_n(g(u)) = u^{2k} - \frac{1}{\sqrt{n}}\, u^k \xi_n(u),$$

where u^k is a normal crossing function and $\xi_n(u)$ is an empirical process which converges to a Gaussian process in law.

Secondly, we prove that, under a natural condition, the stochastic complexity of an arbitrary statistical model can be asymptotically expanded as

$$F_n = -\log \int \prod_{i=1}^{n} p(X_i|w)^{\beta}\, \varphi(w) dw$$

$$= n\beta S_n + \lambda \log n - (m-1) \log\log n + F_n^R,$$

where S_n is the empirical entropy of the true distribution, F_n^R is a random variable which converges to a random variable in law, and $(-\lambda)$ and m are respectively equal to the largest pole and its order of the zeta function of a statistical model,

$$\zeta(z) = \int K(w)^z \varphi(w) dw.$$

In regular statistical models $\lambda = d/2$ and $m = 1$ where d is the dimension of the parameter space, whereas in singular learning machines $\lambda \neq d/2$ and $m \geq 1$

in general. The constant λ, the learning coefficient, is equal to the real log canonical threshold of the set of true parameters if $\varphi(w) > 0$ on singularities.

Thirdly, we prove that the means of Bayes generalization error B_{g}, Bayes training error B_{t}, Gibbs generalization error G_{g}, and Gibbs training error G_{t} are respectively given by

$$E[B_{\mathrm{g}}] = \Big(\frac{\lambda + \nu\beta - \nu}{\beta}\Big)\frac{1}{n} + o\Big(\frac{1}{n}\Big), \tag{6.1}$$

$$E[B_{\mathrm{t}}] = \Big(\frac{\lambda - \nu\beta - \nu}{\beta}\Big)\frac{1}{n} + o\Big(\frac{1}{n}\Big), \tag{6.2}$$

$$E[G_{\mathrm{g}}] = \Big(\frac{\lambda + \nu\beta}{\beta}\Big)\frac{1}{n} + o\Big(\frac{1}{n}\Big), \tag{6.3}$$

$$E[G_{\mathrm{t}}] = \Big(\frac{\lambda - \nu\beta}{\beta}\Big)\frac{1}{n} + o\Big(\frac{1}{n}\Big), \tag{6.4}$$

where n is the number of random samples, $\beta > 0$ is the inverse temperature of the *a posteriori* distribution, and $\nu = \nu(\beta) > 0$ is the singular fluctuation. From these equations, the equations of states in statistical estimation are derived:

$$E[B_{\mathrm{g}}] - E[B_{\mathrm{t}}] = 2\beta(E[G_{\mathrm{t}}] - E[B_{\mathrm{t}}]) + o\Big(\frac{1}{n}\Big),$$

$$E[G_{\mathrm{g}}] - E[G_{\mathrm{t}}] = 2\beta(E[G_{\mathrm{t}}] - E[B_{\mathrm{t}}]) + o\Big(\frac{1}{n}\Big).$$

Although both λ and $\nu(\beta)$ strongly depend on the set of a true distribution, a statistical model, and an *a priori* distribution, the equations of states always hold independent of them. Therefore, based on the equations of states, we can predict Bayes and Gibbs generalization errors from Bayes and Gibbs training errors without any knowledge of the true distribution. Based on random samples, we can evaluate how appropriate a model and an *a priori* distribution are.

And, lastly, we show the symmetry of the generalization and training errors in the maximum likelihood or the maximum *a posteriori* estimation,

$$E[R_{\mathrm{g}}] = \frac{C}{n} + o\Big(\frac{1}{n}\Big),$$

$$E[R_{\mathrm{t}}] = -\frac{C}{n} + o\Big(\frac{1}{n}\Big),$$

where R_{g} and R_{t} are the generalization error and the training error of the maximum likelihood or *a posteriori* estimator respectively. In singular statistical models, the constant $C > 0$ is given by the maximum values of a Gaussian process on the set of true parameters, hence C is far larger than $d/2$ in general. In order to make C small, we need a strong improvement by some penalty term, which is not appropriate in singular statistical estimation.

From a historical point of view, the concepts and proofs of this chapter were found by the author of this book.

6.1 Standard form of likelihood ratio function

To establish singular learning theory we need some fundamental conditions.

Definition 6.1 (Fundamental condition (I)) Let $q(x)$ and $p(x|w)$ be probability density functions on $\mathbb{R}^N$ which have the same support. Here $p(x|w)$ is a parametric probability density function for a given parameter $w \in W \subset \mathbb{R}^d$. The set of all parameters W is a compact set in $\mathbb{R}^d$. The Kullback–Leibler distance is defined by

$$K(w) = \int q(x) \log \frac{q(x)}{p(x|w)} dx.$$

We assume

$$W_0 = \{w \in W;\ K(w) = 0\}$$

is not the empty set. It is said that $q(x)$ and $p(x|w)$ satisfy the fundamental condition (I) with index s ($s \geq 2$) if the following conditions are satisfied.
(1) For $f(x, w) = \log(q(x)/p(x|w))$, there exists an open set $W^{(C)} \subset \mathbb{C}^d$ such that:
(1-a) $W \subset W^{(C)}$,
(1-b) $W^{(C)} \ni w \mapsto f(\cdot, w)$ is an $L^s(q)$-valued complex analytic function,
(1-c) $M(x) \equiv \sup_{w \in W^{(C)}} |f(x, w)|$ is contained in $L^s(q)$.
(2) There exists $\epsilon > 0$ such that, for

$$Q(x) \equiv \sup_{K(w) \leq \epsilon} p(x|w),$$

the following integral is finite,

$$\int M(x)^2 Q(x) dx < \infty.$$

Remark 6.1 These are remarks about the fundamental condition (I).
(1) By definition, there exists a real open set $W^{(R)} \subset \mathbb{R}^d$ such that

$$W \subset W^{(R)} \subset W^{(C)}.$$

The log density ratio function $f(x, w)$ is an $L^s(q)$-valued real analytic function on $W^{(R)}$ and an $L^s(q)$-valued complex analytic function on $W^{(C)}$. For a sufficiently small constant $\epsilon > 0$, we define

$$W_\epsilon = \{w \in W; K(w) \leq \epsilon\}.$$

Based on the resolution theorem in Chapters 2 and 3, there exist open sets $W_\epsilon^{(R)} \subset W^{(R)}$, $W_\epsilon^{(C)} \subset W^{(C)}$ and subsets of manifolds $\mathcal{M}$, $\mathcal{M}^{(R)}$, and $\mathcal{M}^{(C)}$ such that

$$W_\epsilon \subset W_\epsilon^{(R)} \subset W_\epsilon^{(C)},$$
$$\mathcal{M} \subset \mathcal{M}^{(R)} \subset \mathcal{M}^{(C)},$$

and that

$$\mathcal{M} \equiv g^{-1}(W_\epsilon),$$
$$\mathcal{M}^{(R)} \equiv g^{-1}\big(W_\epsilon^{(R)}\big),$$
$$\mathcal{M}^{(C)} \equiv g^{-1}\big(W_\epsilon^{(C)}\big),$$

where $w = g(u)$ is the resolution map and its complexification. In the following, we use this notation.

(2) In the fundamental condition (I), we mainly study the case in which

$$W_0 \equiv \{w \in W; K(w) = 0\}$$

is not one point but a set with singularities. In other words, the Fisher information matrix of $p(x|w)$ is degenerate on W_0 in general. However, this condition contains the regular statistical model as a special case.

(3) From conditions (1-b) and (1-c), $f(x, w)$ is represented by the absolutely convergent power series in the neighborhood of arbitrary $w_0 \in W$:

$$f(x, w) = \sum_\alpha a_\alpha(x)(w - w_0)^\alpha.$$

The function $a_\alpha(x) \in L^s(q)$ is bounded by

$$|a_\alpha(x)| \leq \frac{M(x)}{R^\alpha},$$

where R is the associated convergence radii.

(4) If $M(x)$ satisfies

$$\int M(x)^2 e^{M(x)} q(x)dx < \infty,$$

then condition (2) is satisfied, because

$$\begin{aligned}\int M(x)^2 Q(x)dx &= \int M(x)^2 \sup_{K(w)\leq\epsilon} p(x|w)dx \\ &= \int M(x)^2 \sup_{K(w)\leq\epsilon} e^{-f(x,w)}q(x)dx \\ &\leq \int M(x)^2 e^{M(x)} q(x)dx.\end{aligned}$$

(5) Let $w = g(u)$ be a real proper analytic function which makes $K(g(u))$ normal crossing. If $q(x)$ and $p(x|w)$ satisfy the fundamental condition (I) with index s, then $q(x)$ and $p(x|g(u))$ also satisfy the same condition, where $\mathbb{R}^d$ and $\mathbb{C}^d$ are replaced by real and complex manifolds respectively.
(6) The condition that W is compact is necessary because, even if the log density ratio function is a real analytic function of the parameter, $|w| = \infty$ is an analytic singularity in general. For this reason, if W is not compact and W_0 contains $|w| = \infty$, the maximum likelihood estimator does not exist in general. For example, if $x = (x_1, x_2)$, $w = (a, b)$, and $p(x_2|x_1, w) \propto \exp(-(x_2 - a\sin(bx_1))^2/2)$, then the maximum likelihood estimator never exists. On the other hand, if $|w| = \infty$ is not a singularity, $\mathbb{R}^d \cup \{|w| = \infty\}$ can be understood as a compact set and the same theorems as this chapter hold. From the mathematical point of view, it is still difficult to construct singular learning theory of a general statistical model in the case when W_0 contains analytic singularities. From the viewpoint of practical applications, we can select W as a sufficiently large compact set.
(7) If a model satisfies the fundamental condition (I) with index $s + t$, $(s \geq 2, t > 0)$, then it automatically satisfies the fundamental condition (I) with index s. The condition with index $s = 6$ is needed to ensure the existence of the asymptotic expansion of the Bayes generalization error in the proof. (See the proof of Theorem 6.8.)
(8) Some non-analytic statistical models can be made analytic. For example, in a simple mixture model $p(x|a) = ap_1(x) + (1 - a)p_2(x)$ with some probability densities $p_1(x)$ and $p_2(x)$, the log density ratio function $f(x, a)$ is not analytic at $a = 0$, but it can be made analytic by the representation $p(x|\theta) = \alpha^2 p_1(x) + \beta^2 p_2(x)$, on the manifold $\theta \in \{\alpha^2 + \beta^2 = 1\}$. As is shown in the proofs, if W is contained in a real analytic manifold, then the same theorems as this chapter hold.
(9) The same results as this chapter can be proven based on the weaker conditions. However, to describe clearly the mathematical structure of singular

learning theory, we adopt the condition (I). For the case of the weaker condition, see Section 7.8.

Theorem 6.1 *Assume that $q(x)$ and $p(x|w)$ satisfy the fundamental condition (I) with index $s = 2$. There exist a constant $\epsilon > 0$, a real analytic manifold $\mathcal{M}^{(R)}$, and a real analytic function $g : \mathcal{M}^{(R)} \to W_\epsilon^{(R)}$ such that, in every local coordinate U of $\mathcal{M}^{(R)}$,*

$$K(g(u)) = u^{2k} = u_1^{2k_1} u_2^{2k_2} \cdots u_d^{2k_d},$$

where $k_1, k_2, \ldots, k_d$ are nonnegative integers. Moreover, there exists an $L^s(q)$-valued real analytic function $a(x, u)$ such that

$$f(x, g(u)) = a(x, u)\, u^k \quad (u \in U), \tag{6.5}$$

and

$$\int a(x, u) q(x) dx = u^k \quad (u \in U). \tag{6.6}$$

Remark 6.2 (1) In this chapter, $\epsilon > 0$ is taken so that Theorem 6.1 holds. The set of parameters W is represented as the union of two subsets. The former is W_ϵ which includes $\{w; K(w) = 0\}$ and the latter is $W \setminus W_\epsilon$ in which $K(w) > \epsilon$. To W_ϵ, we can apply resolution of singularities.
(2) A typical example of $K(w)$, $w = g(u)$, $f(x, g(u))$, and $a(x, u)$ is shown in Example 7.1.

Proof of Theorem 6.1 Existence of $\epsilon > 0$, $\mathcal{M}^{(R)}$, and g is shown by resolution of singularities, Theorem 2.3. Let us prove eq.(6.5). For arbitrary $u \in U$,

$$\begin{aligned} K(g(u)) &= \int f(x, g(u)) q(x) dx \\ &= \int (e^{-f(x,g(u))} + f(x, g(u))) - 1) q(x) dx \\ &= \int \frac{f(x, g(u))^2}{2} e^{-t^* f(x,g(u))} q(x) dx, \end{aligned}$$

where $0 < t^* < 1$. Let $U' \subset U$ be a neighborhood of $u = 0$. For arbitrary $L > 0$ the set D_L is defined by

$$D_L \equiv \Big\{ x \in \mathbb{R}^N; \sup_{u \in U'} |f(x, g(u))| \le L \Big\}.$$

Then for any $u \in U'$,

$$u^{2k} \ge \int_{D_L} \frac{f(x, g(u))^2}{2} e^{-L} q(x) dx,$$

giving the result that, for any $u^k \neq 0$ $(u \in U')$,

$$1 \geq e^{-L} \int_{D_L} \frac{f(x, g(u))^2}{2u^{2k}} q(x)dx. \tag{6.7}$$

Since $f(x, g(u))$ is an $L^s(q)$-valued real analytic function, it is given by an absolutely convergent power series,

$$\begin{aligned} f(x, g(u)) &= \sum_{\alpha} a_\alpha(x) u^\alpha \\ &= a(x, u)u^k + b(x, u)u^k, \end{aligned}$$

where

$$\begin{aligned} a(x, u) &= \sum_{\alpha \geq k} a_\alpha(x) u^{\alpha - k}, \\ b(x, u) &= \sum_{\alpha < k} a_\alpha(x) u^{\alpha - k}, \end{aligned}$$

and $\sum_{\alpha \geq k}$ shows the sum of indices that satisfy

$$\alpha_i \geq k_i \quad (i = 1, 2, \ldots, d) \tag{6.8}$$

and $\sum_{\alpha < k}$ shows the sum of indeces that do not satisfy at least one of eq.(6.8). Here $a(x, u)$ is an $L^s(q)$-valued real analytic function. From eq.(6.7), for an arbitrary $u^k \neq 0$ $(u \in U')$,

$$\begin{aligned} 1 &\geq e^{-L} \int_{D_L} (a(x, u) + b(x, u))^2 q(x)dx \\ &\geq \frac{e^{-L}}{2} \int_{D_L} b(x, u)^2 q(x)dx - e^{-L} \int_{D_L} a(x, u)^2 q(x)dx. \end{aligned}$$

Here $|a(x, u)|$ is a bounded function of $u \in U'$. If $b(x, u) \equiv 0$ does not hold, then $|b(x, u)| \to \infty$ $(u \to 0)$, hence we can choose u and D_L so that the above inequality does not hold. Therefore, we have $b(x, u) \equiv 0$, which shows eq.(6.5). From

$$u^{2k} = \int f(x, g(u))q(x)dx = \int a(x, u)u^k q(x)dx,$$

we obtain eq.(6.6). □

Let $X_1, X_2, \ldots, X_n$ be a set of random variables which are independently subject to the probability distribution $q(x)dx$. The log likelihood ratio function is defined by

$$K_n(w) = \frac{1}{n}\sum_{i=1}^{n} f(X_i, w).$$

The expectation of $K_n(w)$ is equal to the Kullback–Leibler distance, $E[K_n(w)] = K(w)$. For w satisfying $K(w) > 0$, the log likelihood ratio function is given as

$$K_n(w) = K(w) - \sqrt{K(w)/n}\ \psi_n(w),$$

where $\psi_n(w)$ is defined by

$$\psi_n(w) = \frac{1}{\sqrt{n}}\sum_{i=1}^{n} \frac{K(w) - f(X_i, w)}{\sqrt{K(w)}}. \tag{6.9}$$

Here $\psi_n(w)$ is an empirical process on $\{w \in W; K(w) \geq \epsilon\}$ and converges to a Gaussian process $\psi(w)$ in law. However, if $K(w) = 0$, $\psi_n(w)$ is ill-defined. For the set $W_\epsilon = \{w \in W; K(w) \leq \epsilon\}$, by Theorem 2.3, there exists a manifold such that, in each local coordinate $(u_1, u_2, \ldots, u_d)$, the Kullback–Leibler distance is given by $K(g(u)) = u^{2k}$. Then

$$\sqrt{K(g(u))} = |u^k|,$$

which is not real analytic at $u^k = 0$. Therefore, eq.(6.9) should be replaced by choosing an appropriate branch so that it is a real analytic function at $u^2 = 0$. The following representation is the theoretical foundation on which singular learning theory is constructed.

Main Theorem 6.1 (Standard form of log likelihood ratio function) *Assume that the fundamental condition (I) holds with index $s = 2$. There exist a real analytic manifold $\mathcal{M}^{(R)}$ and a real analytic and proper map $g : \mathcal{M}^{(R)} \to W_\epsilon^{(R)}$ such that*

$$K(g(u)) = u^{2k}$$

on each local coordinate U of $\mathcal{M}^{(R)}$. The log likelihood ratio function is represented in U by

$$K_n(g(u)) = u^{2k} - \frac{1}{\sqrt{n}}\, u^k\, \xi_n(u), \tag{6.10}$$

where

$$\xi_n(u) = \frac{1}{\sqrt{n}}\sum_{i=1}^{n}\{a(X_i, u) - E_X[a(X, u)]\} \tag{6.11}$$

is an empirical process. Equation (6.10) is called the standard form of the log likelihood ratio function.

Proof of Main Theorem 6.1 From Theorems 2.3, 5.9, and 6.1, this Main Theorem is obtained. □

Remark 6.3 (1) From the definition, the empirical process $\xi_n(u)$ satisfies

$$E[\xi_n(u)] = 0$$

and

$$\begin{aligned} E[\xi_n(u)\xi_n(v)] &= E_X[a(X,u)a(X,v)] - E_X[a(X,u)]E_X[a(X,v)] \\ &= E_X[a(X,u)a(X,v)] - u^k v^k. \end{aligned}$$

In particular, if $K(g(u)) = K(g(v)) = 0$, then

$$E[\xi_n(u)\xi_n(v)] = E_X[a(X,u)a(X,v)].$$

(2) The empirical process $\xi_n(u)$ is well defined on the manifold even for $K(g(u)) = 0$, whereas the empirical process $\xi_n(g^{-1}(w))$ is ill-defined if $K(w) = 0$ in general, which is one of the reasons why algebraic geometry is necessary in statistical learning theory.

Theorem 6.2 *(1) Assume that the fundamental condition (I) holds with $s = 2$. Then the empirical processes $\psi_n(w)$ on $\{w; K(w) > \epsilon\}$ and $\xi_n(u)$ on $\mathcal{M}$ converge in law to the Gaussian processes $\psi(w)$ and $\xi(u)$, respectively.*
(2) Assume that the fundamental condition (I) holds with $s = 4$. Then the empirical processes satisfy

$$\lim_{n\to\infty} E\Big[\sup_{K(w)>\epsilon} |\psi_n(w)|^{s-2}\Big] = E\Big[\sup_{K(w)>\epsilon} |\psi(w)|^{s-2}\Big] < \infty,$$

$$\lim_{n\to\infty} E\Big[\sup_{u\in\mathcal{M}} |\xi_n(u)|^{s-2}\Big] = E\Big[\sup_{u\in\mathcal{M}} |\xi(u)|^{s-2}\Big] < \infty.$$

Proof of Theorem 6.2 For $\psi_n(w)$, this theorem is immediately derived from Theorems 5.9 and 5.10. Let us prove the theorem for $\xi_n(u)$. The subset $\mathcal{M}$ is compact because W_ϵ is compact and the resolution map $g : \mathcal{M}^{(R)} \to W_\epsilon^{(R)}$ is proper. Therefore $\mathcal{M}$ can be covered by a finite union of local coordinates. From Theorems 5.9 and 5.10, we immediately obtain the theorem. □

Remark 6.4 By the above theorem, the limiting process $\xi(u)$ is a Gaussian process on $\mathcal{M}$ which satisfies

$$E[\xi(u)] = 0$$

and

$$E[\xi(u)\xi(v)] = E_X[a(X,u)a(X,v)] - u^k v^k.$$

In particular, if $K(g(u)) = K(g(v)) = 0$, then

$$E[\xi(u)\xi(v)] = E_X[a(X, u)a(X, v)].$$

In other words, the Gaussian process $\xi(u)$ has the same mean and covariance function as $\xi_n(u)$. Note that the tight Gaussian process is uniquely determined by its mean and covariance.

Theorem 6.3 *Assume that $q(x)$ and $p(x|w)$ satisfy the fundamental condition (I) with index $s = 4$. If $K(g(u)) = 0$, then*

$$E_X[a(x, u)^2] = E[|\xi_n(u)|^2] = E[|\xi(u)|^2] = 2.$$

Proof of Theorem 6.3 It is sufficient to prove $E_X[a(X, u)^2] = 2$ when $K(g(u)) = 0$. Let the Taylor expansion of $f(x, g(u))$ be

$$f(x, g(u)) = \sum_{\alpha} a_\alpha(x)u^\alpha.$$

Then

$$|a_\alpha(x)| \leq \frac{M(x)}{R^\alpha},$$

where R are associated convergence radii and

$$a(x, u) = \sum_{\alpha \geq k} a_\alpha(x)u^{\alpha - k}.$$

Hence

$$\begin{aligned} |a(x, u)| &\leq \sum_{\alpha \geq k} \frac{M(x)}{R^\alpha} r^{\alpha - k} \\ &= c_1 \frac{M(x)}{R^k}, \end{aligned}$$

where $c_1 > 0$ is a constant. In the same way as in the proof of Theorem 6.1 and with $f(x, g(u)) = a(x, u)u^k$, for arbitrary u ($u^k \neq 0$), we have

$$1 = \int \frac{a(x, u)^2}{2} e^{-t^* a(x,u)u^k} q(x)dx,$$

where $0 < t^* < 1$. Put

$$S(x, u) = \frac{a(x, u)^2}{2} e^{-t^* a(x,u)u^k} q(x).$$

Then

$$\begin{aligned}
S(x,u) &\leq c_1 \frac{M(x)^2}{R^{2k}} \max_u \left\{1, e^{-a(x,u)u^k}\right\} q(x) \\
&= c_1 \frac{M(x)^2}{R^{2k}} \max_w \{q(x), p(x|w)\} \\
&\leq c_1 \frac{M(x)^2}{R^{2k}} Q(x).
\end{aligned}$$

By the fundamental condition (I), $M(x)^2 Q(x)$ is an integrable function, hence $S(x,u)$ is bounded by the integrable function. By using Lebesgue's convergence theorem for $u^k \to 0$, we obtain

$$1 = \int \frac{a(x,u)^2}{2} q(x) dx$$

for any u that satisfies $u^{2k} = 0$. □

6.2 Evidence and stochastic complexity

Definition 6.2 (Evidence) Let $q(x)$ and $p(x|w)$ be probability distributions which satisfy the fundamental condition (I) with $s = 2$. The set $D_n = \{X_1, X_2, \ldots, X_n\}$ consists of random variables which are independently subject to $q(x)dx$. Let $\varphi(w)$ be a probability density function on $\mathbb{R}^d$. The evidence of a pair $p(x|w)$ and $\varphi(w)$ for D_n is defined by

$$Z_n = \int \Big(\prod_{i=1}^{n} p(X_i|w)^\beta\Big) \varphi(w) dw.$$

Also the stochastic complexity is defined by

$$F_n = -\log Z_n.$$

The normalized evidence and the normalized stochastic complexity are respectively defined by

$$\begin{aligned}
Z_n^0 &= \frac{Z_n}{\displaystyle\prod_{i=1}^{n} q(x_i)^\beta} \\
&= \int \exp(-n\beta K_n(w)) \varphi(w) dw, \qquad (6.12) \\
F_n^0 &= -\log Z_n^0. \qquad (6.13)
\end{aligned}$$

Theorem 6.4 *For arbitrary natural number n, the normalized stochastic complexity satisfies*

$$E\big[F_n^0\big] \le -\log \int \exp(-n\beta K(w))\varphi(w)dw.$$

Proof of Theorem 6.4

$$\begin{aligned} F_n^0 &= -\log \int \exp(-n\beta K_n(w))\varphi(w)dw \\ &= -\log \int \exp(-n\beta(K_n(w) - K(w)) - n\beta K(w))\varphi(w)dw \\ &= -\log \int \exp(-n\beta(K_n(w) - K(w)))\rho(w)dw \\ &\quad -\log \int \exp(-n\beta K(w))\varphi(w)dw, \end{aligned}$$

where

$$\rho(w) = \frac{\exp(-n\beta K(w))\varphi(w)}{\int \exp(-n\beta K(w'))\varphi(w')dw'}.$$

By Jensen's inequality and a definition $K^*(w) \equiv K_n(w) - K(w)$,

$$\int \exp(-n\beta K^*(w))\rho(w)dw \ge \exp\Big(-\int n\beta K^*(w)\rho(w)dw\Big).$$

Using $E[K^*(w)] = 0$, we obtain the theorem. □

Remark 6.5 In the proof of Theorem 6.4, the convergence in law of $\xi_n(u) \to \xi(u)$ is not needed. Therefore, the upper bound of the stochastic complexity can be shown by the weaker condition.

Definition 6.3 (Fundamental condition (II)) Assume that the set of parameters W is a compact set defined by

$$W = \{w \in \mathbb{R}^d; \pi_1(w) \ge 0, \pi_2(w) \ge 0, \ldots, \pi_k(w) \ge 0\},$$

where $\pi_1(w), \pi_2(w), \ldots, \pi_k(w)$ are real analytic functions on some real open set $W^{(R)} \subset \mathbb{R}^d$. The *a priori* probability density function $\varphi(w)$ is given by $\varphi(w) = \varphi_1(w)\varphi_2(w)$ where $\varphi_1(w) > 0$ is a function of class C^∞ and $\varphi_2(w) \ge 0$ is a real analytic function.

Remark 6.6 In singular statistical models, the set of parameters and the *a priori* distribution should be carefully prepared from the theoretical point of view. In particular, their behavior in the neighborhood of $K(w) = 0$ and the boundary of W have to be set naturally. The condition that $\pi_1(w), \pi_2(w), \ldots, \pi_k(w)$

and $\varphi_1(w)$ are real analytic functions is necessary because, if at least one of them is a function of class C^∞, there is a pathological example. In fact, if $\varphi_1(w) = \exp(-1/w^2)$ $(w \in \mathbb{R}^1)$ and $K(w) = w^2$ in a neighborhood of the origin, then $(d/dw)^k \varphi_1(0) = 0$ for an arbitrary $k \geq 0$, and

$$\int_{-1}^{1} (w^2)^z \exp\left(-\frac{1}{w^2}\right) dw$$

has no pole.

Theorem 6.5 (Partition of parameter space) *Assume the fundamental conditions (I) and (II) with index $s = 2$. Let $\epsilon > 0$ be a constant. By applying Hironaka's resolution theorem (Theorem 2.3) to a real analytic function,*

$$K(w)(\epsilon - K(w))\varphi_2(w) \prod_{j=1}^{k} \pi_j(w),$$

we can find a real analytic manifold $\mathcal{M}^{(R)}$ and a proper and real analytic map $g : \mathcal{M}^{(R)} \to W_\epsilon^{(R)}$ such that all functions

$$K(g(u)), \quad \epsilon - K(g(u)), \quad \varphi_2(g(u)), \quad \pi_1(g(u)), \ldots, \pi_k(g(u))$$

have only normal crossing singularities. By using Remark 2.14 and Theorem 2.11, we can divide the set $W_\epsilon = \{w \in W; K(w) \leq \epsilon\}$ such that the following conditions (1), (2), (3), and (4) are satisfied.
(1) The set of parameters $\mathcal{M} = g^{-1}(W_\epsilon)$ is covered by a finite set

$$\mathcal{M} = \cup_\alpha M_\alpha,$$

where M_α is given by a local coordinate,

$$M_\alpha = [0, b]^d = \{(u_1, u_2, \ldots, u_d)\ ;\ 0 \leq u_1, u_2, \ldots, u_d \leq b\}.$$

(2) In each M_α,

$$K(g(u)) = u^{2k} = u_1^{2k_1} u_2^{2k_2} \cdots u_d^{2k_d},$$

where $k_1, k_2, \ldots, k_d$ are nonnegative integers.
(3) There exists a function $\phi(u)$ of class C^∞ such that

$$\varphi(g(u))|g'(u)| = \phi(u)u^h = \phi(u)u_1^{h_1} u_2^{h_2} \cdots h_d^{h_d},$$

where $|g'(u)|$ is the absolute value of the Jacobian determinant and

$$\phi(u) > c > 0 \quad (u \in [0, b]^d)$$

is a function of class C^∞, where $c > 0$ is a positive constant.

(4) There exists a set of functions $\{\sigma_\alpha(u)\}$ of class C^∞ which satisfy

$$\sigma_\alpha(u) \geq 0,$$
$$\sum_\alpha \sigma_\alpha(u) = 1,$$
$$\sigma_\alpha(u) > 0 \quad (u \in [0, b)^d),$$
$$supp\ \sigma_\alpha(u) = [0, b]^d,$$

such that, for an arbitrary integrable function $H(w)$,

$$\int_{W_\epsilon} H(w)\varphi(w)dw = \int_{\mathcal{M}} H(g(u))\varphi(g(u))|g'(u)|du$$
$$= \sum_\alpha \int_{M_\alpha} H(g(u))\phi^*(u)u^h du,$$

where we defined $\phi^*(u)$ by omitting local coordinate α,

$$\phi^*(u) \equiv \sigma_\alpha(u)\phi(u).$$

Moreover there exist constants $C_1 > 0$ such that

$$C_1 \sum_\alpha \int_{M_\alpha} H(g(u))\phi(u)u^h du \leq \int_{W_\epsilon} H(w)\varphi(w)dw$$
$$\leq \sum_\alpha \int_{M_\alpha} H(g(u))\phi(u)u^h du. \tag{6.14}$$

Proof of Theorem 6.5 This theorem is obtained by the resolution theorem (Theorem 2.3), Theorem 2.11, and Remark 2.14. □

Theorem 6.6 *Assume the fundamental conditions (I) and (II) with index $s = 2$. The holomorphic function of $z \in \mathbb{C}$,*

$$\zeta(z) = \int K(w)^z \varphi(w)dw \quad (\mathrm{Re}(z) > 0), \tag{6.15}$$

can be analytically continued to the unique meromorphic function on the entire complex plane whose poles are all real, negative, and rational numbers.

Proof of Theorem 6.6 Let us define

$$\zeta_1(z) = \int_{K(w)<\epsilon} K(w)^z \varphi(w)dw,$$

$$\zeta_2(z) = \int_{K(w)\geq\epsilon} K(w)^z \varphi(w)dw.$$

In an arbitrary neighborhood of $z \in \mathbb{C}$, $|(\partial K(w)^z/\partial z)\varphi(w)|$ is a bounded function on the compact set $\{w \in W; K(w) \geq \epsilon\}$, hence $\zeta_2(z)$ is a holomorphic function on the entire complex plane. Let us study $\zeta_1(z)$. The Kullback–Leibler distance and the *a priori* distribution can be represented as in Theorem 6.5.

$$\zeta_1(z) = \sum_\alpha \int_{M_\alpha} u^{2kz}\, u^h\, \phi^*(u)du.$$

Since $\phi^*(u)$ has the finite-order Taylor expansion,

$$\phi^*(u) = \sum_{|j|\leq n} a_j u^j + R_n(u),$$

where n can be taken as large as necessary. In the region $\mathrm{Re}(z) > 0$,

$$\int_{[0,b]^d} u^{2kz+h+j} du = \prod_{p=1}^{d} \frac{b^{2k_p z + h_p + j_p + 1}}{(2k_p z + h_p + j_p + 1)}.$$

Hence the function $\zeta_1(z)$ can be analytically continued to the unique meromorphic function and all poles are real, negative, and rational numbers. □

Definition 6.4 (Zeta function and learning coefficient) The meromorphic function $\zeta(z)$ that is analytically continued from eq.(6.15) is called the zeta function of a statistical model. The largest pole and its order are denoted by $(-\lambda)$ and m, respectively, where λ and m are respectively called the learning coefficient and its order. If the Kullback–Leibler distance and the *a priori* distribution are represented as in Theorem 6.5, then the learning coefficient is given by

$$\lambda = \min_\alpha \min_{1\leq j\leq d} \left(\frac{h_j + 1}{2k_j}\right), \tag{6.16}$$

and its order m is

$$m = \max_\alpha \sharp\{j; \lambda = (h_j + 1)/(2k_j)\}, \tag{6.17}$$

where $\sharp$ shows the number of elements of the set S. Let $\{\alpha^*\}$ be a set of all local coordinates in which both the minimization in eq.(6.16) and the maximization in eq.(6.17) are attained. Such a set of local coordinates $\{\alpha^*\}$ is said to be the essential family of local coordinates. For each local coordinate α^* in the essential family of local coordinates, we can assume without loss of generality that u is represented as $u = (x, y)$ such that

$$x = (u_1, u_2, \ldots, u_m),$$
$$y = (u_{m+1}, u_{m+2}, \ldots, u_d),$$

and that

$$\lambda = \frac{h_j + 1}{2k_j} \quad (1 \le j \le m),$$

$$\lambda < \frac{h_j + 1}{2k_j} \quad (m + 1 \le j \le d).$$

For a given function $f(u) = f(x, y)$, we use the notation $f_0(y) \equiv f(0, y)$.

Theorem 6.7 (Convergence in law of evidence) *Assume the fundamental conditions (I) and (II) with index $s = 2$. The constants λ and m are the learning coefficient and its order respectively. Let Z_n^0 be the normalized evidence defined in eq.(6.12). When $n \to \infty$, the following convergence in law holds,*

$$\frac{n^\lambda Z_n^0}{(\log n)^{m-1}} \to \sum_{\alpha^*} \gamma_b \int_0^\infty dt \int_{M_{\alpha^*}} t^{\lambda-1}\, e^{-\beta t + \sqrt{t}\beta \xi_0(y)}\, \phi_0^*(y) dy,$$

where $\gamma_b > 0$ is a constant defined by eq.(4.16).

Proof of Theorem 6.7 The normalized evidence can be divided as

$$Z_n^0 = Z_n^{(1)} + Z_n^{(2)},$$

where

$$Z_n^{(1)} = \int_{K(w) \le \epsilon} e^{-n\beta K_n(w)} \varphi(w) dw,$$

$$Z_n^{(2)} = \int_{K(w) > \epsilon} e^{-n\beta K_n(w)} \varphi(w) dw.$$

Firstly, let us study $Z_n^{(2)}$. If $K(w) > \epsilon$, by using the Cauchy–Schwarz inequality,

$$\begin{aligned} nK_n(w) &= nK(w) - \sqrt{K(w)}\psi_n(w) \\ &\ge \frac{nK(w) - \psi_n(w)^2}{2} \\ &\ge \frac{1}{2}\Big(n\epsilon - \sup_{K(w)>\epsilon} |\psi_n(w)|^2\Big). \end{aligned}$$

Since $\psi_n(w)$ is an empirical process which converges to a Gaussian process with supremum norm in law, $\sup_{K(w)>\epsilon} |\psi_n(w)|^2$ converges in law, and therefore

$$0 \le \frac{n^\lambda}{(\log n)^{m-1}} Z_n^{(2)} \le \frac{n^\lambda e^{-n\beta\epsilon/2}}{(\log n)^{m-1}} \exp\Big(\frac{\beta}{2} \sup_{K(w)>\epsilon} |\psi_n(w)|^2\Big) \tag{6.18}$$

converges to zero in probability by Theorem 5.2. Secondly, $Z_n^{(1)}$ is given by

$$Z_n^{(1)} = \sum_{\alpha} \int_0^\infty dt \int_{M_\alpha} \exp(-n\beta u^{2k} + \beta\sqrt{n} u^k \xi_n(u)) u^h \phi^*(u) du.$$

Let us define

$$Y^{(1)}(\xi_n) \equiv \gamma_b \sum_{\alpha^*} \int_0^\infty dt \int dy\, t^{\lambda-1} y^\mu e^{-\beta t + \beta\sqrt{t}\xi_{n,0}(y)} \phi_0^*(y).$$

Then $Y^{(1)}(\xi)$ is a continuous function of ξ with respect to the norm $\|\cdot\|$, hence the convergence in law $Y^{(1)}(\xi_n) \to Y^{(1)}(\xi)$ holds. Let us apply Theorem 4.9 to the coordinates α^* with $p = 0$, $r = m$, $f = \xi_n$. Also we apply Theorem 4.8 to the other coordinates with $r = m-1$ and $f = \xi_n$. Then there exists a constant $C_1 > 0$ such that

$$\left| \frac{n^\lambda Z_n^{(1)}}{(\log n)^{m-1}} - Y^{(1)}(\xi_n) \right| \leq \frac{C_1}{\log n} \sum_\alpha e^{\beta\|\xi_n\|^2/2} \{\beta\|\xi_n\| \|\phi^*\| + \|\nabla\phi^*\| + \|\phi^*\|\}.$$

Since $\|\xi_n\|$ converges in law, the right-hand side of this equation converges to zero in probability. Therefore, the convergence in law

$$\frac{n^\lambda}{(\log n)^{m-1}} Z_n^{(1)} \to Y^{(1)}(\xi)$$

holds, which completes the theorem. □

Main Theorem 6.2 (Convergence of stochastic complexity)
(1) Assume that $q(x)$, $p(x|w)$ and $\varphi(w)$ satisfy the fundamental conditions (I) and (II) with index $s = 2$. Then the following convergence in law holds:

$$F_n^0 - \lambda \log n + (m-1) \log\log n$$
$$\to -\log \sum_{\alpha^*} \gamma_b \int_0^\infty dt \int t^{\lambda-1}\, e^{-\beta t + \beta\sqrt{t}\xi_0(y)}\, \phi_0^*(y) dy.$$

(2) Assume that $q(x)$, $p(x|w)$ and $\varphi(w)$ satisfy the fundamental conditions (I) and (II) with index $s = 4$. Then the following convergence of expectation holds:

$$E\big[F_n^0\big] - \lambda \log n + (m-1) \log\log n$$
$$\to -E\Big[\log \sum_{\alpha^*} \gamma_b \int_0^\infty dt \int t^{\lambda-1}\, e^{-\beta t + \beta\sqrt{t}\xi_0(y)}\, \phi_0^*(y) dy\Big].$$

Proof of Main Theorem 6.2 (1) From Theorem 6.7 and the fact that $-\log(\cdot)$ is a continuous function, the first part is proved by Theorem 5.1.

(2) For the second part, it is sufficient to prove that

$$A_n \equiv -\log \frac{Z_n^0 n^\lambda}{(\log n)^{m-1}}$$

is asymptotically uniformly integrable. By using the same decomposition of Z_n^0 as in the proof of Theorem 6.7,

$$A_n = -\log\Big(\frac{Z_n^{(1)} n^\lambda}{(\log n)^{m-1}} + \frac{Z_n^{(2)} n^\lambda}{(\log n)^{m-1}}\Big).$$

By eq.(6.18) and Theorem 4.8 with $p = 0$ and $r = m$,

$$A_n \geq -\log\Big\{\exp(\beta\|\xi_n\|^2/2)\|\varphi\| + \exp\Big(\frac{\beta}{2}\sup_{K(w)>\epsilon}|\psi_n(w)|^2\Big)\Big\}$$

$$\geq -(\beta/2)\max\Big\{\|\xi_n\|^2, \sup_{K(w)>\epsilon}|\psi_n(w)|^2\Big\} + C_2,$$

where C_2 is a constant and we used the fact that $\log(e^p + e^q) \leq \max\{p, q\} + \log 2$ for arbitrary p, q. On the other hand, by $\phi(u) > 0$ and Theorem 6.5 (4),

$$Z_n^{(2)} \geq C_1 \sum_\alpha \int_0^\infty dt \int du \exp(-n\beta u^{2k} + \sqrt{n}\beta u^k \xi_n) u^h du.$$

Hence, by Theorem 4.8, and $A_n \leq -\log(Z_n^{(2)} n^\lambda/(\log n)^{m-1})$,

$$A_n \leq \beta\|\xi_n\|^2/2 - \log\min|\phi| + C_3,$$

where C_3 is a constant. By Theorem 5.8, $E[\|\xi_n\|^4] < \infty$, $E[\|\psi_n\|^4] < \infty$. Hence A_n is asymptotically uniformly integrable. By Theorem 5.5, we obtain the theorem. □

Corollary 6.1 *(1) Assume that $q(x)$, $p(x|w)$ and $\varphi(w)$ satisfy the fundamental conditions (I) and (II) with index $s = 2$. Then the following asymptotic expansion holds,*

$$F_n = n\beta S_n + \lambda \log n - (m-1)\log\log n + F_n^R,$$

where F_n^R is a random variable which converges to a random variable in law.
(2) Assume that $q(x)$, $p(x|w)$ and $\varphi(w)$ satisfy the fundamental conditions (I) and (II) with index $s = 4$. Then the following asymptotic expansion of the expectation holds,

$$E[F_n] = n\beta S + \lambda \log n - (m-1)\log\log n + E\big[F_n^R\big],$$

where $E[F_n^R]$ converges to a constant.

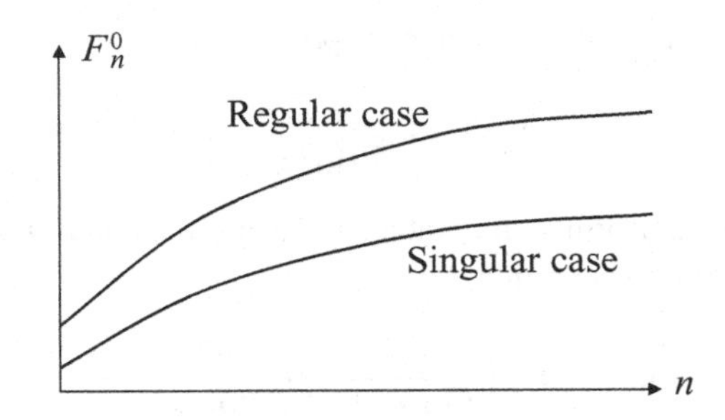

Fig. 6.1. Stochastic complexity

Proof of Corollary 6.1 By the definition

$$F_n^R = -\log \sum_{\alpha^*} \gamma_b \int_0^\infty dt \int du \; t^{\lambda-1} \; e^{-\beta t+\beta\sqrt{t}\xi_{n,0}(y)} \; \phi_0^*(y)dy,$$

this corollary is immediately derived from Main Theorem 6.2. □

Remark 6.7 (1) Figure 6.1 shows the behavior of the normalized stochastic complexity. If the *a priori* distribution is positive on $K(w) = 0$, then the learning coefficient is equal to the real log canonical threshold. Note that the learning coefficient is invariant under a transform

$$p(x|w) \mapsto p(x|g(u)),$$
$$\varphi(w)dw \mapsto \varphi(g(u))|g'(u)|du.$$

(2) If a model is regular, $\lambda = d/2$ and $m = 1$, where d is the dimension of the parameter space. Examples of λ and m in several models are shown in Chapter 7.
(3) Assume that $\beta = 1$. By Main Theorem 6.2 and Theorem 1.2, if the mean of Bayes generalization error B_g has an asymptotic expansion, then

$$E[B_g] = E\big[F_{n+1}^0\big] - E\big[F_n^0\big],$$
$$= \frac{\lambda}{n} + o\Big(\frac{1}{n}\Big).$$

However, in general, even if a function $f(n)$ has an asymptotic expansion, $f(n+1) - f(n)$ may not have an asymptotic expansion. To prove the Bayes generalization error has an asymptotic expansion, we need to show a more precise result as in the following section.

6.3 Bayes and Gibbs estimation

In the previous section, we studied the asymptotic expansion of the stochastic complexity. In this section, mathematical relations in the Bayes quartet are proved, which are called equations of states in learning. Throughout this section, we assume the fundamental conditions (I) and (II) with index $s = 6$. Firstly, the main theorems are introduced without proof. Secondly, basic lemmas are prepared. Finally, the main theorems are proved.

6.3.1 Equations of states

Assume that random samples $X_1, X_2, \ldots, X_n$ are independently taken from the probability distribution $q(x)dx$. For a given set of random samples $D_n = \{X_1, X_2, \ldots, X_n\}$, the generalized *a posteriori* distribution is defined by

$$p(w|D_n) = \frac{1}{Z_n}\varphi(w)\prod_{i=1}^{n} p(X_i|w)^{\beta},$$

which can be rewritten as

$$p(w|D_n) = \frac{1}{Z_n^0}\exp(-n\beta K_n(w))\varphi(w),$$

where $\beta > 0$ is the inverse temperature.

Definition 6.5 (Bayes quartet) Let $E_w[\cdot]$ be the expectation value using $p(w|D_n)$. Four errors are defined.
(1) Bayes generalization error,

$$B_{\mathrm{g}} = E_X\Big[\log\frac{q(X)}{E_w[p(X|w)]}\Big].$$

(2) Bayes training error,

$$B_{\mathrm{t}} = \frac{1}{n}\sum_{i=1}^{n}\log\frac{q(X_i)}{E_w[p(X_i|w)]}.$$

(3) Gibbs generalization error,

$$G_{\mathrm{g}} = E_w\Big[E_X\Big[\log\frac{q(X)}{p(X|w)}\Big]\Big].$$

(4) Gibbs training error,

$$G_{\mathrm{t}} = E_w\Big[\frac{1}{n}\sum_{i=1}^{n}\log\frac{q(X_i)}{p(X_i|w)}\Big].$$

This set of four errors is called the Bayes quartet.

The most important variable in practical applications among them is the Bayes generalization error because it determines the accuracy of the estimation. However, we prove that there are mathematical relations between them. It is shown in Theorem 1.3 that, by using the log density ratio function

$$f(x, w) = \log \frac{q(x)}{p(x|w)},$$

and the log likelihood ratio function

$$K_n(w) = \frac{1}{n} \sum_{i=1}^{n} f(X_i, w),$$

the Bayes quartet can be rewritten as

$$\begin{aligned} B_{\mathrm{g}} &= E_X\Big[-\log E_w[e^{-f(X,w)}]\Big], \\ B_{\mathrm{t}} &= \frac{1}{n} \sum_{i=1}^{n} -\log E_w[e^{-f(X_i,w)}], \\ G_{\mathrm{g}} &= E_w[K(w)], \\ G_{\mathrm{t}} &= E_w[K_n(w)]. \end{aligned}$$

If the true distribution $q(x)$ is contained in the statistical model $p(x|w)$, then the four errors in the Bayes quartet converge to zero in probability when n tends to infinity. In this section, we show how fast random variables in the Bayes quartet tend to zero.

Theorem 6.8 *Assume the fundamental conditions (I) and (II) with $s = 6$.*
(1) There exist random variables B_{g}^*, B_{t}^*, G_{g}^*, and G_{t}^* such that, when $n \to \infty$, the following convergences in law hold:

$$nB_{\mathrm{g}} \to B_{\mathrm{g}}^*, \quad nB_{\mathrm{t}} \to B_{\mathrm{t}}^*, \quad nG_{\mathrm{g}} \to G_{\mathrm{g}}^*, \quad nG_{\mathrm{t}} \to G_{\mathrm{t}}^*.$$

(2) When $n \to \infty$, the following convergence in probability holds:

$$n(B_{\mathrm{g}} - B_{\mathrm{t}} - G_{\mathrm{g}} + G_{\mathrm{t}}) \to 0.$$

(3) Expectation values of the Bayes quartet converge:

$$\begin{aligned} E[nB_{\mathrm{g}}] &\to E[B_{\mathrm{g}}^*], \\ E[nB_{\mathrm{t}}] &\to E[B_{\mathrm{t}}^*], \\ E[nG_{\mathrm{g}}] &\to E[G_{\mathrm{g}}^*], \\ E[nG_{\mathrm{t}}] &\to E[G_{\mathrm{t}}^*]. \end{aligned}$$

Main Theorem 6.3 (Equations of states in statistical estimation) *Assume the fundamental conditions (I) and (II) with $s = 6$. For arbitrary $q(x)$, $p(x|w)$, and $\varphi(w)$, the following equations hold.*

$$E[B_g^*] - E[B_t^*] = 2\beta(E[G_t^*] - E[B_t^*]), \tag{6.19}$$

$$E[G_g^*] - E[G_t^*] = 2\beta(E[G_t^*] - E[B_t^*]). \tag{6.20}$$

Remark 6.8 (1) Main Theorem 6.3 shows that the increases of errors from training to prediction are in proportion to the differences between the Bayes and Gibbs training. We give Main Theorem 6.3 the title **Equations of states in statistical estimation**, because these hold for any true distribution, any statistical model, any *a priori* distribution, and any singularities. If two of these errors are measured by observation, then the other two errors can be estimated without any knowledge of the true distribution.
(2) From the equations of states, widely applicable information criteria (WAIC) are obtained. See Section 8.3.
(3) Although the equations of states hold universally, the four errors themselves strongly depend on a true distribution, a statistical model, an *a priori* distribution, and singularities.

Corollary 6.2 *The two generalization errors can be estimated by the two training errors,*

$$\begin{pmatrix} E[B_g^*] \\ E[G_g^*] \end{pmatrix} = \begin{pmatrix} 1-2\beta & 2\beta \\ -2\beta & 1+2\beta \end{pmatrix} \begin{pmatrix} E[B_t^*] \\ E[G_t^*] \end{pmatrix}. \tag{6.21}$$

Proof of Corollary 6.2 This corollary is directly derived from Main Theorem 6.3. □

Remark 6.9 (1) From eq.(6.21), it follows that

$$\begin{pmatrix} E[G_t^*] \\ E[B_t^*] \end{pmatrix} = \begin{pmatrix} 1-2\beta & 2\beta \\ -2\beta & 1+2\beta \end{pmatrix} \begin{pmatrix} E[G_g^*] \\ E[B_g^*] \end{pmatrix},$$

which shows that there is symmetry in the Bayes quartet.

Theorem 6.9 *Assume the fundamental conditions (I) and (II) with index $s = 6$. When $n \to \infty$, the convergence in probability*

$$nG_g + nG_t - \frac{2\lambda}{\beta} \to 0$$

holds, where λ is the learning coefficient. Moreover,

$$E[G_g^*] + E[G_t^*] = \frac{2\lambda}{\beta}. \tag{6.22}$$

Corollary 6.3 *Assume the fundamental conditions (I) and (II) with $s = 6$. The following convergence in probability holds,*

$$nB_{\mathrm{g}} - nB_{\mathrm{t}} + 2nG_{\mathrm{t}} - \frac{2\lambda}{\beta} \to 0,$$

where λ is the learning coefficient. Moreover,

$$E[B_{\mathrm{g}}^*] - E[B_{\mathrm{t}}^*] + 2E[G_{\mathrm{t}}^*] = \frac{2\lambda}{\beta}.$$

In particular, if $\beta = 1$, $E[B_{\mathrm{g}}^*] = \lambda$.

Proof of Corollary 6.3 This corollary is derived from Theorem 6.8 (1) and 6.9. □

Definition 6.6 (Empirical variance) The empirical variance V of the log likelihood function is defined by

$$V = \sum_{i=1}^{n} \Big\{ E_w[(\log p(X_i|w))^2\,] - (\,E_w[\log p(X_i|w)]\,)^2 \Big\}. \tag{6.23}$$

By using the log density ratio function $f(x, w) = \log(q(x)/p(x|w))$, this can be rewritten as

$$V = \sum_{i=1}^{n} \Big\{ E_w[\,f(X_i|w)^2\,] - E_w[f(X_i|w)]^2 \Big\}. \tag{6.24}$$

Theorem 6.10 *The following convergences in probability hold:*

$$V - 2(nG_{\mathrm{t}} - nB_{\mathrm{t}}) \to 0, \tag{6.25}$$

$$V - 2(nG_{\mathrm{g}} - nB_{\mathrm{g}}) \to 0. \tag{6.26}$$

There exists a constant $\nu = \nu(\beta) > 0$ such that

$$\lim_{n\to\infty} E[V] = \frac{2\nu(\beta)}{\beta}$$

and

$$E[B_{\mathrm{g}}^*] = \frac{\lambda}{\beta} + \Big(1 - \frac{1}{\beta}\Big)\nu(\beta), \tag{6.27}$$

$$E[B_{\mathrm{t}}^*] = \frac{\lambda}{\beta} - \Big(1 + \frac{1}{\beta}\Big)\nu(\beta), \tag{6.28}$$

$$E[G_{\mathrm{g}}^*] = \frac{\lambda}{\beta} + \nu(\beta), \tag{6.29}$$

$$E[G_{\mathrm{t}}^*] = \frac{\lambda}{\beta} - \nu(\beta). \tag{6.30}$$

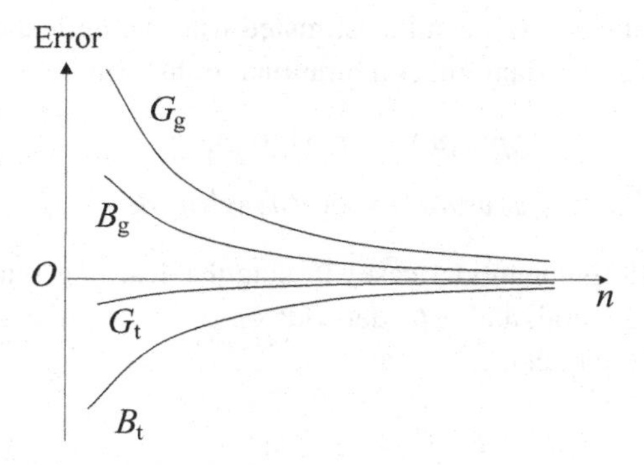

Fig. 6.2. Bayes and Gibbs errors

The constant $\nu(\beta) > 0$ is a singular fluctuation defined in eq.(6.48), which satisfies an inequality,

$$0 \le \nu(\beta) \le \frac{E_\xi[\|\xi\|^2]}{8} + \frac{E_\xi[\|\xi\|^2]^{1/2}}{8}\sqrt{E_\xi[\|\xi\|^2] + 16\lambda/\beta},$$

where $\|\xi\|$ is the maximum value of the random process $\xi(u)$.

Remark 6.10 The behavior of the Bayes quartet is shown in Figure 6.2. From Theorem 6.10, the singular fluctuation $\nu(\beta)$ can be represented in several ways:

$$\begin{aligned}
\nu(\beta) &= (1/2)(E[B_g^*] - E[B_t^*]) \\
&= (1/2)(E[G_g^*] - E[G_t^*]) \\
&= \beta\,(E[G_g^*] - E[B_g^*]) \\
&= \beta\,(E[G_t^*] - E[B_t^*]) \\
&= \lim_{n\to\infty} \frac{\beta}{2} E[V].
\end{aligned}$$

From the equations of states,

$$E[B_g] = E[B_t] + \frac{\beta}{n}E[V] + o(1/n), \tag{6.31}$$

$$E[G_g] = E[G_t] + \frac{\beta}{n}E[V] + o(1/n). \tag{6.32}$$

Using these relations, $\nu(\beta)$ can be estimated from numerical experiments. By definition, $\nu(\beta)$ is invariant under a birational transform

$$p(x|w) \mapsto p(x|g(u)),$$

$$\varphi(w)dw \mapsto \varphi(g(u))|g'(u)|du.$$

Remark 6.11 (Regular model case) In singular learning machines, λ is not equal to $d/2$ in general, and $\nu(\beta)$ depends on β. In a regular statistical model, $\lambda = \nu(\beta) = d/2$, which means that,

$$E[B_{\mathrm{g}}^*] = \frac{d}{2},$$

$$E[B_{\mathrm{t}}^*] = -\frac{d}{2},$$

$$E[G_{\mathrm{g}}^*] = \Big(1 + \frac{1}{\beta}\Big)\frac{d}{2},$$

$$E[G_{\mathrm{t}}^*] = \Big(-1 + \frac{1}{\beta}\Big)\frac{d}{2},$$

which is a special case of Main Theorem 6.3. This result is obtained by the classical asymptotic theory of the maximum likelihood estimator with positive definite Fisher information matrix. Assume that examples are independently taken from $q(x) = p(x|w_0)$. Let $I_n(w)$ and $I(w)$ be the empirical and mean Fisher information matrices respectively. The expectation of a function $f(w)$ under the *a posteriori* distribution is asymptotically given by

$$E_w[f(w)] \cong \frac{\int f(w)\exp\big(-\frac{n\beta}{2}|I_n(\hat{w})^{1/2}(w-\hat{w})|^2\big)dw}{\int \exp\big(-\frac{n\beta}{2}|I_n(\hat{w})^{1/2}(w-\hat{w})|^2\big)dw},$$

where, for a given symmetric matrix I and a vector v, $|I^{1/2}v|^2 = (v, Iv)$. The random variable $\sqrt{n}(\hat{w} - w_0)$ converges in law to the normal distribution with mean zero and covariance matrix $I(w_0)^{-1}$, and the Taylor expansions for w_0 and $\hat{w}$ are given by

$$\begin{aligned} E[nK(\hat{w})] &= E\Big[nK(w_0) + \frac{n}{2}|I(w_0)^{1/2}(\hat{w} - w_0)|^2\Big] \\ &= \frac{d}{2} + o(1) \\ E[nK_n(w_0)] &= E\Big[nK_n(\hat{w}) + \frac{n}{2}|I_n(\hat{w})^{1/2}(w_0 - \hat{w})|^2\Big] \\ &= 0. \end{aligned}$$

Therefore the Gibbs generalization and training errors are given by

$$\begin{aligned}
E[nG_{\mathrm{g}}] &= E[E_w[nK(w)]] \\
&= E\Big[\frac{n}{2}E_w[|I(\hat{w})^{1/2}(w-\hat{w})|^2] + nK(\hat{w})\Big] + o(1) \\
&= \frac{d}{2\beta} + \frac{d}{2} + o(1), \\
E[nG_{\mathrm{t}}] &= E[E_w[nK_n(w)]] \\
&= E\Big[\frac{n}{2}E_w[|I(\hat{w})^{1/2}(w-\hat{w})|^2] + nK_n(\hat{w})\Big] + o(1) \\
&= \frac{d}{2\beta} - \frac{d}{2} + o(1).
\end{aligned}$$

Consequently, in regular statistical models, $\lambda = \nu(\beta) = d/2$.

6.3.2 Basic lemmas

In this subsection, some basic lemmas are prepared which are used in the proofs of the above theorems.

Note that there are three different expectations. The first is the expectation over the parameter space of the *a posteriori* distribution. The second is the expectation over random samples $D_n = \{X_1, X_2, \ldots, X_n\}$. It is denoted by $E[\;]$, which is also used for the expectation over the limit process ξ. The last is the expectation over the test sample X. It is denoted by $E_X[\;]$.

For a given constant $a > 0$, we define an expectation value in the restricted set $\{w; K(w) \leq a\}$ by

$$E_w[f(w)|_{K(w)\leq a}] = \frac{\displaystyle\int_{K(w)\leq a} f(w)e^{-\beta nK_n(w)}\varphi(w)dw}{\displaystyle\int_{K(w)\leq a} e^{-\beta nK_n(w)}\varphi(w)dw}.$$

The four errors of the Bayes quartet in the restricted region are given by

$$\begin{aligned}
B_{\mathrm{g}}(a) &= E_X\Big[-\log E_w[e^{-f(X,w)}|_{K(w)\leq a}]\Big], \\
B_{\mathrm{t}}(a) &= \frac{1}{n}\sum_{j=1}^{n} -\log E_w[e^{-f(X_j,w)}|_{K(w)\leq a}], \\
G_{\mathrm{g}}(a) &= E_w[K(w)|_{K(w)\leq a}], \\
G_{\mathrm{t}}(a) &= E_w[K_n(w)|_{K(w)\leq a}].
\end{aligned}$$

Since W is compact and $K(w)$ is a real analytic function,

$$\overline{K} = \sup_{w \in W} K(w)$$

is finite, therefore

$$\begin{aligned} B_{\mathrm{g}}(\overline{K}) &= B_{\mathrm{g}}, \\ B_{\mathrm{t}}(\overline{K}) &= B_{\mathrm{t}}, \\ G_{\mathrm{g}}(\overline{K}) &= G_{\mathrm{g}}, \\ G_{\mathrm{t}}(\overline{K}) &= G_{\mathrm{t}}. \end{aligned}$$

Also we define $\eta_n(w)$ for w such that $K(w) > 0$ by

$$\eta_n(w) = \frac{K(w) - K_n(w)}{\sqrt{K(w)}}, \tag{6.33}$$

and

$$H_{\mathrm{t}}(a) = \sup_{0<K(w)\leq a} |\eta_n(w)|^2.$$

Let H_{t} denote $H_{\mathrm{t}}(\overline{K})$.

Lemma 6.1 *(1) For an arbitrary $a > 0$, the following inequalities hold.*

$$\begin{aligned} & B_{\mathrm{t}}(a) \leq G_{\mathrm{t}}(a) \leq \tfrac{3}{2} G_{\mathrm{g}}(a) + \tfrac{1}{2} H_{\mathrm{t}}(a), \\ & 0 \leq B_{\mathrm{g}}(a) \leq G_{\mathrm{g}}(a), \\ & -\tfrac{1}{4} H_{\mathrm{t}}(a) \leq G_{\mathrm{t}}(a). \end{aligned}$$

(2) In particular, by putting $a = \overline{K}$,

$$\begin{aligned} & B_{\mathrm{t}} \leq G_{\mathrm{t}} \leq \tfrac{3}{2} G_{\mathrm{g}} + \tfrac{1}{2} H_{\mathrm{t}}, \\ & 0 \leq B_{\mathrm{g}} \leq G_{\mathrm{g}}, \\ & -\tfrac{1}{4} H_{\mathrm{t}} \leq G_{\mathrm{t}}. \end{aligned}$$

Proof of Lemma 6.1 (1) Because $B_{\mathrm{g}}(a)$ is the Kullback–Leibler distance from $q(x)$ to the Bayes predictive distribution using the restricted *a priori* distribution on $K(w) \leq a$, it follows that $B_{\mathrm{g}}(a) \geq 0$. Using Jensen's inequality,

$$E_w[e^{-f(x,w)}|_{K(w)\leq a}] \geq \exp(-E_w[f(x,w)|_{K(w)\leq a}]) \quad (\forall x),$$

hence $B_g(a) \leq G_g(a)$ and $B_t(a) \leq G_t(a)$. By using the Cauchy–Schwarz inequality,

$$\begin{aligned} K_n(w) &= K(w) - \sqrt{K(w)}\, \eta_n(w) \\ &\leq K(w) + \tfrac{1}{2}\{K(w) + \eta_n(w)^2\}. \end{aligned} \tag{6.34}$$

Therefore $G_t(a) \leq (3G_g(a) + H_t(a))/2$. Also

$$\begin{aligned} K_n(w) &= K(w) - \sqrt{K(w)}\eta_n(w) \\ &\geq \left(\sqrt{K(w)} - \frac{\eta_n(w)}{2}\right)^2 - \frac{\eta_n(w)^2}{4} \\ &\geq -\frac{\eta_n(w)^2}{4}. \end{aligned} \tag{6.35}$$

Hence we have $G_t(a) \geq -H_t(a)/4$. (2) is immediately derived from (1). □

Remark 6.12 (1) By Lemma 6.1, if $nH_t(a)$, $nG_g(a)$, and $nB_t(a)$ are asymptotically uniformly integrable (AUI), then $nG_t(a)$ and $nB_g(a)$ are also AUI, for an arbitrary $a > 0$.
(2) In Lemma 6.4, we prove $nH_t(\epsilon)$ is AUI. In Lemma 6.5, we prove $nG_g(\epsilon)$ and $nB_t(\epsilon)$ are AUI. In Lemma 6.2, we show nH_t is AUI using Lemma 6.4. In Lemma 6.3, we show nG_g and nB_t are AUI using Lemma 6.5. Then the four errors in the Bayes quartet are all AUI.
(3) Based on Theorem 5.4 (3), if $E[|X_n|^s] < \infty$, then $X_n^{s-\delta}$ ($\delta > 0$) is AUI.

Lemma 6.2 *(1) There exists a constant $C_H > 0$ such that*

$$E[(nH_t)^3] = C_H < \infty.$$

(2) For an arbitrary $\delta > 0$,

$$P(nH_t > n^\delta) \leq \frac{C_H}{n^{3\delta}}. \tag{6.36}$$

(3) nH_t is asymptotically uniformly integrable.

Proof of Lemma 6.2 (1) For any $\epsilon > 0$ and $a > 0$,

$$\sqrt{n}\, \eta_n(w) = \frac{1}{\sqrt{K(w)}} \cdot \frac{1}{\sqrt{n}} \sum_{j=1}^{n} (E_X[f(X, w)] - f(X_j, w))$$

is an empirical process and $f(x, w)$ is a real analytic function of w. Therefore

$$E\Big[\sup_{\epsilon < K(w) < a} |\sqrt{n}\, \eta_n(w)|^6\Big] < \text{const.}$$

It is proved in Lemma 6.4 that

$$E\Big[\sup_{K(w)\le\epsilon} |\sqrt{n}\,\eta_n(w)|^6\Big] < \text{const.}$$

Therefore, by the definition of H_t, (1) is obtained.
(2) Let S be a random variable defined by

$$S = \begin{cases} 1 & \text{if } nH_t > n^\delta \\ 0 & \text{otherwise.} \end{cases}$$

Then $E[S] = P(nH_t > n^\delta)$ and

$$C_H = E[(nH_t)^3] \ge E[(nH_t)^3\, S] \ge E[S]\, n^{3\delta}.$$

(3) is immediately derived from (1). □

Lemma 6.3 *(1) For an arbitrary $\epsilon > 0$, the following convergences in probability hold:*

$$n(B_g - B_g(\epsilon)) \to 0,$$

$$n(B_t - B_t(\epsilon)) \to 0,$$

$$n(G_g - G_g(\epsilon)) \to 0,$$

$$n(G_t - G_t(\epsilon)) \to 0.$$

(2) The four errors of the Bayes quartet nB_g, nB_t, nG_g, and nG_t are all asymptotically uniformly integrable.

Proof of Lemma 6.3 We use the notation,

$$S_0(f(w)) = \int_{K(w)<\epsilon} f(w)\, e^{-n\beta K_n(w)}\, \varphi(w)dw,$$

$$S_1(f(w)) = \int_{K(w)\ge\epsilon} f(w)\, e^{-n\beta K_n(w)}\, \varphi(w)dw.$$

By using the Cauchy–Schwarz inequality,

$$\tfrac{1}{2}K(w) - \tfrac{1}{2}\eta_n(w)^2 \le K_n(w) \le \tfrac{3}{2}K(w) + \tfrac{1}{2}\eta_n(w)^2,$$

we have inequalities for arbitrary $f(w), g(w) > 0$,

$$S_1(f(w)) \le \big(\sup_w f(w)\big)\, e^{-n\beta\epsilon/2} \exp\Big(\frac{\beta}{2} nH_t\Big),$$

$$S_0(g(w)) \ge c_0\, \big(\inf_w g(w)\big)\, n^{-d/2} \exp\Big(-\frac{\beta}{2} nH_t\Big),$$

where $c_0 > 0$ is a constant and d is the dimension of the parameter space. Hence

$$\frac{S_1(f(w))}{S_0(g(w))} \le \frac{\sup_w f(w)}{\inf_w g(w)}\, s(n),$$

where, by using Theorem 7.2,

$$s(n) = \frac{n^{d/2}}{c_0}\, e^{-n\beta\epsilon/2+n\beta H_t},$$

Then $|\log s(n)| \le n\beta\epsilon/2 + n\beta H_t + o(n)$. Let $M_n \equiv \sum_{j=1}^n M(X_j)/n$. Then $E[M_n^3] \le E_X[M(X)^3]$, $E_X[M(X)^k]_{\{M(X)>n\}} \le E_X[M(X)^3]/n^{3-k}$.
(1) Firstly, we study the Bayes generalization error,

$$\begin{aligned} n(B_g - B_g(\epsilon)) &= nE_X\Big[-\log \frac{E_w[e^{-f(X,w)}]}{E_w[e^{-f(X,w)}|_{K(w)\le\epsilon}]}\Big] \\ &= nE_X\Big[-\log\Big(1+\frac{S_1(e^{-f(X,w)})}{S_0(e^{-f(X,w)})}\Big) + \log\Big(1+\frac{S_1(1)}{S_0(1)}\Big)\Big]. \end{aligned}$$

Thus

$$\begin{aligned} n|B_g - B_g(\epsilon)| &\le nE_X\Big[\log\Big(1+\frac{S_1(e^{-f(X,w)})}{S_0(e^{-f(X,w)})}\Big) + \log\Big(1+\frac{S_1(1)}{S_0(1)}\Big)\Big] \\ &\le nE_X[\log(1+s(n)\,e^{2M(X)})] + ns(n) \\ &= ns(n) + nE_X[\log(1+s(n)\,e^{2M(X)})]_{\{2M(X)\le n\beta\epsilon/4\}} \\ &\quad + nE_X[\log(1+s(n)\,e^{2M(X)})]_{\{2M(X)>n\beta\epsilon/4\}} \\ &\le ns(n) + ns(n)\exp(n\beta\epsilon/4) \\ &\quad + nE_X[|2M(X)| + |\log s(n)|]_{\{2M(X)>n\beta\epsilon/4\}}. \end{aligned}$$

It follows that $n(B_g - B_g(\epsilon)) \to 0$. Secondly, in the same way, the Bayes training error satisfies

$$n|B_t - B_t(\epsilon)| \le \sum_{j=1}^n \log(1+s(n)\,e^{2M(X_j)}) + n\log(1+s(n)) \equiv L_n. \quad (6.37)$$

We can prove the convergence in mean $E[L_n] \to 0$ because

$$\begin{aligned} E[L_n] &= E[L_n]_{\{H_t\le\beta\epsilon/4\}} + E[L_n]_{\{H_t>\beta\epsilon/4\}} \\ &\le nE_X[\log(1+(n^d/c_0)\,e^{2M(X)-n\beta\epsilon/4})] \\ &\quad + \frac{n^{d+1}}{c_0}\exp(-n\beta\epsilon/4) + 2nE[M_n + |\log s(n)|]_{\{H_t>\beta\epsilon/4\}}. \end{aligned}$$

Thus we obtain $n(B_g - B_g(\epsilon)) \to 0$. Thirdly, the Gibbs generalization error can be estimated as

$$\begin{aligned} n|G_g - G_g(\epsilon)| &\le \left| n\frac{S_0(K(w)) + S_1(K(w))}{S_0(1) + S_1(1)} - \frac{nS_0(K(w))}{S_0(1)} \right| \\ &\le \frac{nS_1(K(w))}{S_0(1)} + \frac{nS_0(K(w))S_1(1)}{S_0(1)^2} \\ &\le 2n\,\overline{K}\,s(n), \end{aligned} \tag{6.38}$$

which converges to zero in probability. Lastly, in the same way, the Gibbs training error satisfies

$$\begin{aligned} n|G_t - G_t(\epsilon)| &\le 2n\ s(n)\ \sup_w |K_n(w)| \\ &\le 2n\ s(n)\, M_n, \end{aligned}$$

which converges to zero in probability.
(2) Firstly, from Lemma 6.2, nH_t is AUI. Secondly, let us prove nB_t is AUI. From eq.(6.37),

$$|nB_t| \le |nB_t(\epsilon)| + L_n.$$

Moreover, by employing a function

$$b(s) = -\frac{1}{n}\sum_{j=1}^{n} \log E_w[e^{-sf(X_j,w)}],$$

there exists $0 < s^* < 1$ such that

$$nB_t = nb(1) = \sum_{j=1}^{n} \frac{E_w[f(X_j,w)e^{-s^* f(X_j,w)}]}{E_w[e^{-s^* f(X_j,w)}]}.$$

Hence the following always holds:

$$|nB_t| \le \sum_{j=1}^{n} \sup_w |f(X_j,w)| \le nM_n.$$

Therefore

$$|nB_t| \le |nB_t(\epsilon)| + B^*,$$

where

$$B^* \equiv \begin{cases} nM_n & (nH_t > \epsilon\beta n/4) \\ L_n & (nH_t \le \epsilon\beta n/4). \end{cases}$$

By summing up the above equations,

$$E[|nB_t|^{3/2}] \leq E[2|nB_t(\epsilon)|^{3/2}] + E[2(B^*)^{3/2}].$$

In Lemma 6.5, we prove $E[|nB_t(\epsilon)|^{3/2}] < \infty$. By Lemma 6.2 (2) with δ such that $n^\delta = \epsilon\beta n/4$, we have $P(H_t > \epsilon\beta/4) \leq C'_H/n^3$, hence

$$\begin{aligned} E[(B^*)^{3/2}] &\leq E[(B^*)^{3/2}]_{\{H_t > \epsilon\beta/4\}} + E[(B^*)^{3/2}]_{\{H_t \leq \epsilon\beta/4\}} \\ &\leq E[(nM_n)^3]^{1/2} E[1]^{1/2}_{\{H_t > \epsilon\beta/4\}} \\ &\quad + E[(L_n)^3]^{1/2}_{\{H_t \leq \epsilon\beta/4\}} < \infty. \end{aligned}$$

Hence $|nB_t|$ is AUI. Lastly, we show nG_g is AUI. From eq.(6.38),

$$0 \leq nG_g \leq nG_g(\epsilon) + 2n\, s(n)\, \overline{K}.$$

Moreover, $nG_g \leq n\overline{K}$ always, by definition. Therefore

$$nG_g \leq nG_g(\epsilon) + K^*$$

where

$$\begin{aligned} K^* &\equiv \begin{cases} n\overline{K} & (nH_t > n^{2/3}) \\ \overline{K}\, n\, s(n) & (nH_t \leq n^{2/3}) \end{cases} \\ &\leq \begin{cases} n\overline{K} & (nH_t > n^{2/3}) \\ \overline{K}\, e^{-n\beta\epsilon/3} & (nH_t \leq n^{2/3}). \end{cases} \end{aligned}$$

Then

$$0 \leq E[(nG_g)^{3/2}] \leq E[2(nG_g(\epsilon))^{3/2}] + E[2(K^*)^{3/2}].$$

It is proven in Lemma 6.5 that $E[(nG_g(\epsilon))^{3/2}] < \infty$. By Lemma 6.2 (2) with $\delta = 2/3$, we have $P(nH_t > n^{2/3}) \leq C_H/n^2$, hence

$$E[(K^*)^{3/2}] \leq n^{3/2}\overline{K}^{3/2}\frac{C_H}{n^2} + \overline{K}e^{-n\beta\epsilon/2} < \infty.$$

Hence nG_g is AUI. Since $E[(nH_t)^3] < \infty$, $E[(nB_t)^{3/2}] < \infty$, and $E[(nG_g)^{3/2}] < \infty$, all four errors are also AUI by Lemma 6.1. □

Based on Lemma 6.3, $B_g(\epsilon)$, $B_t(\epsilon)$, $G_g(\epsilon)$, and $G_t(\epsilon)$ are the major parts of the four errors when $n \to \infty$. The region in the parameter set to be studied is

$$W_\epsilon = \{w \in W;\ K(w) \leq \epsilon\}$$

for a sufficiently small $\epsilon > 0$. Since W_ϵ contains singularities of $K(w) = 0$, we need Theorems 6.1 and 6.5. Let us define the supremum norm by

$$\|f\| = \sup_{u \in \mathcal{M}} |f(u)|.$$

There exists an $L^s(q)$-valued analytic function $\mathcal{M} \ni u \mapsto a(x, u) \in L^s(q)$ such that, in each local coordinate,

$$\begin{aligned} f(x, g(u)) &= a(x, u)\, u^k, \\ E_X[a(X, u)] &= u^k, \\ K(g(u)) = 0 \Rightarrow E_X[a(X, u)^2] &= 2, \\ E_X[\|a(X)\|^s] &< \infty. \end{aligned}$$

We define $\|a(X)\| = \sup_{u \in \mathcal{M}} |a(X, u)|$. An empirical process $\xi_n(u)$ is defined by eq.(6.11). Then the empirical process satisfies the following lemma.

Lemma 6.4 *(1) Let $s = 6$. The empirical process $\xi_n(u)$ satisfies*

$$E[\|\xi_n\|^s] < \text{const.} < \infty$$

$$E[\|\nabla \xi_n\|^s] < \text{const.} < \infty$$

where the constant does not depend on n, and $\|\nabla \xi_n\| = \sum_{j=1}^{d} \|\partial_j \xi_n\|$.
(2) The random variable $n H_t(\epsilon)$ is asymptotically uniformly integrable.

Proof of Lemma 6.4 (1) This lemma is derived from Theorem 5.8 and fundamental condition (I). (2) is immediately derived from (1). □

Let the Banach space of uniformly bounded and continuous functions on $\mathcal{M}$ be

$$B(\mathcal{M}) = \{f(u)\ ;\ \|f\| < \infty\}.$$

Since $\mathcal{M}$ is compact, $B(\mathcal{M})$ is a separable norm space. The empirical process $\xi_n(u)$ defined on $B(\mathcal{M})$ weakly converges to the tight Gaussian process $\xi(u)$.

Definition 6.7 (Integral over parameters) Let $\xi(u)$ be an arbitrary function on $\mathcal{M}$ of class C^1. We define the mean of $f(u)$ over $\mathcal{M}$ for a given $\xi(u)$ by

$$E_u^\sigma[f(u)|\xi] = \frac{\displaystyle\sum_\alpha \int_{[0,b]^d} f(u)\, Z(u, \xi)\, du}{\displaystyle\sum_\alpha \int_{[0,b]^d} Z(u, \xi)\, du},$$

where $\sum_\alpha$ is the summation over all coordinates of $\mathcal{M}$, $0 \leq \sigma \leq 1$, and

$$Z(u, \xi) = u^h \ \phi^*(u) \ e^{-\beta n u^{2k} + \beta \sqrt{n} u^k \xi(u) - \sigma u^k a(X,u)}.$$

Based on this definition of $E_u^\sigma[\ |\xi]$ and the standard form of the log likelihood ratio function, the major parts of the four errors are given by the case $\sigma = 0$,

$$B_g(\epsilon) = E_X\Big[-\log E_u^0[e^{-a(X,u)u^k}|\xi_n]\Big], \tag{6.39}$$

$$B_t(\epsilon) = \frac{1}{n}\sum_{j=1}^{n} -\log E_u^0[e^{-a(X_j,u)u^k}|\xi_n], \tag{6.40}$$

$$G_g(\epsilon) = E_u^0[u^{2k}|\xi_n], \tag{6.41}$$

$$G_t(\epsilon) = E_u^0\Big[u^{2k} - \frac{1}{\sqrt{n}}\ u^k \xi_n(u)\Big|\xi_n\Big]. \tag{6.42}$$

Lemma 6.5 *Assume that $k_1 > 0$, where k_1 is the first coefficient of the multi-index $k = (k_1, k_2, \ldots, k_d)$, and that $0 \leq \sigma \leq 1$.*
(1) For an arbitrary real analytic function $\xi(u)$ and $a(x, u)$,

$$E_u^\sigma[u^{2k}|\xi] \leq \frac{c_1}{n}\{1 + \|\xi\|^2 + \|\partial_1 \xi\|^2 + \|a(X)\| + \|\partial_1 a(X)\|\},$$

$$E_u^\sigma[u^{3k}|\xi] \leq \frac{c_2}{n^{3/2}}\{1 + \|\xi\|^3 + \|\partial_1 \xi\|^3 + \|a(X)\|^{3/2} + \|\partial_1 a(X)\|^{3/2}\},$$

where $\partial_1 = (\partial/\partial u_1)$, $c_1, c_2 > 0$ are constants.
(2) For the empirical process $\xi_n(u)$,

$$E\big[E_u^\sigma[n\ u^{2k}|\xi_n]\big] < \infty,$$
$$E\big[E_u^\sigma[n^{3/2}\ u^{3k}|\xi_n]\big] < \infty.$$

(3) Random variables $nG_g(\epsilon)$ ard $nB_t(\epsilon)$ are asymptotically uniformly integrable.

Proof of Lemma 6.5 (1) Let $0 \leq p \leq 3$. We use the notation $g(u) = u_2^{k_2} \cdots u_d^{k_d}$ and $h(u) = u_2^{h_2} \cdots u_d^{h_d}$, which do not depend on u_1. Then

$$u^k = u_1^{k_1} g(u),$$
$$u^h = u_1^{h_1} h(u),$$
$$N_p = \sum_\alpha \int_{[0,b]^d} (u^k)^p \ u^h \ e^{-\beta n u^{2k} + f(u)} du,$$
$$f(u) = \beta\sqrt{n} u^k \xi(u) - \sigma u^k a(X, u).$$

By eq.(6.14) and given $\phi(u) > 0$, for each $0 \leq p \leq 3$, there exists a constant $c_p > 0$ such that

$$0 \leq E_u^\sigma[u^{pk}|\xi] \leq c_p \frac{N_p}{N_0}.$$

By applying partial integration to N_p and using $q = (p-2)k_1 + h_1 + 1$,

$$\begin{aligned}
N_p &= \sum_\alpha \int_{[0,b]^d} g(u)^p h(u) u_1^{2k_1-1+q} e^{-\beta n u^{2k} + f(u)} du \\
&= -\sum_\alpha \int_{[0,b]^d} \frac{g(u)^{p-2}h(u)}{2\beta n k_1} u_1^q e^{f(u)} \, \partial_1(e^{-\beta n u^{2k}}) \, du \\
&\leq \sum_\alpha \int_{[0,b]^d} \frac{g(u)^{p-2}h(u)}{2\beta n k_1} \partial_1(u_1^q e^{f(u)}) \, e^{-\beta n u^{2k}} \, du \\
&= \sum_\alpha \int_{[0,b]^d} \frac{(u^k)^{p-2}u^h}{2\beta n k_1} e^{-\beta n u^{2k} + f(u)} (q + u_1 \partial_1 f(u)) \, du.
\end{aligned}$$

By the relation

$$\begin{aligned}
u_1 \partial_1 f(u) = \beta\sqrt{n} u^k (k_1 \xi(u) + u_1 \partial_1 \xi(u)) \\
- \sigma u^k (k_1 a(X,u) + u_1 \partial_1 a(X,u)),
\end{aligned}$$

and the Cauchy–Schwarz inequality, since $u \in [0,b]^d$, there exists $B > 0$ such that

$$\begin{aligned}
|u_1 \partial_1 f(u)| \leq \frac{\beta n k_1 u^{2k}}{2} + B(\|\xi\|^2 + \|\partial_1 \xi\|^2 \\
+ \|a\| + \|\partial_1 a\|).
\end{aligned}$$

Hence, by $B' = \max\{B, q\}$,

$$\begin{aligned}
\frac{N_p}{N_0} \leq \frac{N_p}{4N_0} + B'(1 + \|\xi\|^2 + \|\partial_1 \xi\|^2 \\
+ \|a\| + \|\partial_1 a\|) \frac{N_{p-2}}{N_0}.
\end{aligned} \tag{6.43}$$

The case $p = 2$ shows the first half of (1). For the latter half, By eq.(6.43) with $p = 3$, using $B'' = 4B'/3$,

$$\frac{N_3}{N_0} \leq B''(1 + \|\xi\|^2 + \|\partial_1 \xi\|^2 + \|a\| + \|\partial_1 a\|) \frac{N_1}{N_0}.$$

Since $N_1/N_0 \leq (N_2/N_0)^{1/2}$ by the Cauchy–Schwarz inequality, there exists $B''' > 0$ such that

$$\frac{N_3}{N_0} \leq B'''(1 + \|\xi\|^2 + \|\partial_1 \xi\|^2 + \|a\| + \|\partial_1 a\|)^{3/2}.$$

In general $(\sum_{i=1}^n |a_i|^2/n)^{1/2} \leq (\sum_{i=1}^n |a_i|^3/n)^{1/3}$, so the latter half of (1) is obtained.
(2) By Lemma 6.4 and the result of (1) of this lemma, part (2) is immediately derived.
(3) By the definition, $nG_g(\epsilon) = E_u^0[n\, u^{2k}|\xi_n]$. Then from (2) of this lemma, $nG_g(\epsilon)$ is asymptotically uniformly integrable (AUI). Let us prove $nB_t(\epsilon)$ is AUI. By using the notation

$$b(s) = -\sum_{j=1}^n \log E_u^0[e^{-s\, a(X_j,u)u^k}],$$

there exists $0 < s^* < 1$ such that

$$\begin{aligned} nB_t(\epsilon) &= b(1) = b(0) + b'(0) + \tfrac{1}{2} b''(s^*) \\ &= B_1 + B_2, \end{aligned}$$

where

$$B_1 = \sum_{j=1}^n E_u^0[a(X_j,u)u^k|\xi_n],$$

$$B_2 = \frac{1}{2n}\sum_{j=1}^n E_u^{s^*}[a(X_j,u)^2\, n\, u^{2k}|\xi_n]\Big|_{X=X_j}.$$

The first term $B_1 = nG_t(\epsilon)$. From Lemma 6.1,

$$-\tfrac{1}{4} nH_t(\epsilon) \leq nG_t(\epsilon) \leq \tfrac{1}{2}(3nG_g(\epsilon) + nH_t(\epsilon)).$$

Therefore $E[|B_1|^{3/2}] < \infty$, because $E[(nG_g(\epsilon))^{3/2}] < \infty$ and $E[(nH_t(\epsilon))^3] < \infty$. Moreover,

$$|B_2|^{3/2} \leq \frac{1}{n}\sum_{j=1}^n \|a(X_j)^3\| \Big(E_u^{s^*}[n\, u^{2k}|\xi_n]\Big|_{X=X_j}\Big)^{3/2}.$$

By the statements (1) and (2) of this lemma, $E[|B_2|^{3/2}] < \infty$, therefore $nB_t(\epsilon)$ is AUI. □

Without loss of generality, for each local coordinate, we can assume $u = (x, y)$ $x \in \mathbb{R}^r$, $y \in \mathbb{R}^{r'}$ $(r' = d - r)$, $k = (k, k')$, $h = (h, h')$, and

$$\frac{h_1 + 1}{2k_1} = \cdots = \frac{h_r + 1}{2k_r} = \lambda_\alpha < \frac{h'_1 + 1}{2k'_1} \leq \cdots .$$

We define $\mu = h' - 2k'\lambda_\alpha \in \mathbb{R}^{r'}$; then

$$\mu_i > h'_i - 2k'_i \frac{h'_i + 1}{2k'_i} = -1,$$

hence y^μ is integrable in $[0, b]^{r'}$. Both λ_α and r depend on the local coordinate. Let λ be the smallest λ_α and m be the largest r among the coordinates in which $\lambda = \lambda_\alpha$. Then $(-\lambda)$ and m are respectively equal to the largest pole and its order of the zeta function, as is shown in Definition 6.4. Let α^* be the index of the set of coordinates which satisfy $\lambda_\alpha = \lambda$ and $r = m$. As is shown in Lemma 6.6, only coordinates M_{α^*} affect the four errors. Let $\sum_{\alpha^*}$ be the sum of such coordinates. For a given function $f(u)$, we use the notation $f_0(y) = f(0, y)$. Also $a_0(X, y) = a(X, 0, y)$.

Definition 6.8 The expectation of a function $f(y, t)$ for a given function $\xi(u)$ on the essential family of local coordinates is defined by

$$E_{y,t}[f(y,t)|\xi] = \frac{\displaystyle\sum_{\alpha^*} \int_0^\infty dt \int dy\ f(y,t)\ Z_0(y,t,\xi)}{\displaystyle\sum_{\alpha^*} \int_0^\infty dt \int dy\ Z_0(y,t,\xi)},$$

where $\int dy$ stands for $\int_{[0,b]^{d-m}} dy$ and

$$Z_0(y, t, \xi) = \gamma_b y^\mu\ t^{\lambda-1} e^{-\beta t + \beta\sqrt{t}\,\xi_0(y)} \phi_0^*(y).$$

Here $\gamma_b > 0$ is a constant defined by eq.(4.16).

Lemma 6.6 *Let $p \geq 0$ be a constant. There exists $c_1 > 0$ such that, for an arbitrary C^1-class function $f(u)$ and analytic function $\xi(u)$, the following inequality holds:*

$$\left| E_u^0[(n\ u^{2k})^p\ f(u)|\xi] - E_{y,t}[t^p f_0(y)|\xi] \right| \leq \frac{D(\xi, f, \phi^*)}{\log n},$$

where

$$D(\xi, f, \phi^*) \equiv \frac{c_1 e^{2\beta\|\xi\|^2} \|\phi^*\|}{(\min \phi^*)^2} \{\beta \|\nabla\xi\| \|f\phi^*\| + \|\nabla(f\phi^*)\| + \|f\phi^*\|\}$$

and $\|\nabla f\| = \sum_j \|\partial_j f\|$.

Proof of Lemma 6.6 Using Z^p and Y^p in Definition 4.10 and eq.(4.19), we define A, B, and C by

$$A \equiv E_u^0[(n\, u^{2k})^p\ f(u)|\xi] = \frac{\sum_\alpha n^p Z^p(n,\xi,f\phi^*)}{\sum_\alpha Z^0(n,\xi,f\phi^*)},$$

$$B \equiv E_{y,t}[t^p f_0(y)|\xi] = \frac{\sum_{\alpha^*} n^p Y^p(n,\xi,f\phi^*)}{\sum_{\alpha^*} Y^0(n,\xi,f\phi^*)},$$

$$C \equiv \frac{\sum_\alpha n^p Y^p(n,\xi,f\phi^*)}{\sum_\alpha Y^0(n,\xi,f\phi^*)},$$

where $\sum_\alpha$ and $\sum_{\alpha^*}$ denote the sum of all local coordinates and the sum of coordinates in the essential family respectively. To prove the lemma, it is sufficient to show $|A-B| \leq D(\xi,f,\phi^*)/\log n$. Since

$$|A-B| \leq |A-C| + |C-B|,$$

we show the inequalities for $|A-C|$ and $|C-B|$ respectively. The set $(n,\xi,f\phi^*)$ is omitted for simplicity. Firstly, $|A-C|$ is bounded by

$$\begin{aligned}
|A-C| &= n^p\left|\frac{\sum_\alpha Z^p}{\sum_\alpha Z^0} - \frac{\sum_\alpha Y^p}{\sum_\alpha Y^0}\right| \\
&= n^p\left|\frac{\sum_\alpha Z^p - \sum_\alpha Y^p}{\sum_\alpha Z^0} + \frac{\sum_\alpha Y^p\{\sum_\alpha Y^0 - \sum_\alpha Z^0\}}{\sum_\alpha Z^0 \sum_\alpha Y^0}\right| \\
&\leq n^p \frac{\sum_\alpha |Z^p - Y^p|}{\sum_\alpha Z^0} + n^p \frac{\sum_\alpha Y^p}{\sum_\alpha Y^0} \times \frac{\sum_\alpha |Y^0 - Z^0|}{\sum_\alpha Z^0}.
\end{aligned}$$

For general $a_i, b_i > 0$,

$$\frac{\sum a_i}{\sum b_i} \leq \sum \frac{a_i}{b_i}.$$

Therefore,

$$|A-C| \leq \sum_\alpha \frac{n^p|Z^p - Y^p|}{Z^0} + \sum_\alpha \frac{n^p Y^p}{Y^0} \times \sum_\alpha \frac{|Y^0 - Z^0|}{Z^0}.$$

Then by using Theorems 4.8, 4.9, 4.10, there exist constants $C_1, C_2 > 0$ such that

$$\begin{aligned}
|A-C| \leq &\ \frac{C_1 e^{\beta\|\xi\|^2}}{\log n} \frac{\{\beta\|\nabla\xi\|\|f\phi^*\| + \|\nabla(f\phi^*)\| + \|f\phi^*\|\}}{\min\phi^*} \\
&+ \frac{C_2 e^{2\beta\|\xi\|^2}}{\log n} \frac{\|\phi^*\|\{\beta\|\nabla\xi\|\|f\phi^*\| + \|\nabla(f\phi^*)\| + \|f\phi^*\|\}}{(\min\phi^*)^2}.
\end{aligned}$$

Secondly,

$$|C - B| = n^p \left| \frac{\sum_\alpha Y^p}{\sum_\alpha Y^0} - \frac{\sum_{\alpha^*} Y^p}{\sum_{\alpha^*} Y^0} \right|.$$

Let us use the simplified notation,

$$T^p = \sum_{\alpha^*} Y^p,$$

$$U^p = \sum_{\alpha \backslash \alpha^*} Y^p.$$

Then, by $\sum_\alpha = \sum_{\alpha^*} + \sum_{\alpha\backslash\alpha^*}$,

$$|C - B| = n^p \left| \frac{T^p + U^p}{T^0 + U^0} - \frac{T^p}{T^0} \right|$$

$$\leq \frac{n^p U^p}{T^0} + \frac{n^p U^0 T^p}{(T^0)^2}.$$

By Theorem 4.10, there exists $C_3 > 0$ such that

$$|C - B| \leq \frac{C_3 e^{\beta \|\xi\|^2} \|\phi^*\|}{\min \phi^*} + \frac{C_4 e^{2\beta \|\xi\|^2} \|\phi^*\|^2}{(\min \phi^*)^2}.$$

By combining two results, we obtain the lemma. □

6.3.3 Proof of the theorems

In this subsection, we prove the theorems.

Definition 6.9 (Explicit representation of the Bayes quartet) Four functionals of a given function $\xi(u)$ are defined by

$$B_{\mathrm{g}}^*(\xi) \equiv \tfrac{1}{2} E_X[\ E_{y,t}[a_0(X, y) t^{1/2} | \xi]^2\ ], \tag{6.44}$$

$$B_{\mathrm{t}}^*(\xi) \equiv G_{\mathrm{t}}^*(\xi) - G_{\mathrm{g}}^*(\xi) + B_{\mathrm{g}}^*(\xi), \tag{6.45}$$

$$G_{\mathrm{g}}^*(\xi) \equiv E_{y,t}[t | \xi], \tag{6.46}$$

$$G_{\mathrm{t}}^*(\xi) \equiv E_{y,t}[t - t^{1/2} \xi_0(y) | \xi]. \tag{6.47}$$

Note that these four functionals do not depend on n. If $\xi(u)$ is a random process, then the four functionals are random variables. The singular fluctuation is defined by

$$\nu(\beta) = \tfrac{1}{2} E_\xi \Big[E_{y,t}[t^{1/2} \xi_0(y) | \xi] \Big]. \tag{6.48}$$

Proof of Theorems 6.8 In this proof, we use the simplified notation

$$E_u^\sigma[f(u)] = E_u^\sigma[f(u)|\xi_n],$$
$$E_{y,t}[f(y,t)] = E_{y,t}[f(y,t)|\xi_n],$$

in other words, '$|\xi_n$' is omitted. Firstly we prove the following convergences in probability.

$$nB_{\rm g}(\epsilon) - B_{\rm g}^*(\xi_n) \to 0, \tag{6.49}$$
$$nB_{\rm t}(\epsilon) - B_{\rm t}^*(\xi_n) \to 0, \tag{6.50}$$
$$nG_{\rm g}(\epsilon) - G_{\rm g}^*(\xi_n) \to 0, \tag{6.51}$$
$$nG_{\rm t}(\epsilon) - G_{\rm t}^*(\xi_n) \to 0. \tag{6.52}$$

Based on eq.(6.41), eq.(6.46), and Lemma 6.6 with $p = 1$,

$$|nG_{\rm g}(\epsilon) - G_{\rm g}^*(\xi_n)| = \left|E_u^0[nu^{2k}] - E_{y,t}[t]\right| \leq \frac{D(\xi_n, 1, \phi^*)}{\log n}.$$

Because the convergence in law $\xi_n \to \xi$ holds, eq.(6.51) is obtained. Also, based on eq.(6.42), eq. (6.47), and Lemma 6.6 with $p = 1, \frac{1}{2}$,

$$|nG_{\rm t}(\epsilon) - G_{\rm t}^*(\xi_n)| = \left|E_u^0[nu^{2k} - \sqrt{n}u^k\xi_n] - E_{y,t}[t - t^{1/2}\xi_0]\right| \leq \frac{D(\xi_n, 1, \phi^*)}{\log n} + \frac{D(\xi_n, \xi_n, \phi^*)}{\log n}.$$

Because the convergence in law $\xi_n \to \xi$ holds, eq.(6.52) is obtained. Let us prove eq.(6.49); we define

$$b_{\rm g}(\sigma) \equiv E_X\Big[-\log E_u^0[e^{-\sigma a(X,u)u^k}]\Big],$$

then it follows that $nB_{\rm g}(\epsilon) = nb_{\rm g}(1)$ and there exists $0 < \sigma^* < 1$ such that

$$nB_{\rm g}(\epsilon) = nb_{\rm g}(0) + nb_{\rm g}'(0) + \frac{n}{2}b_{\rm g}''(0) + \frac{n}{6}b_{\rm g}^{(3)}(\sigma^*) \tag{6.53}$$
$$= nE_u^0[u^{2k}] - \frac{n}{2}E_X E_u^0[a(X,u)^2u^{2k}]$$
$$+\frac{n}{2}E_X E_u^0[a(X,u)u^k]^2 + \tfrac{1}{6}nb_{\rm g}^{(3)}(\sigma^*), \tag{6.54}$$

where we used $b_g(0) = 0$, and $E_X[a(X, u)] = u^k$ hence $b'_g(0) = E_u^0[u^{2k}]$. The first term on the right-hand side of eq.(6.54) is equal to $nG_g(\epsilon)$. By Lemma 6.6, the following convergence in probability

$$\begin{aligned} &\left|nE_X E_u^0[a(X,u)^2 u^{2k}] - E_X E_{y,t}[a_0(X,y)^2 t]\right| \\ &\leq \frac{E_X[D(\xi, a(X,u)^2, \phi^*)]}{\log n} \to 0 \end{aligned} \tag{6.55}$$

holds, where $D(\beta, \xi, a(X, u), \phi^*)$ is defined as in Lemma 6.6. Since $E_X[a_0(X, y)^2] = 2$, the sum of the first two terms on the right-hand side of eq.(6.54) converges to zero in probability. For the third term, by using the notation

$$\begin{aligned} \rho(u, v) &= E_X[a(X,u)a(X,v)], \\ \rho_0(u, y) &= \rho(u, (0, y)), \\ \rho_{00}(y', y) &= \rho((0, y'), (0, y)), \end{aligned}$$

and applying Lemma 6.6,

$$\begin{aligned} &\left|nE_X E_u^0[a(X,u)u^k]^2 - E_{y,t}[a_0(X,y)t^{1/2}]^2\right| \\ &\leq \left|\sqrt{n}E_u^0\Big[u^k\big(\sqrt{n}E_v^0[\rho(u,v)v^k] - E_{y,t}[\rho_0(u,y)t^{1/2}]\big)\Big]\right| \\ &\quad + \left|E_{y,t}\Big[t^{1/2}\big(\sqrt{n}E_u^0[\rho_0(u,y)u^k] - E_{y',t'}[\rho_{00}(y',y)(t't)^{1/2}]\big)\Big]\right| \\ &\leq \frac{c_1\sqrt{n}}{\log n}E_u^0[u^k]\; D(\xi_n, \rho(\cdot,\cdot), \phi^*) \\ &\quad + \frac{c_1}{\log n}E_{y,t}[t^{1/2}D(\xi_n, \rho(\cdot, y), \phi^*)]. \end{aligned} \tag{6.56}$$

Equation (6.56) converges to zero in probability by Lemma 6.5. Therefore the difference between the third term and $B_g^*(\xi_n)$ converges to zero in probability. For the last term, we have

$$\begin{aligned} \left|nb_g^{(3)}(\sigma^*)\right| &= n\left|E_X\Big\{E_u^{\sigma^*}[a(X,u)^3 u^{3k}] + 2E_u^{\sigma^*}[a(X,u)u^k]^3\right. \\ &\qquad \left. -3E_u^{\sigma^*}[a(X,u)^2 u^{2k}]E_u^{\sigma^*}[a(X,u)u]\Big\}\right| \\ &\leq 6nE_X\Big[\|a(X)\|^3\; E_u^{\sigma^*}[u^{3k}]\Big], \end{aligned}$$

where we used Hölder's inequality. By applying Lemma 6.5,

$$\left|nb_{\rm g}^{(3)}(\sigma^*)\right| \le \frac{6c_2}{n^{1/2}} E_X\Big[\|a(X)\|^3 \{1 + \|\xi_n\|^3 + \|\partial\xi_n\|^3 + \|a(X)\|^{3/2} + \|\partial a(X)\|^{3/2}\}\Big], \tag{6.57}$$

which shows $nb_{\rm g}^{(3)}(\sigma^*)$ converges to zero in probability, because the fundamental condition (I) with index $s = 6$ is assumed. Hence eq.(6.49) is proved. We proceed to the proof of eq.(6.50). An empirical expectation $E_j^*[\ ]$ is simply denoted by

$$E_j^*[f(X_j)] = \frac{1}{n}\sum_{j=1}^{n} f(X_j).$$

By defining

$$b_{\rm t}(\sigma) = E_j^*[-\log E_u^0[e^{-\sigma a(X_j,u)u^k}]],$$

it follows that $nB_{\rm t}(\epsilon) = nb_{\rm t}(1)$ and there exists $0 < \sigma^* < 1$ such that

$$nB_{\rm t}(\epsilon) = nG_{\rm t}(\epsilon) - \frac{n}{2}E_j^* E_u^0[a(X_j,u)^2u^{2k}] + \frac{n}{2}E_j^* E_u^0[a(X_j,u)u^k]^2 + \tfrac{1}{6}nb_{\rm t}^{(3)}(\sigma^*). \tag{6.58}$$

Then, by applying Lemma 6.5, $nb_{\rm t}^{(3)}(\sigma^*)$ converges to zero in probability in the same way as eq.(6.57). In fact,

$$\begin{aligned}\left|nb_{\rm t}^{(3)}(\sigma^*)\right| &= \Big|E_j^*\Big\{E_u^{\sigma^*}[a(X_j,u)^3u^{3k}] + 2E_u^{\sigma^*}[a(X_j,u)u^k]^3 \\ &\quad -3E_u^{\sigma^*}[a(X_j,u)^2u^{2k}]E_u^{\sigma^*}[a(X_j,u)u]\Big\}\Big| \\ &\le 6nE_j^*\Big[\|a(X_j)\|^3\ E_u^{\sigma^*}[u^{3k}]\Big].\end{aligned}$$

By applying Lemma 6.5, using the fundamental condition (7) with $5 = 6$,

$$\left|nb_{\rm t}^{(3)}(\sigma^*)\right| \le \frac{6c_2}{n^{1/2}} E_j^*\Big[\|a(X_j)\|^3 \{1 + \|\xi_n\|^3 + \|\partial\xi_n\|^3 + \|a(X_j)\|^{3/2} + \|\partial a(X_j)\|^{3/2}\}\Big], \tag{6.59}$$

which converges to zero in probability. By the same methods as eq.(6.55) and eq.(6.56), replacing respectively $E_X[\|a(X)^2\|]$ with $E_j^*\|a(X_j)^2\|$ and $\rho(u,v)$

with $\rho_n(u, v) = E_j^* a(X_j, u) a(X_j, v)$,

$$\begin{aligned}
&\left|\frac{n}{2} E_j^* E_u^0[a(X_j, u)^2 u^{2k}] - G_g^*(\xi_n)\right| \\
&\leq \frac{n}{2}\left|E_j^* E_u^0[a(X_j, u)^2 u^{2k}] - E_X E_u^0[a(X, u)^2 u^{2k}]\right| \\
&\quad + \left|\frac{n}{2} E_X E_u^0[a(X, u)^2 u^{2k}] - G_g^*(\xi_n)\right| \\
&\leq \left(\sup_u |E_j^* a(X_j, u) - E_X a(X, u)|\right) \frac{n}{2} E_u^0[u^{2k}] \\
&\quad + \left|\frac{n}{2} E_X E_u^0[a(X, u)^2 u^{2k}] - G_g^*(\xi_n)\right|,
\end{aligned}$$

which converges to zero by Lemma 6.5 and eq.(6.55). In the same way, the following convergence in probability holds,

$$\tfrac{1}{2} E_j^* E_u^0[a(X_j, u) u^k]^2 - B_g^*(\xi_n) \to 0,$$

and therefore the following convergence in probability also holds:

$$n B_t(\epsilon) - n G_t(\epsilon) + n G_g(\epsilon) - n B_g(\epsilon) \to 0. \tag{6.60}$$

Therefore eq.(6.50) is obtained. By combining eq.(6.49)–eq.(6.52) with Lemma 6.3 (2), we obtain the following convergences in probability:

$$n B_g - B_g^*(\xi_n) \to 0, \tag{6.61}$$

$$n B_t - B_t^*(\xi_n) \to 0, \tag{6.62}$$

$$n G_g - G_g^*(\xi_n) \to 0, \tag{6.63}$$

$$n G_t - G_t^*(\xi_n) \to 0. \tag{6.64}$$

The four functionals $B_g^*(\xi)$, $B_t^*(\xi)$, $G_g^*(\xi)$, and $G_t^*(\xi)$ are continuous functions of $\xi \in B(\mathcal{M})$. From the convergence in law of the empirical process $\xi_n \to \xi$, these convergences in law

$$B_g^*(\xi_n) \to B_g^*(\xi), \quad B_t^*(\xi_n) \to B_t^*(\xi),$$
$$G_g^*(\xi_n) \to G_g^*(\xi), \quad G_t^*(\xi_n) \to G_t^*(\xi),$$

are derived. Therefore Theorem 6.8 (1) and (2) are obtained. Theorem 6.8 (3) is shown because the four errors are asymptotically uniformly integrable by Lemma 6.3. □

Proof of Main Theorem 6.3 Before proving the theorem, we introduce a property of a Gaussian process. We use the notation

$$S_\lambda(a) = \int_0^\infty dt\, t^{\lambda-1}\, e^{-\beta t + a\beta\sqrt{t}},$$

$$\int du^* = \sum_{\alpha^*} \gamma_b \int dx\, dy\, \delta(x)\, y^\mu,$$

$$Z(\xi) = \int du^*\, S_\lambda(\xi(u)),$$

where $u = (x, y)$. Then, by Definition 6.9,

$$E[B_{\rm g}^*] = \frac{1}{2\beta^2} E\Big[E_X\Big[\Big(\frac{\int du^* a(X,u) S_\lambda'(\xi(u))}{Z(\xi)}\Big)^2\Big]\Big],$$

$$E[B_{\rm t}^*] = E[B_{\rm g}^*] + E[G_{\rm t}^*] - E[G_{\rm g}^*],$$

$$E[G_{\rm g}^*] = \frac{1}{\beta^2} E\Big[\frac{\int du^* S_\lambda''(\xi(u))}{Z(\xi)}\Big],$$

$$E[G_{\rm t}^*] = \frac{1}{\beta^2} E\Big[\frac{\int du^* S_\lambda''(\xi(u))}{Z(\xi)}\Big] - \frac{1}{\beta} E\Big[\frac{\int du^*\, \xi(u) S_\lambda'(\xi(u))}{Z(\xi)}\Big].$$

Let $\nu = \nu(\beta)$ be the singular fluctuation in eq.(6.48) and A be a constant,

$$\nu = \frac{1}{2\beta} E\Big[\frac{\int du^*\, \xi(u) S_\lambda'(\xi(u))}{Z(\xi)}\Big], \tag{6.65}$$

$$A = \frac{1}{\beta^2} E\Big[\frac{\int du^* S_\lambda''(\xi(u))}{Z(\xi)}\Big]. \tag{6.66}$$

By eq.(5.24) in Theorem 5.11 and $\rho(u,u) = E_X[a(X,u)^2] = 2$ with $u = (0, y)$, we have

$$E[B_g^*] = A - \frac{1}{\beta}\nu, \tag{6.67}$$

$$E[B_t^*] = A - \Big(2 + \frac{1}{\beta}\Big)\nu, \tag{6.68}$$

$$E[G_g^*] = A, \tag{6.69}$$

$$E[G_t^*] = A - 2\nu. \tag{6.70}$$

By combining these equations to eliminate A and ν, we obtain two equations which do not contain either A or ν, giving Main Theorem 6.3. □

Proof of Theorem 6.9 By eq.(5.22) with $a = \xi_n(u)$

$$\frac{2}{\beta^2}\frac{S''_\lambda(\xi_n(u))}{S_\lambda(\xi_n(u))} - \frac{1}{\beta}\frac{\xi_n(u)S'_\lambda(\xi_n(u))}{S_\lambda(\xi_n(u))} - \frac{2\lambda}{\beta} = 0.$$

By the definitions of $G^*_{\rm g}(\xi_n)$ and $G^*_{\rm t}(\xi_n)$,

$$G^*_{\rm g}(\xi_n) + G^*_{\rm t}(\xi_n) - \frac{2\lambda}{\beta} = 0.$$

By the convergences in law $nG_{\rm g} \to G^*_{\rm g}(\xi)$ and $nG_{\rm t} \to G^*_{\rm t}(\xi)$, Theorem 6.9 is obtained. □

Proof of Theorem 6.10 Let $\nu = \nu(\beta)$ be the singular fluctuation in eq.(6.48). By eq.(5.23) in Theorem 5.11, eq.(6.66), and eq.(6.65),

$$A = \frac{\lambda}{\beta} + \nu(\beta).$$

Hence from eqs.(6.67)–(6.70), we obtain eqs.(6.27)–(6.30). From the definition in eq.(6.24),

$$V = nE^*_j E^0_u[a(X_j, u)^2 u^{2k}] - nE^*_j E^0_u[a(X_j, u)u^k]^2 + o_p(1),$$

where $o_p(1)$ is a random variable which converges to zero in probability. Then, based on eq.(6.58) in the proof of Theorem 6.8, the following convergence in probability holds,

$$V - 2(G^*_{\rm t}(\xi_n) - B^*_{\rm t}(\xi_n)) \to 0,$$

which gives eqs.(6.25) and (6.26). Therefore V converges in law and is asymptotically uniformly integrable, so, when $n \to \infty$,

$$E[V] \to \frac{2\nu(\beta)}{\beta}.$$

Let us introduce an expectation $\langle\ \rangle$ defined by

$$\langle f(u, t)\rangle = \frac{\int du^*\ f(u, t)\, t^{\lambda-1}\ e^{-\beta t + \beta\sqrt{t}\xi(u)}}{\int du^*\ t^{\lambda-1}\ e^{-\beta t+\beta\sqrt{t}\xi(u)}}. \tag{6.71}$$

Then

$$A = E_\xi[\langle t\rangle],$$
$$\nu(\beta) = \tfrac{1}{2}E_\xi[\langle\sqrt{t}\xi(u)\rangle].$$

By using the Cauchy–Schwarz inequality,

$$E_\xi[\langle\sqrt{t}\xi(u)\rangle] \le E_\xi[\langle t\rangle]^{1/2} E_\xi[\langle\xi(u)^2\rangle]^{1/2} \le \sqrt{A}\, E_\xi[\|\xi\|^2]^{1/2}.$$

By combining this inequality with

$$A = \frac{\lambda}{\beta} + \nu(\beta) \le \frac{\lambda}{\beta} + \frac{\sqrt{A}}{2} E_\xi[\|\xi\|^2]^{1/2},$$

we obtain

$$\sqrt{A} \le \frac{E_\xi[\|\xi\|^2]^{1/2}}{4} + \sqrt{E_\xi[\|\xi\|^2]/16 + \lambda/\beta},$$

which completes the proof. □

Remark 6.13 (Singular fluctuation) By using the expectation notation defined in eq.(6.71) we can represent the variance of $a_0(X, y)$,

$$\nu(\beta) = \tfrac{1}{2} E_\xi[\langle\sqrt{t}\xi_0(y)\rangle] = \frac{\beta}{2} E_\xi E_X\Big[\Big\langle \big(\sqrt{t}a_0(X, y)\big)^2 \Big\rangle - \big\langle\sqrt{t}a_0(X, y)\big\rangle^2\Big].$$

Note that

$$a(x, w) = \frac{\log(q(x)/p(x|w)) - K(w)}{\sqrt{K(w)}}.$$

Although $a(x, w)$ is not well defined at a singularity in the original parameter space, it can be made well-defined on the manifold by resolution of singularities. Both the real log canonical threshold λ and the singular fluctuation $\nu(\beta)$ determine the asymptotic behavior of a statistical model. In regular statistical models, $\lambda = \nu(\beta) = d/2$, where d is the dimension of the parameter space, whereas, in singular statistical models, λ and $\nu(\beta)$ are different from $d/2$ in general.

6.4 Maximum likelihood and *a posteriori*

In this section, we study the estimator $\hat{w}$ which minimizes

$$-\sum_{i}^{n} \log p(X_i|w) + a_n\sigma(w)$$

in a compact set W, which is equal to the parameter that minimizes

$$R_n^0(w) = nK_n(w) + a_n\sigma(w).$$

We assume that $\sigma(w)$ is a C^2-class function of w in an open set which contains W, and that $\{a_n \geq 0\}$ is a nondecreasing sequence. We can assume $\sigma(w) \geq 0$ without loss of generality. If $a_n = 0$ for arbitrary n, then $\hat{w}$ is called the maximum likelihood estimator (MLE) and if $a_n = 1$ for arbitrary n and $\sigma(w) = -\log\varphi(w)$, where $\varphi(w)$ is an *a priori* probability density function, then $\hat{w}$ is called the maximum *a posteriori* estimator (MAP). The generalization and training errors are respectively defined by

$$R_{\rm g} = K(\hat{w}),$$
$$R_{\rm t} = K_n(\hat{w}).$$

Although the MAP employs an *a priori* distribution, its generalization error is quite different from that of Bayes estimation.

To study the ML or MAP method, we have to analyze the geometry of the parameter space. Let us assume the fundamental conditions (I) and (II) with index $s = 4$. We prove that, for arbitrary $\epsilon > 0$, $P(K(\hat{w}) > \epsilon)$ is sufficiently small that it does not affect the asymptotic generalization and training errors. To study the event $K(\hat{w}) \leq \epsilon$, we use the resolution of singularities and the standard form of the log likelihood ratio function. Then the Kulback–Leibler distance becomes a normal crossing function defined on a local coordinate $[0, b]^d$ of a manifold

$$u^{2k} = u_1^{2k_1} u_2^{2k_2} \cdots u_r^{2k_r},$$

where r is an integer which satisfies $1 \leq r \leq d$ and $k_1, k_2, \ldots, k_r > 0$ are natural numbers. Without loss of generality, we can assume that $b = 1$. For a given $u \in [0, 1]^d$, the integer a is defined by the number $(1 \leq a \leq r)$ which satisfies

$$\frac{u_a^2}{k_a} \leq \frac{u_i^2}{k_i} \quad (i = 1, 2, \ldots, r). \tag{6.72}$$

Intuitively, a is the number on the axis which is farthest from the given point u. A map $[0, 1]^d \ni u \to (t, v)$, where $t \in \mathbb{R}^1$, $v = (v_1, v_2, \ldots, v_d) \in \mathbb{R}^d$, is defined by

$$t = u^{2k},$$

$$v_i = \begin{cases} \sqrt{u_i^2 - (k_i/k_a)u_a^2} & (1 \leq i \leq r) \\ u_i & (r < i \leq d). \end{cases}$$

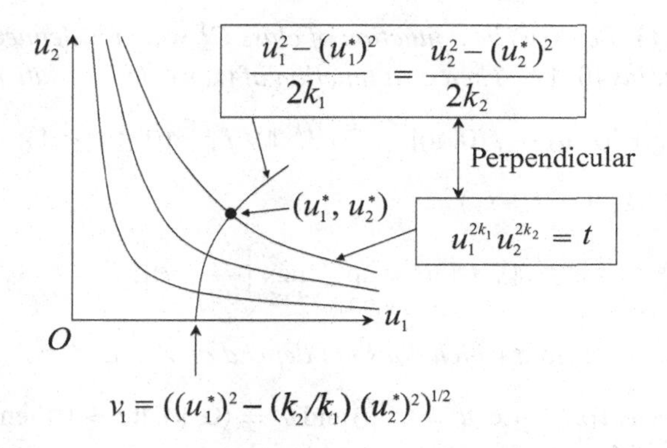

Fig. 6.3. Normal crossing coordinate

Then, by definition, $v_a = 0$. The set V is defined by

$$V \equiv \{v = (v_1, v_2, \ldots, v_d) \in [0, 1]^d;\ v_1 v_2 \cdots v_r = 0\}.$$

Then $[0, 1]^d \ni u \mapsto (t, v) \in T \times V$ is a one-to-one map as in Figure 6.3. Under this correspondence u is identified with (t, v).

Remark 6.14 Note that, if a surface parameterized by t,

$$u_1^{2k_1} u_2^{2k_2} \cdots u_r^{2k_r} = t, \tag{6.73}$$

and the curve parameterized by $(u_1^*, u_2^*, \ldots, u_r^*)$,

$$\frac{u_1^2 - (u_1^*)^2}{2k_1} = \frac{u_2^2 - (u_2^*)^2}{2k_2} = \cdots = \frac{u_r^2 - (u_r^*)^2}{2k_r}, \tag{6.74}$$

have a crossing point, then they are perpendicular to each other in the restricted space $\mathbb{R}^r$ at the crossing point. In fact, the perpendicular vector of eq.(6.73) and the tangent vector at $(u_1, u_2, \ldots, u_r)$ of eq.(6.74) are equal to each other,

$$\left(\frac{k_1}{u_1}, \frac{k_1}{u_2}, \ldots, \frac{k_r}{u_r}\right).$$

The map $u \mapsto (t, v)$ can be understood as a function from a point $(u_1^*, u_2^*, \ldots, u_r^*)$ to the crossing point of eq.(6.74) and $u^{2k} = 0$. It is equal to the limit point of steepest descent dynamics, eq.(1.35).

Theorem 6.11 *Let $f(u)$ be a function of class C^1 which is defined on an open set that contains $[0, 1]^d$. Then as a function of (t, v), $f(t, v)$ satisfies*

$$|f(t, v) - f(0, v)| \leq C\, t^{1/k^*} \|\nabla f\| \quad (0 \leq t < 1),$$

where $k^* = 2(k_1 + \cdots + k_r)$,

$$\|\nabla f\| = \sup_{u \in [0,1]^d} \max_{1 \leq j \leq d} \left| \frac{\partial f}{\partial u_j}(u) \right|,$$

and $C > 0$ is a constant which does not depend on t, s, and f.

Proof of Theorem 6.11 Let $u = (t, v)$ and $u' = (0, v)$. It $t > 0$ then the Jacobian determinant of the map $u \mapsto (t, v)$ is not equal to zero. There exists $u^* \in [0, 1]^d$ such that

$$|f(t, v) - f(0, v)| \leq \sum_{j=1}^{r} |u_j - u'_j| \left| \frac{\partial f}{\partial u_j}(u^*) \right|.$$

Hence

$$|f(t, v) - f(0, v)| \leq \|\nabla f\| \sum_{j=1}^{r} |u_j - u'_j|,$$

where

$$\|\nabla f\| = \sum_{j=1}^{d} \max_{u \in [0,1]^d} \left| \frac{\partial f}{\partial u_j} \right|.$$

If $j = a$ then $|u_j - u'_j| = u_a$. If $j \neq a$, then $u_a^2/k_a \leq u_j^2/k_j$,

$$\begin{aligned}
|u_j - u'_j| &= \left| u_j - \left(u_j^2 - (k_j/k_a)u_a^2\right)^{1/2} \right| \\
&= \frac{(k_j/k_a)u_a^2}{u_j + \left(u_j^2 - (k_j/k_a)u_a^2\right)^{1/2}} \\
&\leq \sqrt{k_j/k_a}\; u_a.
\end{aligned}$$

Hence there exists $C' > 0$ such that

$$|f(t, v) - f(0, v)| \leq C' \|\nabla f\| u_a.$$

On the other hand by eq.(6.72) there exists $C'' > 0$ such that

$$t = u^{2k} \geq C''(u_a)^{2k^*},$$

which completes the theorem. □

Theorem 6.12 *Assume the fundamental conditions (I) and (II) with index s ($s \geq 6$) and that $a_n/n \to 0$ ($n \to \infty$). Let $\psi_n(w)$ be an empirical process on $\{w; K(w) > \epsilon\}$,*

$$\psi_n(w) = \sum_{i=1}^{n} \frac{K(w) - f(X_i, w)}{\sqrt{nK(w)}},$$

and

$$\|\psi_n\| = \sup_{K(w)>\epsilon} |\psi_n(w)|.$$

Then the following hold.
(1) For a given $\epsilon > 0$, there exists a constant $C > 0$, such that, for arbitrary $n \geq 1$,

$$P(\|\psi_n\|^2 > n\epsilon) \leq \frac{C}{n^{s/2}},$$

$$P(K(\hat{w}) > \epsilon) \leq \frac{C}{n^{s/2}}.$$

(2) For a given $\epsilon > 0$, there exists a constant $C' > 0$, such that, for arbitrary $n \geq 1$,

$$E[\|\psi_n\|^2]_{\{\|\psi_n\|^2 > n\epsilon\}} \leq \frac{C'}{n^{s/2-1}},$$

$$E[nK(\hat{w})]_{\{K(\hat{w})>\epsilon\}} \leq \frac{C'}{n^{s/2-1}}.$$

Proof of Theorem 6.12 (1) From Theorem 5.8,

$$E[\|\psi_n\|^s] = C < \infty,$$

and it follows that

$$\begin{aligned} C &\geq E[\|\psi_n\|^s]_{\{\|\psi_n\|^2 > n\epsilon\}} \\ &\geq (n\epsilon)^{s/2} P(\|\psi_n\|^2 > n\epsilon). \end{aligned}$$

Therefore

$$P(\|\psi_n\|^2 > n\epsilon) \leq \frac{C}{(n\epsilon)^{s/2}}. \tag{6.75}$$

By the Cauchy–Schwarz inequality,

$$\begin{aligned} R_n^0(w) &= nK(w) - \sqrt{nK(w)}\,\psi_n(w) + a_n\sigma(w) \\ &\geq \tfrac{1}{2}(nK(w) - \|\psi_n\|^2) + a_n\sigma(w). \end{aligned}$$

A parameter w_0 in the set of true parameters satisfies $K(w_0) = 0$, hence

$$R_n^0(w_0) = a_n \sigma(w_0).$$

Therefore, by the definition of $\hat{w}$, $R_n^0(\hat{w}) \leq R_n^0(w_0)$, and consequently

$$\tfrac{1}{2}(nK(\hat{w}) - \|\psi_n\|^2) + a_n \sigma(\hat{w}) \leq a_n \sigma(w_0).$$

Hence, if $\|\psi_n\|^2 \geq n\epsilon$, there exists a constant $c_1 > 0$ such that

$$nK(\hat{w}) \leq \|\psi_n\|^2 + 4a_n \|\sigma\| \leq c_1 \|\psi_n\|^2.$$

Because $a_n/n \to 0$,

$$\begin{aligned} P(K(\hat{w}) > \epsilon) &\leq P(c_1 \|\psi_n\|^2 > n\,\epsilon) \\ &\leq \frac{c_2}{(n\epsilon)^{s/2}}. \end{aligned}$$

(2) By using the above results,

$$\begin{aligned} E[\|\psi_n\|^2]_{\{\|\psi_n\|^2 > n\epsilon\}} &\leq \sum_{j=1}^{\infty} E[\|\psi_n\|^2]_{\{jn\epsilon < \|\psi_n\|^2 \leq (j+1)n\epsilon\}} \\ &\leq \sum_{j=1}^{\infty} (j+1)\epsilon n \times \frac{C}{(n\epsilon j)^{s/2}} \\ &\leq \frac{c_3}{(n\epsilon)^{s/2-1}}. \end{aligned}$$

In the same way,

$$\begin{aligned} E[nK(\hat{w})]_{\{K(w) > \epsilon\}} &\leq \sum_{j=1}^{\infty} E[nK(\hat{w})]_{\{jn\epsilon < nK(\hat{w}) \leq (j+1)n\epsilon\}} \\ &\leq \sum_{j=1}^{\infty} (j+1)\epsilon n \times \frac{c_2}{(n\epsilon j)^{s/2}} \\ &\leq \frac{c_4}{(n\epsilon)^{s/2-1}}. \end{aligned}$$

□

Remark 6.15 (Consistency of estimation) By Theorem 6.12, if the fundamental conditions (I) and (II) are satisfied, then both the maximum likelihood estimator and the maximum *a posteriori* estimator converge to the true set of parameters in probability. This property is called consistency of estimation. If the fundamental conditions are not satisfied, then such a model may not have consistency.

Theorem 6.13 *Assume the fundamental conditions (I) and (II) with index s ($s \geq 6$) and that, for an arbitrary $p > 0$, $a_n/n^p \to 0$ ($n \to \infty$). Let $\xi_n(w)$ be an empirical process on $\{w; K(w) < \epsilon\}$,*

$$\xi_n(w) = \frac{1}{\sqrt{n}} \sum_{i=1}^{n} \{a(X_i, u) - E_X[a(X, u)]\},$$

and

$$\|\xi_n\| = \sup_{u \in \mathcal{M}} |\xi_n(u)|.$$

Then the following hold.
(1) For a given $0 < \delta < 1$, there exists a constant $C > 0$, such that, for arbitrary $n \geq 1$,

$$P(\|\xi_n\|^2 > n^\delta) \leq \frac{C}{n^{s\delta/2}},$$

$$P(nK(g(\hat{u})) > n^\delta) \leq \frac{C}{n^{s\delta/2}},$$

where $\hat{u}$ is defined by $\hat{w} = K(g(\hat{u}))$).
(2) For a given $0 < \delta < 1$, there exists a constant $C' > 0$, such that, for arbitrary $n \geq 1$,

$$E[\|\xi_n\|^2]_{\{\|\psi_n\|^2 > n^\delta\}} \leq \frac{C'}{n^{\delta(s/2-1)}},$$

$$E[nK(g(\hat{u}))]_{\{nK(g(\hat{u})) > n^\delta\}} \leq \frac{C'}{n^{\delta(s/2-1)}}.$$

Proof of Theorem 6.13 This theorem is proved in the same way as the previous theorem. Let $\sigma(u) \equiv \sigma(g(u))$.
(1) From Theorem 5.8,

$$E[\|\xi_n\|^s] = C < \infty,$$

and it follows that

$$\begin{aligned} C &\geq E[\|\xi_n\|^s]_{\{\|\xi_n\|^2 > n^\delta\}} \\ &\geq n^{s\delta/2} P(\|\xi_n\|^2 > n^\delta). \end{aligned}$$

Therefore

$$P(\|\xi_n\|^2 > n^\delta) \leq \frac{C}{n^{s\delta/2}}. \tag{6.76}$$

By using $K(g(u)) = u^{2k}$ and the Cauchy–Schwarz inequality,

$$R_n^0(g(u)) = nu^{2k} - u^k\, \xi_n(u) + a_n\sigma(u)$$
$$\geq \tfrac{1}{2}(nu^{2k} - \|\xi_n\|^2) + a_n\sigma(u).$$

A parameter u_0 in the set of true parameters satisfies $K(g(u_0)) = u_0^{2k} = 0$, hence

$$R_n^0(g(u_0)) = a_n\sigma(u_0).$$

Therefore, by the definition of $\hat{u}$, $R_n^0(\hat{u}) \leq R_n^0(w_0)$,

$$\tfrac{1}{2}(n\hat{u}^{2k} - \|\xi_n\|^2) + a_n\sigma(\hat{u}) \leq a_n\sigma(u_0).$$

Hence, if $\|\xi_n\|^2 \geq n^\delta$, since a_n is smaller than any power of n^p $(p > 0)$, there exists a constant $c_1 > 0$ such that

$$n\hat{u}^{2k} \leq \|\xi_n\|^2 + 4a_n\|\sigma\| \leq c_1\|\xi_n\|^2.$$

Therefore

$$P(nK(g(\hat{u})) > n^\delta) \leq P(c_1\|\xi_n\|^2 > n^\delta)$$
$$\leq \frac{c_2}{n^{s\delta/2}}.$$

(2) By using the above results,

$$E[\|\xi_n\|^2]_{\{\|\xi_n\|^2 > n^\delta\}} \leq \sum_{j=1}^{\infty} E[\|\xi_n\|^2]_{\{jn^\delta < \|\xi_n\|^2 \leq (j+1)n^\delta\}}$$
$$\leq \sum_{j=1}^{\infty} (j+1)n^\delta \times \frac{C}{(jn^\delta)^{s/2}}$$
$$\leq \frac{c_3}{n^{\delta(s/2-1)}}.$$

In the same way,

$$E[nK(g(\hat{u}))]_{\{nK(g(\hat{u})) > n^\delta\}} \leq \sum_{j=1}^{\infty} E[nK(g(\hat{u}))]_{\{jn^\delta < nK(g(\hat{u})) \leq (j+1)n^\delta\}}$$
$$\leq \sum_{j=1}^{\infty} (j+1)n \times \frac{c_2}{(jn^\delta)^{s/2}}$$
$$\leq \frac{c_4}{n^{\delta(s/2-1)}}.$$

□

Main Theorem 6.4 *Assume that $q(x)$ and $p(x|w)$ satisfy the fundamental conditions (I) and (II) with index s ($s \geq 6$). Let $\{a_n \geq 0\}$ be a nondecreasing sequence that satisfies the condition that, for arbitrary $p > 0$,*

$$\lim_{n\to\infty} \frac{a_n}{n^p} = 0.$$

Let $M = \{M_\alpha\}$ be the manifold found by resolution of singularities and its local coordinate.
(1) If $a_n \equiv 0$, then

$$\lim_{n\to\infty} nE[R_{\mathrm{g}}] = \tfrac{1}{4}E\Big[\max_{\alpha}\max_{u\in M_{\alpha 0}}\Big(\max\{0,\xi(u)\}\Big)^2\Big],$$

$$\lim_{n\to\infty} nE[R_{\mathrm{t}}] = -\tfrac{1}{4}E\Big[\max_{\alpha}\max_{u\in M_{\alpha 0}}\Big(\max\{0,\xi(u)\}\Big)^2\Big],$$

where $\max_\alpha$ shows the maximization for local coordinates and

$$M_{\alpha 0} = \{u \in M_\alpha; K(g(u)) = 0\}.$$

(2) If $\lim_{n\to\infty} a_n = a^*$, then

$$\lim_{n\to\infty} nE[R_{\mathrm{g}}] = \tfrac{1}{4}E\Big[\max_{\alpha}\Big(\max\{0,\xi(u^*)\}\Big)^2\Big],$$

$$\lim_{n\to\infty} nE[R_{\mathrm{t}}] = -\tfrac{1}{4}E\Big[\max_{\alpha}\Big(\max\{0,\xi(u^*)\}\Big)^2\Big],$$

where u^* is the parameter in $M_{\alpha 0}$ that maximizes

$$\tfrac{1}{4}\max\{0,\xi(u)\}^2 - a^*\sigma(g(u)).$$

(3) If $\lim_{n\to\infty} a_n = \infty$, then

$$\lim_{n\to\infty} nE[R_{\mathrm{g}}] = \tfrac{1}{4}E\Big[\max_{\alpha}\max_{u\in M_{\alpha 00}}\Big(\max\{0,\xi(u)\}\Big)^2\Big],$$

$$\lim_{n\to\infty} nE[R_{\mathrm{t}}] = -\tfrac{1}{4}E\Big[\max_{\alpha}\max_{u\in M_{\alpha 00}}\Big(\max\{0,\xi(u)\}\Big)^2\Big],$$

where $M_{\alpha 00}$ is the set of parameters which minimizes $\sigma(g(u))$ in the set $M_{\alpha 0}$.

Proof of Main Theorem 6.4 Let $\epsilon > 0$ be a sufficiently small constant. The proof is divided into several cases.
Case (A), $\|\psi_n\|^2 > n\epsilon$. By the proof of Theorems 6.12, $nK(\hat{w}) \leq c_1\|\psi_n\|^2$. The generalization error of the partial expectation is bounded by Theorem 6.12,

$$E[nK(\hat{w})]_{\{\|\psi_n\|^2 > n\epsilon\}} \leq \frac{C'}{n^2}, \tag{6.77}$$

Also the training error of the partial expectation is

$$E[nK_n(\hat{w})]_{\{\|\psi_n\|^2>n\epsilon\}} \leq \tfrac{1}{2}E[3nK(\hat{w}) + \|\psi_n\|^2]_{\{\|\psi_n\|^2>n\epsilon\}} \leq \tfrac{C''}{n^2}. \qquad (6.78)$$

Therefore the event $\|\psi_n\|^2 > n\epsilon$ does not affect the generalization and training errors asymptotically.
Case (B), $\|\psi_n\|^2 \leq n\epsilon$. As is shown by the proofs of Theorems 6.12 and 6.13, if $\|\psi_n\|^2 \leq n\epsilon$, then $K(\hat{w}) \leq c_1 n\epsilon$. Let us use resolution of singularities and Main Theorem 6.1. The function to be minimized, which is called a loss function, is

$$R_n^0(g(u)) = nu^{2k} - \sqrt{n}\; u^k \xi_n(u) + a_n\sigma(g(u)),$$

where $u \in [0, 1]^d$. By using parameterization $(t, v) \in T \times S$ of each local coordinate, the loss function is given by

$$R_n^0(g(t, v)) = nt^2 - \sqrt{n}\; t\; \xi_n(t, v) + a_n\sigma(t, v),$$

where we use the notation

$$\begin{aligned} K(t, v) &= K(g(t, v)) = t^2, \\ K_n(t, v) &= K_n(g(t, v)), \\ R_n^0(t, v) &\equiv R_n^0(g(t, v)), \\ \sigma(t, v) &\equiv \sigma(g(t, v)). \end{aligned}$$

Note that, even if the optimal parameter is on the boundary of $[0, 1]^d$, it asymptotically does not affect the value t because, sufficiently near the point $(0, v)$, the surface $v =$ const. can be taken perpendicular to the boundary. The loss function is rewritten as

$$R_n^0(t, v) = nt^2 - \sqrt{n}\; t\; \xi_n(0, v) + a_n\sigma(0, v) + R_1(t, v),$$

where

$$R_1(t, v) = -\sqrt{n}\; t\; \Big(\xi_n(t, v) - \xi_n(0, v)\Big) + a_n(\sigma(t, v) - \sigma(0, v)).$$

Case (B1), $\|\xi_n\|^2 > n^\delta$ $(0 < \delta \leq 1)$. By Theorem 6.13 and $nK(\hat{u}) < c_1\|\xi_n\|^2$,

$$E[nK(\hat{u})]_{\{\|\xi_n\|^2>n^\delta\}} \leq \frac{c_5}{n^{\delta(s/2-1)}},$$

and

$$\begin{aligned} E[nK_n(\hat{u})]_{\{\|\xi_n\|^2>n^\delta\}} &\leq (1/2)E[3nK(\hat{w}) + \|\xi_n\|^2]_{\{\|\xi_n\|^2>n^\delta\}} && (6.79) \\ &\leq \frac{c_6}{n^{\delta(s/2-1)}}. && (6.80) \end{aligned}$$

Therefore the event $\|\xi_n\|^2 > n^\delta$ $(\delta > 0)$ does not affect the generalization and training errors asymptotically.

Case (B2), $\|\xi_n\|^2 \leq n^\delta$ $(\delta > 0)$. We know $nK(\hat{u}) = n\hat{t}^2$ is not larger than $c_7 n^\delta$, hence $\hat{t} \leq c_8 n^{(\delta-1)/2}$. Thus we can restrict t in the region,

$$T_\delta \equiv \{0 \leq t \leq c_8 n^{(\delta-1)/2}\}.$$

By using Theorem 6.11,

$$\begin{aligned} \|R_1\| &\equiv \sup_{t\in T_\delta} |R_1(t, v)| \\ &\leq \sup_{t\in T_\delta} \{n^{1/2} t^{1+k_0} \|\nabla \xi_n\| + a_n t^{k_0} \|\nabla \sigma\|\} \\ &\leq c_9 n^{-\delta/2} \|\nabla \xi_n\| + c_{10} a_n n^{k_0(\delta-1)/2} \|\nabla \sigma\|, \end{aligned} \tag{6.81}$$

where $k_0 = 1/k^*$. Therefore $\|R_1\| \to 0$ in probability. We need to minimize

$$R_n^0(t, v) = nt^2 - \sqrt{n}\; t\; \xi_n(0, v) + a_n \sigma(0, v) + R_1(t, v)$$

in $T_\delta \times V$. For a given v, the parameter t that minimizes $R_n^0(t, v)$ is denoted by $t(v)$.

Case (B2-1), $\xi_n(0, v) \leq 0$. If $\xi_n(0, v) \leq 0$ then $R_n^0(t(v), v)$ is not larger than the special case $t = 0$,

$$R_n^0(t(v), v) \leq a_n \sigma(0, v) + \|R_1\|.$$

On the other hand, by removing the nonnegative term,

$$R_n^0(t(v), v) \geq a_n \sigma(0, v) - \|R_1\|.$$

Therefore

$$|nt(v)^2 - \sqrt{n}\; t(v)\; \xi_n(0, v)| \leq 2\|R_1\|.$$

Moreover, since $\xi_n(0, v) \leq 0$,

$$|nR_{\mathrm{g}}| \leq 2\|R_1\|, \tag{6.82}$$

$$|nR_{\mathrm{t}}| \leq 2\|R_1\|. \tag{6.83}$$

Case (B2-2), $\xi(0, v) > 0$. We have

$$R_n^0(t, v) = (\sqrt{n}t - \xi_n(0, v)/2)^2 - \tfrac{1}{4}\xi_n(0, v)^2 + a_n \sigma(0, v) + R_1(t, v).$$

Then $R_n^0(t(v), v)$ is not larger than the special case $t = \xi_n(0, v)/(2\sqrt{n})$,

$$R_n^0(t(v), v) \leq -\tfrac{1}{4}\xi_n(0, v)^2 + a_n \sigma(0, v) + \|R_1\|. \tag{6.84}$$

On the other hand, by removing the nonnegative term,

$$R_n^0(t(v), v) \geq -\tfrac{1}{4}\xi_n(0, v)^2 + a_n \sigma(0, v) - \|R_1\|. \tag{6.85}$$

Therefore

$$(\sqrt{n}\,t - \xi_n(0,v)/2)^2 \leq 2\|R_1\|,$$

which means that

$$\xi_n(0,v)^2/4 - 2\|R_1\| \leq nR_{\mathrm{g}} \leq \xi_n(0,v)^2/4 + 2\|R_1\|, \tag{6.86}$$

$$-\xi_n(0,v)^2/4 - 2\|R_1\| \leq nR_{\mathrm{t}} \leq -\xi_n(0,v)^2/4 + 2\|R_1\|. \tag{6.87}$$

Then by using the convergence in law $\xi_n(u) \to \xi_n(u)$, and eq.(6.84) and eq.(6.85), the minimizing procedure for v is divided into three cases. By comparing $-\xi(0,v)^2/4$ with $a_n\sigma(0,v)$, we have following results.
(1) If $a_n \equiv 0$, the following convergences in probability hold:

$$nR_{\mathrm{g}} \to (1/4) \max_{\alpha} \max_{u \in M_{\alpha 0}} \max\{0, \xi(u)^2\},$$

$$nR_{\mathrm{t}} \to -(1/4) \max_{\alpha} \max_{u \in M_{\alpha 0}} \max\{0, \xi(u)^2\}.$$

(3) If $a_n \to \infty$, the following convergences in probability hold:

$$nR_{\mathrm{g}} \to (1/4) \max_{\alpha} \max_{u \in M_{\alpha 00}} \max\{0, \xi(u)^2\},$$

$$nR_{\mathrm{t}} \to -(1/4) \max_{\alpha} \max_{u \in M_{\alpha 00}} \max\{0, \xi(u)^2\}.$$

(2) If $\lim_n a_n = a^*$, the following covergences in probability hold:

$$nR_{\mathrm{g}} \to (1/4) \max_{\alpha} \overset{*}{\max_{u \in M_{\alpha 0}}} \max\{0, \xi(u)^2\},$$

$$nR_{\mathrm{t}} \to -(1/4) \max_{\alpha} \overset{*}{\max_{u \in M_{\alpha 0}}} \max\{0, \xi(u)^2\},$$

where $\overset{*}{\max_{u \in M_{\alpha 0}}}$ shows the maximization of $(1/4)\xi(u)^2 - a^*\sigma(u)$ in the set $M_{\alpha 0}$.
Lastly, from eqs.(6.77), (6.78), (6.79), (6.80), (6.82), (6.83), (6.86), (6.87), and $E[\|R_1\|^2] < \infty$, both nR_{g} and nR_{t} are asymptotically uniformly integrable, which completes Main Theorem 6.4. □

Corollary 6.4 *Let $\hat{w}$ be the maximum likelihood or a posteriori estimator.*
(1) Assume that $q(x)$, $p(x|w)$ and $\varphi(w)$ satisfy the fundamental conditions (I) and (II) with index s ($s \geq 6$). Then

$$F_n = -\sum_{i=1}^{n} \log p(X_i|\hat{w}) + \lambda \log n - (m-1)\log\log n + F_n^{MR},$$

where F_n^{MR} is a random variable which converges to a random variable F^{MR} in law.
(2) Assume that $q(x)$, $p(x|w)$ and $\varphi(w)$ satisfy the fundamental conditions (I) and (II) with index s ($s \geq 6$). Then the following convergence of expectation holds,

$$E[F_n] = nE_X[E[\log p(X|\hat{w})]] + \lambda \log n - (m-1)\log\log n + E[F_n^{MR}],$$

where $E[R_n^{MR}] \to E[F^{MR}]$.

Proof of Corollary 6.4 From Main Theorem 6.4,

$$\sum_{i=1}^{n} \log \frac{q(X_i)}{p(X_i|\hat{w})}$$

converges in law. Its expectation also converges. This corollary is immediately derived from Main Theorem 6.2. □

Remark 6.16 (1) In the equation

$$E[R_g] = -E[R_t] + o\Big(\frac{1}{n}\Big)$$

the generalization error is represented by the training error; however, the left-hand side contains the entropy $+1$, whereas the right-hand side has entropy -1. Therefore an information criterion cannot be directly derived from this equation. To construct an information criterion, we need a constant $C > 0$ such that

$$E[R_g] = E[R_t] + \frac{2C}{n} + o\Big(\frac{1}{n}\Big).$$

In a regular statistical model $C = d/2$; however, in a singular model, it depends on the true distribution and a statistical model.
(2) If the set of parameters is not compact, then $|w| = \infty$ may be an analytic singularity. For the behavior of the maximum value of the random process and its application to the maximum likelihood training errors, see, for example, [20, 36, 113]. Although several results were obtained on the asymptotic behavior of the training errors, it is still difficult to know their generalization errors. It is conjectured that the asymptotic generalization error of the maximum likelihood is very large.

Remark 6.17 (Phase transition) It seems that, when $\beta \to \infty$, the that of Bayes and Gibbs generalization errors converge to the maximum likelihood estimation. However, such convergence does not hold in general, even if the set of parameters is compact. In Gibbs and Bayes estimations, the main part of the *a posteriori* distribution is contained in the essential coordinates which minimize λ and maximize its order. However, in the maximum likelihood estimation, such a restriction is not introduced. Therefore, the asymptotic generalization and training errors may not be continuous at $\beta \to \infty$. Such a phenomenon is called a phase transition in statistical physics. From the viewpoint of statistical physics, a singular learning machine have phase transition at $\beta = \infty$, in general.

7

Singular learning machines

Singular learning machines are now being used in artificial intelligence, pattern recognition, robotic control, time series prediction, and bioinformatics. In order to build the foundation on which their learning processes are understood, we need to clarify the effect of singularities. In this chapter, we study the phenomenon caused by singularities in several concrete learning machines.

7.1 Learning coefficient

For a given set of the true probability density function $q(x)$, a learning machine $p(x|w)$, and an *a priori* probability density function $\varphi(w)$, the zeta function is defined by

$$\zeta(z) = \int K(w)^z \, \varphi(w) \, dw,$$

where $K(w)$ is the Kullback–Leibler distance,

$$K(w) = \int q(x) \log \frac{q(x)}{p(x|w)} dx.$$

Let $(-\lambda)$ and m be the largest pole of the meromorphic function $\zeta(z)$ and its order. Then $\lambda > 0$ is called a learning coefficient. If we obtain the learning coefficient, then we can predict the stochastic complexity and the Bayes generalization error for $\beta = 1$ theoretically.

Theorem 7.1 *Assume that the set of true parameters*

$$\{w \in supp\ \varphi; K(w) = 0\}$$

is not an empty set. Then λ and m satisfy the following conditions.

(1) There exists a constant $c_1 > 0$ *such that*

$$\lim_{n\to\infty} \frac{(\log n)^{m-1}}{n^{\lambda}} \int \exp(-nK(w))\, \varphi(w)\, dw = c_1.$$

(2) This equation holds:

$$\lambda = -\lim_{n\to\infty} \frac{\log \int \exp(-nK(w))\, \varphi(w)\, dw}{\log n}.$$

(3) There exists a constant $c_2 > 0$ *such that*

$$\lim_{t\to 0} \frac{1}{t^{\lambda-1}(-\log t)^{m-1}} \int \delta(t - K(w))\, \varphi(w)\, dw = c_2.$$

(4) Let $V(t)$ *be a volume function,*

$$V(t) = \int_{K(w)<t} \varphi(w)\, dw.$$

For an arbitrary $a > 0$ *(*$a \neq 1$*),*

$$\lambda = \lim_{t\to 0} \frac{\log\{V(at)/V(t)\}}{\log a}.$$

Proof of Theorem 7.1 We have already proved (1), (2), and (3) in Chapter 4. Let us prove (4). The state density function is defined by

$$v(t) = \int \delta(t - K(w))\varphi(w)dw.$$

By (3), there exists $c_2 > 0$ such that

$$v(t) = c_2 t^{\lambda-1}(-\log t)^{m-1} + o(t^{\lambda-1}(-\log t)^{m-1}).$$

By the definition,

$$V(t) = \int_0^t v(s)ds.$$

Using partial integration, we obtain

$$V(t) = c_2' t^{\lambda}(-\log t)^{m-1} + o(t^{\lambda}(-\log t)^{m-1}),$$

which shows

$$\log\{V(at)/V(t)\} = \lambda \log a + o(1).$$

□

Remark 7.1 (1) Figure 7.1 shows the shape of $\{w; K(w) < t\}$ near normal crossing singularities. Theorem 7.1 (4) shows that the learning coefficient is equal to the volume dimension of the set of almost correct parameters $\{w; K(w) < t\}$ when $t \to 0$.

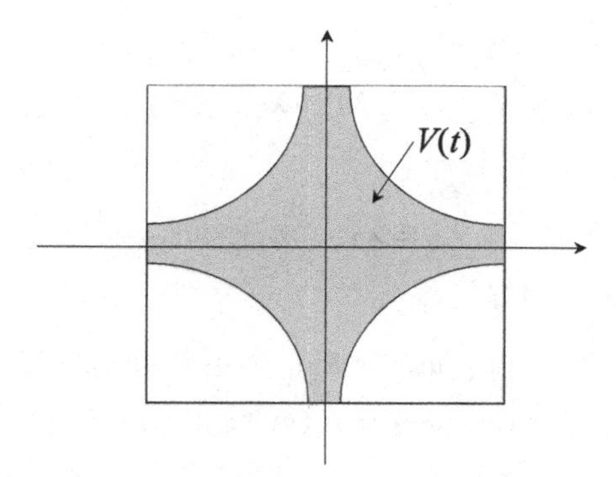

Fig. 7.1. Learning coefficient and volume dimension

(2) If it is difficult to find a complete resolution of singularities for $K(w) = 0$, then we can numerically calculate λ by Theorem 7.1.

Remark 7.2 To calculate the learning coefficient for a given Kullback–Leibler distance, the following properties may be helpful.
(1) If there exist $c_1, c_2 > 0$ such that $K(w)$ and $H(w)$ satisfy

$$c_1 H(w) \le K(w) \le c_2 H(w),$$

then $K(w)$ is said to be equivalent to $H(w)$. If $K(w)$ and $H(w)$ are equivalent, they have the same learning coefficient. For example,

$$K(a, b, c) = a^2 + (a + bc)^2,$$
$$H(a, b, c) = a^2 + b^2c^2$$

are equivalent. In a compact set which contains the origin,

$$K(a, b, c) = a^2 + a^2b^2,$$
$$H(a, b, c) = a^2 + a^2c^2$$

are equivalent.
(2) For pairs $(K_1(w), \varphi_1(w))$ and $(K_2(w), \varphi_2(w))$, two zeta functions are defined by

$$\zeta_i(z) = \int K_i(w)^z \varphi_i(w)\, dw \quad (i = 1, 2).$$

Let (λ_1, m_1) and (λ_2, m_2) be the pairs of learning coefficients and their orders. If, for all $w \in W$,

$$K_1(w) \le K_2(w), \quad \varphi_1(w) \ge \varphi_2(w)$$

then

$$\lambda_1 > \lambda_2$$

or

$$\lambda_1 = \lambda_2, \quad m_1 \leq m_2.$$

(3) If $w = (w_a, w_b)$ and if

$$K(w_a, w_b) = K_a(w_a) + K_b(w_b),$$
$$\varphi(w_a, w_b) = \varphi_a(w_a)\, \varphi_b(w_b),$$

then the coefficient λ and the order m satisfy

$$\lambda = \lambda_a + \lambda_b,$$
$$m = m_a + m_b - 1,$$

where (λ_a, m_a) and (λ_b, m_b) are pairs of learning coefficients and orders of (K_a, φ_a) and (K_b, φ_b) respectively.

Theorem 7.2 *Let $W \subset \mathbb{R}^d$ be the set of parameters. If there exists an open set $U \subset W$ such that*

$$\{w \in U; K(w) = 0, \quad \varphi(w) > 0\}$$

is not the empty set, then

$$\lambda \leq \frac{d}{2}.$$

Proof of Theorem 7.2 Let w_0 be a parameter which satisfies $K(w_0) = 0$ and $\varphi(w_0) > 0$. We can assume $w_0 = 0$ without loss of generality. If $\epsilon > 0$ is a sufficiently small constant,

$$Z(n) = \int \exp(-nK(w))\, \varphi(w)\, dw$$
$$\geq \int_{|w|<\epsilon} \exp(-nK(w))\varphi(w)\, dw.$$

The matrix K_{ij} is defined by

$$K_{ij} = \frac{\partial^2 K(w)}{\partial w_i \partial w_j}\bigg|_{w=0}.$$

For sufficiently small $|w| > 0$

$$K(w) = \frac{1}{2} \sum_{i,j=1}^{d} K_{ij} w_i w_j + o(|w|^2).$$

Hence

$$Z(n) \geq \int_{|w|<\epsilon} \exp\Big\{-\frac{n}{2}\sum_{i,j=1}^{d} K_{ij} w_i w_j - n\, o(|w|^2)\Big\}\, \varphi(w)\, dw.$$

By putting $w' = \sqrt{n}\, w$ ($\epsilon n > 1$),

$$Z(n) n^{d/2} \geq \int_{|w'|<1} \exp\Big\{-\frac{1}{2}\sum_{i,j=1}^{d} K_{ij} w'_i w'_j + \frac{o(|w'|^3)}{\sqrt{n}}\Big\}\, \varphi(\frac{w'}{\sqrt{n}})\, dw',$$

which converges to the positive constant,

$$\int_{|w'|<1} \exp\Big\{-\frac{1}{2}\sum_{i,j=1}^{d} K_{ij} w'_i w'_j\Big\}\, \varphi(0)\, dw'.$$

From Theorem 7.1(1), we obtain the theorem. □

Theorem 7.3 *Assume that a parameter is represented by* $w = (u, v) \in W$ ($u \in \mathbb{R}^{d_1}$, $v \in \mathbb{R}^{d_2}$) *and that* $K(u, v)$ *and* $\varphi(u, v)$ *satisfy the following conditions.*
(1) For arbitrary v, $K(u_0, v) = 0$.
(2) There exists an open set $V \subset \mathbb{R}^{d_2}$ *such that* $\varphi(u_0, v) > 0$ ($v \in V$).
Then

$$\lambda \leq \frac{d_1}{2}.$$

Proof of Theorem 7.3 Without loss of generality, we can assume $u_0 = 0$. By the same method as in the proof of Theorem 7.2, where the rank K_{ij} is not larger than d_1, we obtain the theorem. □

Remark 7.3 (1) In some statistical models, it is easy to show that $K(w) = 0$ when d_1 parameters are equal to a fixed value and the other d_2 parameters are free. Then λ is not larger than $d_1/2 = (d - d_2)/2$. Although this is not a tight bound of the learning coefficient, it might be useful in practical applications.
(2) Let (d_1, d_2) and (d'_1, d'_2) be the numbers of parameters which satisfy Theorem 7.3. Then $\lambda \leq \min(d_1, d'_1)/2$.
(3) If the assumption of Theorem 7.3 is satisfied, then $\{w; K(w) = 0\}$ contains a d_2-dimensional manifold. If the set $\{w; K(w) = 0\}$ contains a d_2-dimensional manifold, then $\lambda \leq (d - d_2)/2$.

Definition 7.1 (Jeffreys' prior) The Fisher information matrix $I(w) = I_{ij}(w)$ is defined by

$$I_{ij}(w) = \int \frac{\partial f(x, w)}{\partial w_i}\frac{\partial f(x, w)}{\partial w_j}\, p(x|w)\, dx,$$

where

$$f(x, w) = \log(q(x)/p(x|w)).$$

A probability density function $\varphi(w)$ on $\mathbb{R}^d$ is called Jeffreys' prior on W if

$$\varphi(w) = \begin{cases} \frac{1}{Z}\sqrt{\det I(w)} & (w \in W) \\ 0 & \text{otherwise.} \end{cases} \tag{7.1}$$

Remark 7.4 (1) At a singularity, $\det I(w) = 0$, hence Jeffreys' prior is equal to zero.
(2) Let $w = g(u)$ be a function from an open set $U \subset \mathbb{R}^d$ to an open set $W \subset \mathbb{R}^d$. The Fisher information matrix of $p(x|g(u))$ is given by

$$\begin{aligned} I_{ij}(u) &= \int \frac{\partial}{\partial u_i} f(x, g(u)) \frac{\partial}{\partial u_j} f(x, g(u))\, p(x|g(u))\, dx \\ &= \sum_k \sum_l \frac{\partial w_k}{\partial u_i} \frac{\partial w_l}{\partial u_j} \int \frac{\partial f}{\partial w_k} \frac{\partial f}{\partial w_l}\, p(x|w)\, dx. \end{aligned}$$

Therefore,

$$\det I(u) = |g'(u)|^2 \det I(w).$$

Let $\sqrt{I(w)}$ and $\sqrt{I(u)}$ be Jeffreys' priors on respective spaces. Then

$$\begin{aligned} \frac{1}{Z}\sqrt{\det I(w)}\, dw &= \frac{1}{Z}\sqrt{\det I(g(u))}\, |g'(u)|\, du \\ &= \frac{1}{Z}\sqrt{\det I(u)}\, du. \end{aligned}$$

In other words, Jeffreys' prior can be defined independently of coordinates. Such a property is called 'coordinate-free'.
(3) In statistical estimation, the pair $(p(x|w), \varphi(w))$ is a statistical model which is optimized for given random samples. Hence, if $p(x|w)$ is fixed and $\varphi(w)$ is made coordinate-free, such a pair $(p(x|w), \varphi(w))$ is not appropriate for statistical estimation in general.

Theorem 7.4 *If Jeffreys' prior is employed, then (1) or (2) holds.*
(1) $\lambda = d/2, \quad m = 1.$
(2) $\lambda > d/2.$

Proof of Theorem 7.4 Let us prove that the largest pole of

$$\zeta(z) = \int_W K(w)^z \sqrt{\det I(w)} dw$$

satisfies '$-\lambda = -d/2$ and $m = 1$' or '$-\lambda < -d/2$'. It is sufficient to show that there exists a function $I_0(w)$ which satisfies

$$\sqrt{\det I(w)} \le I_0(w),$$

and the largest pole of

$$\zeta_0(z) = \int K(w)^z I_0(w) dw$$

is equal to $-d/2$ and its order is equal to $m = 1$. By using resolution of singularities, in each local coordinate, there exists a set of natural numbers $k_1, k_2, \ldots, k_b > 0$

$$K(w) = w_1^{2k_1} w_2^{2k_2} \cdots w_b^{2k_b},$$

where $1 \le b \le d$. The log density ratio function can be written as

$$f(x, w) = a(x, w) w_1^{k_1} w_2^{k_2} \cdots w_b^{k_b}.$$

Let us define

$$r_i(x, w) = \begin{cases} \frac{\partial a}{\partial w_i} w_i + k_i a(x, w) & (1 \le i \le b) \\ \frac{\partial a}{\partial w_i} & (b < i \le d). \end{cases} \tag{7.2}$$

Then

$$\frac{\partial f(x, w)}{\partial w_i} = \begin{cases} r_i(x, w) w_1^{k_1} \cdots w_i^{k_i - 1} \cdots w_b^{k_b} & (1 \le i \le b) \\ r_i(x, w) w_1^{k_1} \cdots w_b^{k_b} & (b < i \le d). \end{cases}$$

By using a matrix $J(w)$ defined by

$$J_{ij}(w) = \int r_i(x, w) r_j(x, w) p(x|w) dx,$$

we have

$$(\det I(w))^{1/2} = \left\{ \prod_{p=1}^{b} w_p^{dk_p - 1} \right\} (\det J(w))^{1/2}.$$

If $b = 1$, there exists $c_1 > 0$ such that

$$K(w) = w_1^{2k_1}$$
$$(\det I(w))^{1/2} \le c_1 w_1^{dk_1 - 1}.$$

Therefore the theorem holds. Let us study the case $b \geq 2$. A transform $w = g(u)$ defined by

$$\begin{aligned} w_1 &= u_1, \\ w_2 &= u_2 u_1, \\ &\vdots \\ w_b &= u_b \cdots u_1 \end{aligned}$$

can be made by blow-ups, where the other coordinates are symmetrically defined. By the definition eq.(7.2), if $u_i = 0$ for some i $(1 \leq i \leq b-1)$, then $\det J(g(u)) = 0$. Because the determinant $\det J(g(u)) \geq 0$ is an analytic function of u, there exists a constant $c_2 > 0$ such that

$$\det J(g(u)) \leq c_2\, u_1^2 \cdots u_{b-1}^2.$$

Let σ_p be a constant defined by

$$\sigma_p = k_1 + k_2 + \cdots + k_p.$$

There exists a constant $c_3 > 0$ such that

$$K(g(u)) = \prod_{p=1}^{b} u_p^{2\sigma_p}$$

$$(\det I(g(u)))^{1/2} \leq c_3 \prod_{p=1}^{b-1} u_p \prod_{p=1}^{b} u_p^{d\sigma_p - b + p - 1}$$

$$|g'(u)| = \prod_{p=1}^{b} u_p^{b-p}.$$

From the integration of u_b, we obtain the pole $(-d/2)$ $(m = 1)$. From the integration of the other parameters $u_1, u_2, \ldots, u_{b-1}$, smaller poles than $(-d/2)$ are obtained, which completes the proof. □

Remark 7.5 (1) If Jeffreys' prior is employed, there exists a case $\lambda > d/2$. For example,

$$p(y|x, a, b) = \tfrac{1}{2}\exp(-\tfrac{1}{2}(y - abx - a^2b^3)^2).$$

Assume that the true distribution is given by $p(y|x, 0, 0)$. In this case, $d = 2$ but $\lambda = 3/2$. Note that, if $ab = c$, $a^2b^3 = d$, this model can be understood as a regular model $\lambda = 2/2$. However, the compact set of (a, b) does not correspond to the compact set of (c, d). Therefore, the model represented by (a, b) with

compact support is not equivalent to that represented by (c, d) with compact support.
(2) Let us study a model of $(x, y) \in \mathbb{R}^M \times \mathbb{R}^N$,

$$p(x, y|w) = \frac{q(x)}{(2\pi\sigma^2)^{N/2}} \exp\Big(-\frac{1}{2\sigma^2}\|y - h(x, w)\|^2\Big),$$

$$h(x, w) = \sum_{k=1}^{K} a_k s(b_k, x),$$

where the parameter w is given by

$$w = \{(a_k, b_k) \in \mathbb{R}^N \times \mathbb{R}^L\}.$$

Assume that $s(b, x)$ satisfies the condition

$$\det I(w) \neq 0 \Longleftrightarrow \{a_k s(b_k, x)\} \text{ are linearly independent.}$$

Then

$$\lambda = \frac{d}{2}, \quad m = 1$$

holds [103].

Example 7.1 In a concrete learning machine, let us calculate the learning coefficient, λ, and its order, m. Let $\mathcal{N}$ be a random variable which is subject to the standard normal distribution. The probability distribution $q(x)$ of X is assumed to have compact support. We study a statistical model in which a random variable Y for a given X is defined by

$$Y = a\sigma(bX) + c\sigma(dX) + \mathcal{N}.$$

This is a layered neural network with one input unit, two hidden units, and one output unit. Let us consider the case $\sigma(x) = e^x - 1$. The true distribution of Y is assumed to be

$$Y = 0 + \mathcal{N}.$$

The *a priori* probability distribution $\varphi(w)$ has compact support and $\varphi(0, 0) > 0$. The Kullback–Leibler distance as a function of $w = (a, b, c, d)$ is given by

$$K(w) = \frac{1}{2}\int (a\sigma(bx) + c\sigma(dx))^2 q(x)dx.$$

By using the Taylor expansion,

$$f(x, w) = a\sigma(bx) + c\sigma(dx) = \sum_{k=1}^{\infty} \frac{x^k}{k!}(ab^k + cd^k).$$

Here $\{x^k\}$ is a set of linearly independent functions. Therefore $K(w) \equiv 0$ is equivalent to

$$p_k = ab^k + cd^k = 0 \quad (\forall k = 1, 2, 3, \ldots).$$

As we have shown in Example 3.2,

$$p_n \in \langle p_1, p_2 \rangle.$$

In Chapter 3, we have already derived the resolution of singularities of this polynomial in Section 3.6.2. In one of the coordinates, the resolution map $w = g(u)$ is given by

$$\begin{aligned} a &= a, \\ b &= b_1 d, \\ c &= a(b_1 - 1)b_1 c_5 d - ab_1, \\ d &= d, \end{aligned}$$

where $u = (a, b_1, c_5, d)$. Hence

$$\begin{aligned} p_1 &= ab_1(b_1 - 1)c_5 d^2, \\ p_2 &= ab_1(b_1 - 1)(1 + c_5)d^2, \\ p_k &= ab_1(b_1 - 1)\big\{b_1^{k-2} + b_1^{k-3} + \cdots + 1 + c_5 d\big\}. \end{aligned}$$

Thus

$$f(x, g(u)) = ab_1(b_1 - 1)d^2 \, a(x, u),$$

where

$$a(x, u) = c_5 x + \tfrac{1}{2}(1 + c_5)x^2 + \sum_{k=3}^{\infty} \frac{x^k}{k!} d^{k-2}\big\{b_1^{k-1} + b_1^{k-2} + \cdots + 1 + c_5 d\big\}.$$

Hence $a(x, 0) \neq 0$, so we have

$$K(g(u)) = \frac{a^2 b_1^2 (b_1 - 1)^2 d^4}{2} \int a(x, u)^2 q(x) dx.$$

The Jacobian determinant of $w = g(u)$ is

$$|g'| = |a(b_1 - 1)b_1 d^2|.$$

The largest pole of the zeta function

$$\zeta(z) = \int \big\{a^2 b_1^2 (b_1 - 1)^2 d^4\big\}^z \, |a(b_1 - 1)b_1 d^2| \, \varphi(g(u)) \, da \, db_1 \, dc_5 \, dd$$

is $-\lambda = -\frac{3}{4}$ and its order is 1. We found that the learning coefficient in this case is equal to $\frac{3}{4}$, hence the normalized stochastic complexity is equal to

$$F_n^0 = \tfrac{3}{4}\log n + \text{random variable}$$

and the mean Bayes generalization error for $\beta = 1$ is equal to

$$E[B_g] = \frac{3}{4n} + o\Big(\frac{1}{n}\Big).$$

Remark 7.6 If a statistical model $p(y|x, w)q(x)$ and a true distribution $q(y|x)q(x)$ are respectively given by

$$p(y|x, w) = \tfrac{1}{2}\exp(-\tfrac{1}{2}(y - f(x, w))^2),$$
$$q(y|x) = \tfrac{1}{2}\exp(-\tfrac{1}{2}(y - f_0(x))^2),$$

then the Kullback–Leibler distance is given by

$$K(w) = \frac{1}{2}\int (f(x, w) - f_0(x))^2 q(x)dx.$$

If

$$f(x, w) - f_0(x) = \sum_{j=1}^{\infty} f_j(w)e_j(x),$$

where $f_k(w)$ is a set of polynomials and $\{e_k(x)\}$ is a set of linearly independent functions on the support of $q(x)$, then by the Hilbert basis theorem, there exists J such that $K(w)$ is equivalent to

$$K_1(w) = \sum_{j=1}^{J} f_j(w)^2.$$

7.2 Three-layered neural networks

There are several kinds of neural networks: three-layered perceptrons, radial basis functions, and reduced rank regressions. They are created by a superposition of parametric functions. A three-layered neural network, which is

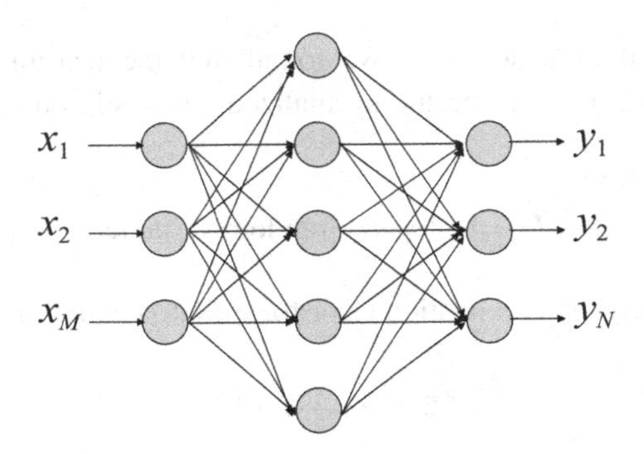

Fig. 7.2. Three-layered neural network

illustrated in Figure 7.2, is defined as a probability density function of $x \in \mathbb{R}^M$ and $y \in \mathbb{R}^N$

$$p(x, y|w) = \frac{q(x)}{\sqrt{2\pi}^N} \exp(-\tfrac{1}{2}|y - h(x, w)|^2),$$

$$h(x, w) = \sum_{k=1}^{H} a_k \sigma(b_k \cdot x + c_k),$$

where $|\ \ |$ is the norm of N-dimensional Euclidean space. The probability density function $q(x)$ is not estimated. Sometimes $\sigma(t) = \tanh(t)$ is employed. The parameter is given by $w = \{(a_k, b_k, c_k)\}$. If the true probability distribution is given by $q(x, y) = p(x, y|w_0)$, the log density ratio function is equal to

$$\begin{aligned} f(x, y, w) &= \log q(x, y) - \log p(x, y|w) \\ &= \tfrac{1}{2}\{|y - h(x, w)|^2 - |y - h(x, w_0)|^2\}. \end{aligned}$$

The convergence radius of one variable $\tanh(x)$ is $\pi/2$. Hence if the support of $q(x)$ is compact then f is an $L^s(q(x, y))$-valued real analytic function. If the true distribution is given by $H = H_0$, then the learning coefficient satisfies

$$\lambda \leq \tfrac{1}{2}\Big[H_0(M + N_1) + \min\Big\{N_1 H_1, M H_1, \frac{M H_1 + 2M N_1}{3}\Big\}\Big],$$

where $H_1 = H - H_0$ $N_1 = N + 1$ [102]. In particular, if $M = N = 1$ and $H_0 = 0$, then

$$\lambda = \frac{[\sqrt{H}]^2 + [\sqrt{H}] + H}{4\sqrt{H} + 2},$$

where $[\sqrt{H}]$ is the largest integer that is not larger than $\sqrt{H}$ [11]. In the case $\sigma(x) = x$, $c_h = 0$, this model is called the reduced rank regression. If the dimensions of input and output are equal to M and N respectively, and if the ranks of the statistical model and the true distribution are H and r respectively, then the following hold [10].

(1) If $N+r < M+H$, $M+r < N+H$, $H+r < M+N$ and $M+H+N+r$ is an even integer, then $m=1$ and

$$\lambda = \tfrac{1}{8}\{2(H+r)(M+N) - (M-N)^2 - (H+r)^2\}.$$

(2) If $N+r < M+H$, $M+r < N+H$, $H+r < M+N$ and $M+H+N+r$ is an odd integer, then $m=2$ and

$$\lambda = \tfrac{1}{8}\{2(H+r)(M+N) - (M-N)^2 - (H+r)^2 + 1\}.$$

(3) If $M+H < N+r$ then $m=1$ and

$$\lambda = \tfrac{1}{2}\{HM - Hr + Nr\}.$$

(4) If $N+H < M+r$ then $m=1$ and

$$\lambda = \tfrac{1}{2}\{HN - Hr + Mr\}.$$

(5) If $M+N < H+r$ then $m=1$ and

$$\lambda = \frac{MN}{2}.$$

Remark 7.7 (1) The reduced rank regression is a statistical model defined by

$$p(y|x,w) = \frac{1}{(2\pi\sigma^2)^{N/2}} \exp\Big(-\frac{1}{2\sigma^2}|y - BAx|^2\Big),$$

where $x \in \mathbb{R}^M$, $y \in \mathbb{R}^N$, A is an $M \times H$ matrix, B is an $H \times N$ matrix, and $\sigma > 0$ is a constant. The parameter is $w = (A, B)$. It the true distribution is given by (A_0, B_0), then the Kullback–Leibler distance is equivalent to

$$K(w) = \tfrac{1}{2}\|BA - B_0A_0\|^2,$$

where $\|\ \ \|$ is the norm of the matrix. For example, if $M = N = H = 2$ and $r = 0$, then

$$\begin{aligned} 2K(w) &= \left\| \begin{pmatrix} a & b \\ c & d \end{pmatrix} \begin{pmatrix} e & f \\ g & h \end{pmatrix} \right\|^2 \\ &= (ae+bg)^2 + (af+bh)^2 + (ce+dg)^2 + (cf+dh)^2. \end{aligned}$$

By using a blow-up and isomorphism,

$$a = a, \quad b = ab', \quad c = ac', \quad d = ad',$$

$$e' = e + b'g, \quad f' = f + b'g, \quad d'' = d' - bc',$$

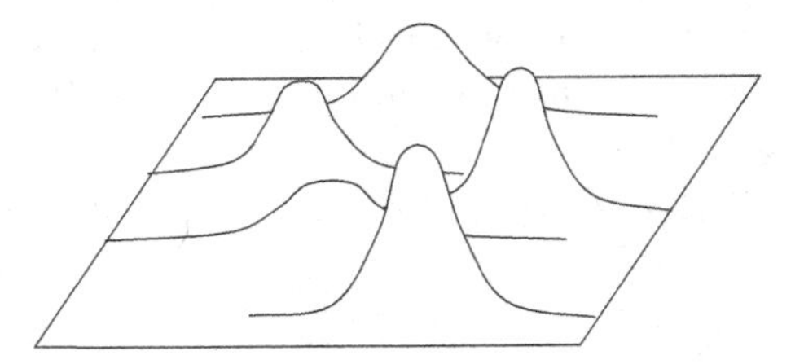

Fig. 7.3. Mixture model

it follows that

$$2K(w) = a^2(e'^2 + f'^2 + (ce' + d''g)^2 + (cf' + d''h)^2),$$

which is equivalent to $a^2(e'^2 + f'^2 + d''^2g^2 + d''^2h^2)$. Furthermore, using blow-up with the center $\langle e', f', d'' \rangle$, we obtain $\lambda = 3/2$.
(2) In statistics, there is a problem of how we can numerically approximate the Bayes *a posteriori* distributions. For example, several Markov chain Monte Carlo methods are studied in Chapter 8. To evaluate the accuracy of such methods, we can compare the numerical stochastic complexity or Bayes generalization error with theoretical values. The reduced rank regression is appropriate for such a purpose because its learning coefficients are completely determined as above.

7.3 Mixture models

Mixture models have very wide applications in automatic clustering of data and density estimation of unknown distributions. A mixture model in Figure 7.3 is defined by

$$p(x|w) = \sum_{h=1}^{H} a_h s(x, b_h),$$

where $x \in \mathbb{R}^N$, $s(x, b_h)$ is itself a probability density function of x for a given $b_h \in \mathbb{R}^M$, and $\{a_h\}$ is a set of nonnegative real numbers which satisfy

$$\sum_{h=1}^{H} a_h = 1.$$

The set of parameters of a mixture model is given by

$$w = \{(a, b) \equiv (a_h, b_h); h = 1, 2, \ldots, H\}.$$

The dimension of the parameter space is $MH+H-1$. For example, if $s(x, b_h)$ is a normal distribution, this model is called a normal mixture, or if $s(x, b_h)$ is a binomial distribution, a binomial mixture.

There are several mathematical problems in mixture models. The singularity caused by the condition that a_k should be nonnegative can be overcome by $a_h = \hat{a}_h^2$ then $\{\hat{a}_h\}$ can be understood as an element of the smooth manifold,

$$\sum_{h=1}^{H} \hat{a}_H^2 = 1,$$

where the *a priori* distribution is also transformed. The log density ratio function is given by

$$\begin{aligned} f(x, w) &= \log \frac{q(x)}{p(x|w)} \\ &= -\log\Big(1 + \frac{p(x|w) - q(x)}{q(x)}\Big), \end{aligned}$$

Hence it is an $L^s(q)$-valued analytic function if there exists $\epsilon > 0$ such that

$$\sup_{|w-w_0|<\epsilon} \sup_{x} \left| \frac{p(x|w) - q(x)}{q(x)} \right| < 1$$

for a true parameter w_0. Binomial and multinomial mixtures satisfy this condition. However, a normal mixture does not, hence the log density ratio function is not an $L^s(q)$-valued analytic function in general. However, there exists a real analytic map such that

$$K(g(u)) = a(u)\, u_1^{2k_1} \cdots u_d^{2k_d},$$

where $a(u) > 0$ is a C^∞-class function but not an analytic function. See Section 7.8. Hence we can derive the same learning theory for normal mixtures as in Chapter 6.

If a learning machine is a mixture of H normal distributions,

$$p(x|w) = \sum_{h=1}^{H} \frac{a_h}{(2\pi)^{N/2}} \exp(-\tfrac{1}{2}|x - b_h|^2),$$

and if the true distribution is made of H_0 components,

$$p(x|w^*) = \sum_{h=1}^{H_0} \frac{a_h^*}{(2\pi)^{N/2}} \exp(-\tfrac{1}{2}|x - b_h^*|^2),$$

then $M = N$. If the *a priori* distribution is not equal to zero on the set of true parameters,

$$\lambda \leq \tfrac{1}{2}(NH_0 + H - 1). \tag{7.3}$$

This is derived from Theorem 7.3. In fact, $q(x) = p(x|w)$ if

$$a_h = \begin{cases} a_h^* & (1 \leq h \leq H_0) \\ 0 & \text{otherwise}, \end{cases}$$

$$b_h = \begin{cases} b_h^* & (1 \leq h \leq H_0) \\ \text{free} & \text{otherwise}. \end{cases}$$

The dimension of the parameter is $NH + H - 1$ and there are $N(H - H_0)$ free parameters, hence by Theorem 7.3, inequality (7.3) holds.

If the Dirichlet distribution defined in Remark 8.9 is employed for the *a priori* distribution of $\{a_h\}$, then the distribution at $a_h = 0$ can be controlled by the hyperparameter. Let $H = K$ and assume that $\phi_k = \phi_0$ $(1 \leq k \leq H)$ in Remark 8.9, then the zeta function of the normal mixture is

$$\zeta(z) = \int K(a, b)^z \varphi(a)\varphi(b)da\,db,$$

where $\varphi(a)$ is the Dirichlet distribution of a and $\varphi(b)$ is a positive probability distribution on b. Then by studying the neighborhood of

$$\begin{aligned} a_h &= a_h^* \quad (1 \leq h \leq H_0), \\ b_h &= b_h^* \quad (1 \leq h \leq H_0), \\ a_h &= 0, \text{ or free} \quad (H_0 < h \leq H), \\ b_h &= \text{free, or } b_1^* \quad (H_0 < h \leq H), \end{aligned}$$

the learning coefficient is bounded by

$$\lambda \leq \tfrac{1}{2}\min\{H_0(N+1) + (H - H_0 - 1)\phi_0,\ HN + H_0\}. \tag{7.4}$$

The multinomial mixtures are studied in the same way. Let T be a natural number. The set S is defined by

$$S = \{x = (x_1, x_2, x_3)\ ;\ x_1 + x_2 + x_3 = T,\ x_1, x_2, x_3 \geq 0\},$$

where x_1, x_2, x_3 are integers. A trinomial mixture $p(x|w)$ of S trials is a probability distribution on S,

$$p(x|w) = \sum_{h=1}^{H} a_h \left(\frac{S!}{x_1!\, x_2!\, x_3!} (p_{h1})^{x_1} (p_{h2})^{x_2} (p_{h3})^{x_3} \right),$$

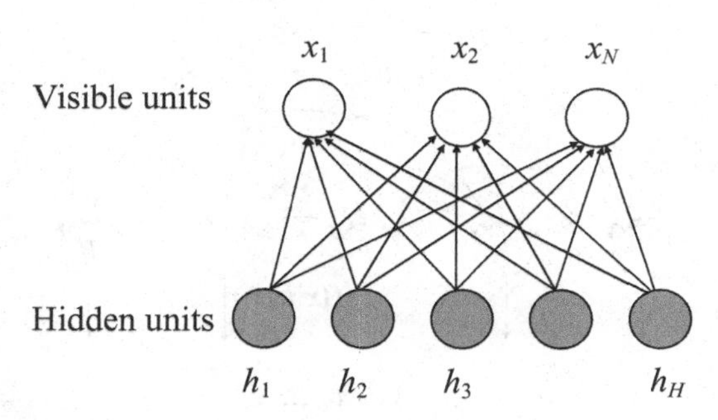

Fig. 7.4. Bayesian network

where $p_h = (p_{h1}, p_{h2}, p_{h3})$ $((p_{h1} + p_{h2} + p_{h3} = 1, 0 \leq p_h \leq 1)$. The parameter is $w = \{(a_h, p_h)\}$. The number of parameters is $3H - 1$. If the true distribution consists of H_0 components, then the number of free parameters is $2(H - H_0)$, hence $\lambda \leq (1/2)(2H_0 + H - 1)$, by Theorem 7.3.

7.4 Bayesian network

Bayesian networks are applied to knowledge discovery, human modeling, and a lot of inference systems in artificial intelligence. Figure 7.4 shows a Bayesian network with visible units and hidden units. Let $x = (x_1, x_2, \ldots, x_N)$ be a set of visible variables, and $h = (h_1, h_2, \ldots, h_H)$ a set of hidden variables, where N and H are the numbers of visible and hidden variables. Assume that each $x_j \in \{1, 2, 3, \ldots, Y\}$ and each $h_j \in \{1, 2, 3, \ldots, S\}$, where Y and S are the numbers of visible and hidden states respectively. The probability that $h_i = k$ is denoted by a_{ik}. Since the sum of probabilities is equal to 1,

$$\sum_{k=1}^{S} a_{ik} = 1 \quad (1 \leq i \leq H).$$

For a given state of hidden variables $(h_1, \ldots, h_H)$, the probability that $x_j = s$ is denoted by $b^{(js)}_{h_1 h_2 \cdots h_H}$, so

$$\sum_{s=1}^{Y} b^{(js)}_{h_1 h_2 \cdots h_H} = 1 \quad (1 \leq j \leq N, (h_1, h_2, \cdots, h_H) \in S^H).$$

The probability distribution of x_j is given by

$$p(x_j | h_1, \ldots, h_H) = \prod_{s=1}^{Y} (b^{(js)}_{h_1 h_2 \ldots h_H})^{\delta(x_j, s)},$$

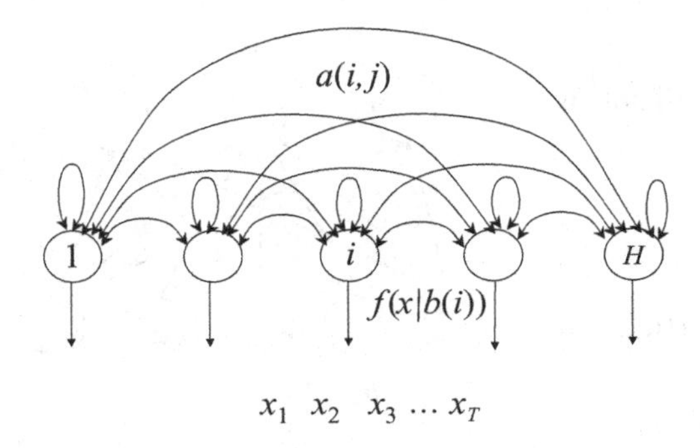

Fig. 7.5. Hidden Markov model

where

$$\delta(x_j, s) = \begin{cases} 1 & (x_j = s) \\ 0 & \text{otherwise.} \end{cases}$$

The learning machine is defined by

$$p(x|a, b) = \sum_{h_1=1}^{S} \cdots \sum_{h_H=1}^{S} a_{1h_1} a_{2h_2} \cdots a_{Hh_H} \prod_{j=1}^{N} p(x_j | h_1, \ldots, h_H).$$

The set of parameters in the learning machine is $a = \{a_{ik}\}$ and $b = \{b^{(js)}_{h_1 h_2 \ldots h_H}\}$. The number of parameters is

$$D_1 = H(S - 1) + (Y - 1)NS^H$$

Assume that the true distribution has H_0 hidden variables and the number of hidden states is equal to S_0. If $a_{ik} = 0$ for $i > H_0$ or $k > S_0$, then $b^{(is)}_{h_1 h_2 \cdots h_H}$ $(i > H_0$ or $h_p > S_0)$ are free parameters, hence the number of free parameters is

$$D_2 = (Y - 1)NS^{H - H_0} + (Y - 1)N(S - S_0)^{H_0},$$

and therefore $\lambda \leq (D_1 - D_2)/2$, by Theorem 7.3.

7.5 Hidden Markov model

Hidden Markov models or probabilistic finite automatons in Figure 7.5 are being applied to speech recognition, motion recognition, and gene analysis. Let H be the number of hidden states. At $t = 1$, the first state is chosen by probability 1. Then at $t = 2, 3, \ldots,$ the new state is chosen with respect to the transition probability $a(i, j)$ from the ith state to the jth state. Since the sum of the probabilities is equal to 1 for each i,

$$\sum_{j=1}^{H} a(i, j) = 1, \quad (i = 1, 2, \ldots, H).$$

When the hidden state is at the jth state, the output y is taken from the probability density function $f(y|b(j))$ where $b(j) \in \mathbb{R}^M$. In some models, y is taken from a discrete set whereas, in other models, y is taken from Euclidean space. By this stochastic process, the model outputs a sequence x of length T,

$$x = (x_1, x_2, \ldots, x_T).$$

The set of parameters is $w = \{(a(i, j), b(j))\}$, where the number of parameters is $H(H - 1) + HM$. The probability distribution of x is

$$p(x|w) = f(x_1|b_1) \prod_{t=2}^{T} \Big(\sum_{k_t=1}^{H} a(k_{t-1}, k_t) f(x_t|b(k_t)) \Big),$$

where $k_1 = 1$. If many sequences of length T are obtained independently, singular learning theory can be applied. Assume that the true distribution has H_0 hidden variables. If the transition probabilities satisfy

$$a(i, j) = 0 \quad (1 \leq i \leq H_0, \quad H_0 < j \leq H),$$

then $a(i, j)$ $(H_0 < i \leq H)$ and $b(i)$ $(H_0 < i \leq H)$ are free parameters. Therefore the numbers of free parameters is $(H - H_0)(H - 1 + M)$, giving $\lambda \leq H_0(H + M - 1)/2$ by Theorem 7.3.

7.6 Singular learning process

In this section, we consider a case when the true distribution is outside of the finite-size parametric model. Even in such a case, singularities of a learning machine affect the learning process.

A three-layered perceptron is defined by a probability distribution of (x, y) where $x \in \mathbb{R}^M$ and $y \in \mathbb{R}^N$,

$$p(x, y|w) = \frac{q(x)}{(2\pi)^{1/2}} \exp(-\tfrac{1}{2}|y - h(x, w)|^2),$$

$$h(x, w) = \sum_{k=1}^{H} a_k \tanh(b_k \cdot x + c_k),$$

and the parameter of this model is

$$w = \{(a_k, b_k, c_k); k = 1, 2, \ldots, H\}.$$

For a given $H_0 < H$, the set of all parameters which satisfies

$$\sum_{k=1}^{H} a_k \tanh(b_k \cdot x + c_k) = \sum_{k=1}^{H_0} a_k^* \tanh(b_k^* \cdot x + c_k^*)$$

contains singularities. If H_0 is smaller, then the singularities are more complicated. In other words,

$$\text{simple function} \Longleftrightarrow \text{complicated singularities,}$$

$$\text{complicated function} \Longleftrightarrow \text{simple singularities.}$$

Assume that the true conditional distribution $q(y|x)$ is outside of the set of the parametric probability distributions, $\{p(x, y|w); w\}$. The set of training samples is denoted by

$$D_n = \{(x_i, y_i); i = 1, 2, \ldots, n\}.$$

The *a posteriori* distribution is

$$p(w|D_n) \propto \varphi(w) \prod_{i=1}^{n} p(y_i|x_i, w).$$

The neighborhood of a parameter w^* is defined by

$$U(w^*) = \{w \in W; |w - w^*| < \epsilon\}.$$

The probability that a parameter is contained in $U(w^*)$ with respect to the *a posteriori* distribution is

$$P(w^*) \equiv \int_{U(w^*)} p(w|D_n) dw.$$

Let us show the following universal phenomenon.

Universal phenomenon. When the number of training samples is small, the parameter that makes $P(w^*)$ largest is a complicated singularity. As the number of training samples increases, the singularity becomes simpler.

Let us explain the above phenomenon. The empirical Kullback–Leibler distance is given by

$$K_n(w) = \frac{1}{n} \sum_{i=1}^{n} \log \frac{q(y_i|x_i)}{p(y_i|x_i, w)}.$$

Then the *a posteriori* distribution is rewritten as

$$p(w|D_n) = \frac{1}{Z_0} \exp(-nK_n(w))\,\varphi(w)$$

and

$$P(w^*) = p(w^*|D_n)\,Z(w^*),$$

where

$$Z(w^*) = \frac{1}{Z_0} \int_{U(w^*)} \exp(-n\{K_n(w) - K_n(w^*)\})\,\varphi(w)\,dw.$$

Note that the probability $P(w^*)$ is determined by not only $p(w^*|D_n)$ but both $p(w^*|D_n)$ and $Z(w^*)$, and that $Z(w^*)$ is larger if w^* is the more complicated singularity. Therefore the learning process is determined as the jump from complicated singularities to simpler singularities. The set of parameters $W \subset \mathbb{R}^d$ can be represented as a union of small subsets,

$$W = \cup_\alpha V_\alpha, \tag{7.5}$$

where V_α is defined by

$$V_\alpha = \{w \in R^d; |w - w_\alpha| \leq \epsilon\}.$$

The empirical Kullback–Leibler distance is

$$nK_n(w) \cong nK(w) + (nK(w))^{1/2}\,\psi(w).$$

Let p_α be the probability that $w \in V_\alpha$, that is to say,

$$p_\alpha = \frac{1}{Z_0} \int_{V_\alpha} e^{-nK_n(w)}\,\varphi(w)\,dw.$$

If n is sufficiently large,

$$p_\alpha \propto e^{-K_\alpha n - \lambda_\alpha \log n}$$

where

$$\begin{aligned} K_\alpha &= K(w_\alpha^*), \\ w_\alpha^* &= \arg\min_{w \in V_\alpha} K(w), \\ K(w) &= \int q(x) \log \frac{q(x)}{p(x|w)} dx, \end{aligned}$$

and $(-\lambda_\alpha)$ is the largest pole of

$$\zeta(z) = \int_{V(w_0)} (K(w) - K(w_\alpha^*))^z \varphi(w) dw.$$

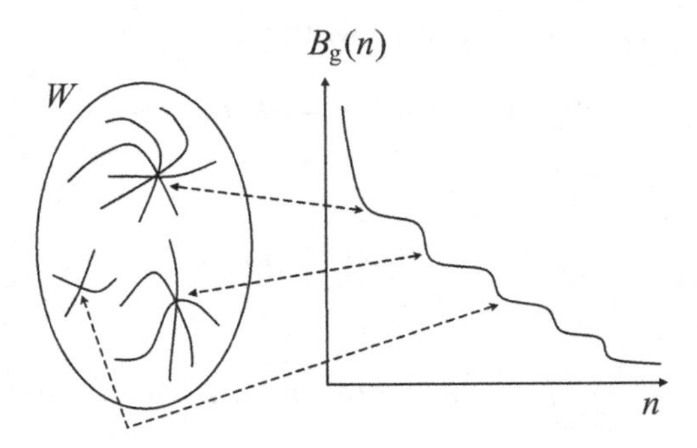

Fig. 7.6. Learning curve with singularities

In general, if the analytic set

$$\{w \in V_\alpha; K(w) - K(w_\alpha^*) = 0\}$$

is more complicated, then λ_α is smaller. Therefore,

$$\begin{aligned} p_\alpha &\propto e^{-f_\alpha} \\ f_\alpha &= K_\alpha n + \lambda_\alpha \log n. \end{aligned}$$

The neighborhood of a parameter w_α which maximizes f_α is realized with the highest probability. As the number of training samples increases, the singularity becomes simpler. In a three-layered perceptron, it is well known that, if V_α is the neighborhood of the parameter that represents k hidden units, K_α and λ_α can be approximated by

$$\begin{aligned} K_\alpha &\approx \frac{C_1}{k}, \\ \lambda_\alpha &\approx C_2 k + C_3, \end{aligned}$$

where $C_1, C_2, C_3 > 0$ are constants. Therefore

$$f_\alpha = \frac{C_1 n}{k} + \{C_2 k + C_3\} \log n.$$

The Bayes generalization error is equal to the increase of the free energy,

$$B_g(n) = \min_k \left\{ \frac{C_1}{k} + \frac{C_2 k + C_3}{2n} \right\}.$$

Figure 7.6 shows the learning process of singular learning machines. The number k which minimizes f_α is

$$k = \sqrt{\frac{Cn}{\log n}}.$$

Hence the number of hidden units which is chosen with high probability by the *a posteriori* distribution is in proportion to $(n/\log n)^{1/2}$.

Remark 7.8 (1) From the statistical physics point of view, the *a posteriori* distribution can be understood as the equilibrium state whose Hamiltonian is equal to the Kullback–Leibler distance. The equilibrium state is determined not by the minimization of not the energy $K_\alpha n$ but by the minimization of the free energy, $K_\alpha n + \lambda \log n$.
(2) Singularities of a model make the learning curve smaller than any nonsingular learning machine. If the true distribution is not contained in the model, then the singularity seems to be the virtual true distribution appropriately with respect to the number of training samples. The most appropriate singularities are selected by minimization of the free energy for the given number of training samples, resulting in the generalization errors being made smaller. It is possible that brain-like systems utilize the effect of singularities in the real world.

7.7 Bias and variance

In the previous section, the true distribution is outside of the finite parametric model but constant compared to the number of training samples. In this section, to analyze the effect of singularities more precisely, we study the balance between bias and variance. Here 'bias' is the function approximation error, whereas 'variance' is the statistical estimation error. If the bias is much larger than the variance, then the statistical model is too simple, hence we should use a more complicated model. If the variance is much larger than the bias, then the model is too complicated, hence it is better to use a simpler model.

If the Kullback–Leibler distance from the singularity to the distribution is in proportion to $1/n$, then both the bias and the variance are of the same order $1/n$, so the bias and variance problem appears. It is still difficult to study such cases based on the general conditions. In this section, we study the problem using a concrete and simple model.

Assume that a set of independent random samples

$$D_n = \{(x_1, y_1), (x_2, y_2), \ldots, (x_n, y_n)\}$$

is taken from the true distribution $q(x)q(y|x)$, where $(x, y) \in \mathbb{R}^N \times \mathbb{R}^1$. Let us study the following conditions.

Learning machine:

$$p(y|x, a, b) = \frac{1}{\sqrt{2\pi}} \exp(-\tfrac{1}{2}(y - H(x, a, b))^2), \tag{7.6}$$

True distribution:

$$q(y|x) = \frac{1}{\sqrt{2\pi}} \exp\Big(-\tfrac{1}{2}\Big(y - \frac{H_0(x)}{\sqrt{n}}\Big)^2\Big), \tag{7.7}$$

where $H_0(x)$ is a general function and

$$H(x, a, b) = \sum_{j=1}^{J} a\, h_j(b)\, e_j(x),$$

which has the set of parameters $\{(a, b) \in R^1 \times R^N\}$. We adopt the set of orthonormal functions, that is to say,

$$\int e_i(x) e_j(x) q(x) dx = \delta_{ij}.$$

We use the notation

$$\|H_0\|^2 = \int H_0(x)^2 q(x) dx,$$

$$H_{0j} = \int H_0(x)\, e_j(x) q(x) dx.$$

Note that the true distribution is outside of the parametric model in general and it depends on the number of random samples. Assume that the *a priori* probability density function $\varphi(a, b)$ is a C^1-class function of a, and that

$$\psi(b) \equiv \varphi(0, b)$$

is a compact support function of b ($\psi(0) > 0$). The *a posteriori* probability distribution with $\beta = 1$ and the predictive distribution are respectively given by

$$p(a, b|D_n) = \frac{1}{C}\, \varphi(a, b) \prod_{i=1}^{n} p(y_i|x_i, a, b),$$

$$p(y|x, D_n) = \int p(y|x, a, b)\, p(a, b|D_n)\, da\, db,$$

where $C > 0$ is a constant. Let $E_{a,b}[\ ]$ be the expectation value using the *a posteriori* distribution.

Theorem 7.5 *Assume that* $\beta = 1$ *and that*

$$\int \frac{\psi(b)\, db}{\sum_{j=1}^{J} h_j(b)^2} < \infty. \tag{7.8}$$

The four errors of the Bayes quartet are given by

$$B_{\mathrm{g}} = \Lambda \frac{1}{n} + o\Big(\frac{1}{n}\Big), \tag{7.9}$$

$$B_{\mathrm{t}} = (\Lambda - 2\nu)\frac{1}{n} + o\Big(\frac{1}{n}\Big), \tag{7.10}$$

$$G_{\mathrm{g}} = (\Lambda + \nu)\frac{1}{n} + o\Big(\frac{1}{n}\Big), \tag{7.11}$$

$$G_{\mathrm{t}} = (\Lambda - \nu)\frac{1}{n} + o\Big(\frac{1}{n}\Big), \tag{7.12}$$

where

$$\Lambda = \tfrac{1}{2}\Big\{1 + \|H_0\|^2 - \sum_{j=1}^{J} H_{0j} E_{\mathrm{g}}\Big[\frac{1}{Z}\frac{\partial Z}{\partial g_j}\Big]\Big\},$$

$$\nu = \sum_{j=1}^{J} E_{\mathrm{g}}\Big[g_j \frac{1}{Z}\frac{\partial Z}{\partial g_j}\Big],$$

and

$$Z(g) = \iint \exp\Big(-\tfrac{1}{2}\sum_{j=1}^{J} \alpha^2 h_j(b)^2 + \sum_{j=1}^{J} \alpha h_j(b)(g_j + H_{0j})\Big)\psi(b) d\alpha\, db. \tag{7.13}$$

Here $g = \{g_j\}$ *is a random variable which is subject to the* J*-dimensional normal distribution with zero mean and identity covariance matrix, and* $E_{\mathrm{g}}[\ ]$ *shows the expectation value over* g.

Remark 7.9 (Universality of equations of states) In this theorem, since the true distribution is outside of the parametric model and it is not fixed but moving as the number of training samples, Theorem 6.10 does not hold. In fact $\Lambda \neq \lambda$. If $H_0(x) \equiv 0$, then this theorem coincides with Theorem 6.10 with $\beta = 1$. However, this theorem shows that, even if $H_0(x) \neq 0$, the equations of states in Main Theorem 6.3 hold. It seems that equations of states are the universal relations among the Bayes quartet.

Proof of Theorem 7.5 From the definition of the true distribution and the statistical model, the log density ratio is

$$f(x, y, a, b) = \tfrac{1}{2}\Big(H(x, a, b) - \frac{H_0(x)}{\sqrt{n}}\Big)^2 + \sigma\Big(H(x, a, b) - \frac{H_0(x)}{\sqrt{n}}\Big),$$

where $\sigma = y - H_0(x)$ is a random variable which is subject to the standard normal distribution. The log likelihood ratio function is

$$n\ K_n(a, b) = \sum_{i=1}^{n} f(x_i, y_i, a, b).$$

By using a rescaling parameter $a = \alpha/\sqrt{n}$, we have

$$nK_n\Big(\frac{\alpha}{\sqrt{n}}, b\Big) = \frac{1}{2n}\sum_{i=1}^{n}(H(x_i, \alpha, b) - H_0(x_i))^2$$

$$-\frac{1}{\sqrt{n}}\sum_{i=1}^{n}\sigma_i(H(x_i, \alpha, b) - H_0(x_i)).$$

By using the central limit theorem, $nK_n(\alpha/\sqrt{n}, b)$ converges to the following function $E(\alpha, b)$ in law,

$$E(\alpha, b) = \frac{1}{2}\sum_{j=1}^{J}\alpha^2 h_j(b)^2 - \sum_{j=1}^{J}\alpha h_j(b) H_{0j}$$

$$+\frac{\|H_0\|^2}{2} - \sum_{j=1}^{J}\alpha h_j(b) g_j + \hat{\sigma}$$

where we used these convergences in law,

$$\frac{1}{\sqrt{n}}\sum_{j=1}^{n}\sigma_i e_j(x) \to g_j,$$

$$\frac{1}{\sqrt{n}}\sum_{j=1}^{n}\sigma_i H_0(x_i) \to \hat{\sigma}.$$

Let us define the renormalized *a posteriori* distribution,

$$E_{\alpha,b}[F(\alpha, b)] = \frac{\int e^{-E(\alpha,b)} F(\alpha, b)\psi(b) d\alpha\, db}{\int e^{-E(\alpha,b)}\psi(b) d\alpha\, db}.$$

Then, for an arbitrary natural number k,

$$E_{a,b}[H(X, a, b)^k] = \frac{1}{n^{k/2}} E_{\alpha,b}[H(X, \alpha, b)^k] + o\Big(\frac{1}{n^{k/2}}\Big).$$

By Theorem 1.3, Remark 1.12, and Subsection 1.4.3,

$$B_{\mathrm{g}} = \frac{1}{2n} E_{\mathrm{g}}[\ E_X[\ E_{\alpha,b}[H(X,\alpha,b) - H_0(X)]^2]] + o\Big(\frac{1}{n}\Big),$$

$$B_{\mathrm{t}} = B_{\mathrm{g}} - G_{\mathrm{g}} + G_{\mathrm{t}} + o(1/n),$$

$$G_{\mathrm{g}} = \frac{1}{2n} E_{\mathrm{g}}[\ E_X[\ E_{\alpha,b}[(H(X,\alpha,b) - H_0(X))^2]]] + o\Big(\frac{1}{n}\Big),$$

$$G_{\mathrm{t}} = G_{\mathrm{g}} - \frac{1}{n} E_{\mathrm{g}}[\ E_{\alpha,b}[\sum_{j=1}^{J} \alpha h_j(b) g_j]] + o(1/n).$$

Then, by the definition of $H(x,\alpha,b)$,

$$B_{\mathrm{g}} = \frac{1}{2n} \sum_{j=1}^{J} \Big(E_{\alpha,b}[\alpha h_j(b)]^2 - 2H_{0j} E_{\alpha,b}[\alpha h_j(b)] \Big) + \frac{\|H_0\|^2}{2n} + o\Big(\frac{1}{n}\Big),$$

$$G_{\mathrm{g}} = \frac{1}{2n} \sum_{j=1}^{J} \Big(E_{\alpha,b}[\alpha^2 h_j(b)^2] - 2H_{0j} E_{\alpha,b}[\alpha h_j(b)] \Big) + \frac{\|H_0\|^2}{2n} + o\Big(\frac{1}{n}\Big).$$

The *a posteriori* distribution satisfies

$$E_{\alpha,b}[\ (\alpha h_j(b))^k\ ] = \frac{1}{Z} \Big(\frac{\partial}{\partial g_j}\Big)^k Z(g),$$

where $Z = Z(g)$ is defined as in eq.(7.13). By using the partial integration for g_j,

$$E_{\mathrm{g}}\Big[g_j \frac{1}{Z}\frac{\partial Z}{\partial g_j}\Big] = E_{\mathrm{g}}\Big[\frac{\partial}{\partial g_j}\Big(\frac{1}{Z}\frac{\partial Z}{\partial g_j}\Big)\Big] = E_{\mathrm{g}}\Big[\frac{1}{Z^2}\frac{\partial^2 Z}{\partial g_j^2}\Big] - E_{\mathrm{g}}\Big[\Big(\frac{1}{Z}\frac{\partial Z}{\partial g_j}\Big)^2\Big].$$

Moreover, by using the partial integration for $\int d\alpha$,

$$\sum_{j=1}^{J} \frac{1}{Z}\frac{\partial^2 Z}{\partial g_j^2} = 1 + \sum_{j=1}^{J} (g_j + H_{0j}) \frac{1}{Z}\Big(\frac{\partial Z}{\partial g_j}\Big).$$

By using these facts, we obtain the theorem. □

If the true distribution is given by $H_0(x) \equiv 0$, then $\Lambda = 1/2$, which corresponds to the fact that the largest pole of the zeta function

$$\zeta(z) = \int a^{2z} |b|^{2z}\ \varphi(a,b)\ da\ db$$

is equal to $z = -1/2$. Let us consider a more specific learning machine,

$$p(y|x,a,b) = \frac{1}{\sqrt{2\pi}} \exp\Big(-\frac{1}{2}(y - \sum_{j=1}^{N} a b_j e_j(x))^2\Big). \tag{7.14}$$

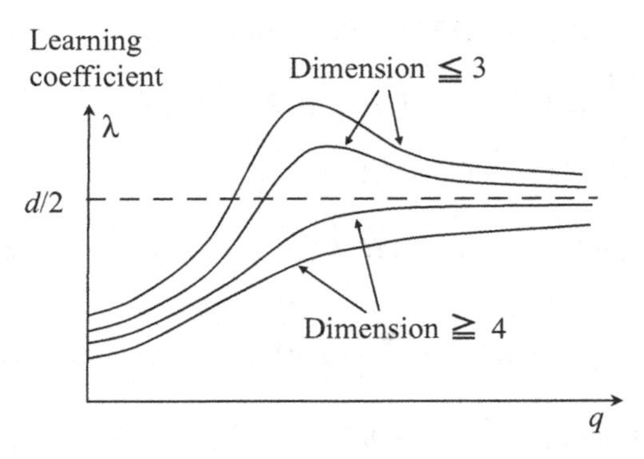

Fig. 7.7. Learning coefficients

where $a \in \mathbb{R}^1$, $b \in \mathbb{R}^N$, $x \in \mathbb{R}^N$ $(N > 1)$ and we assume $\psi(b)$ is symmetric for the direction of b hence it can be written as $\psi(|b|)$. If $N = 1$, eq.(7.8) is not satisfied. If $N \geq 2$, by $w_i = ab_i$, the model is formally equal to the regular model,

$$p(y|x, w) = \frac{1}{\sqrt{2\pi}} \exp\Big(-\frac{1}{2}(y - \sum_{j=1}^{N} w_j e_j(x)))^2\Big).$$

Let us compare the generalization error of the singular model represented by $\{a, b_i\}$ with that of the regular model represented by $\{w_i\}$. In the following we assume $H_0(x) = w_0 \cdot x$. Let $\Lambda(w_0)$ be the value in Theorem 7.5, by which the Bayes generalization error is given as $B_{\rm g} = \Lambda/n + o(1/n)$. For arbitrary w_0, the learning coefficient of the regular statistical model is $N/2$. Therefore, we compare $N/2$ with the learning coefficient of the singular model, $\Lambda(w_0)$. If the true model is given by eq.(7.7) with $H_0(x) = w_0 \cdot x$, then the singular model eq.(7.14) has the following learning coefficient:

$$2\Lambda(w_0) = 1 + E_{\rm g}\Big[(|w_0|^2 + w_0 \cdot g)\frac{Y_N(g)}{Y_{N-2}(g)}\Big] \tag{7.15}$$

where

$$Y_N(g) = \int_0^{\pi/2} d\theta \ \sin^N\theta \ \exp(-\tfrac{1}{2}|w_0 + g|^2 \sin^2\theta).$$

This results can be shown by using polar coordinates. In eq.(7.15) the learning coefficient $\Lambda(w_0)$ cannot be represented by a simple function; however, the numerical result is shown in Figure 7.7 for $N = 2, 3, \ldots, 6$. Here the horizontal and vertical lines show $|w_0|$ and $2\Lambda(w_0)/N$ respectively. If $2\Lambda(w_0)/N < 1$, then the learning coefficient is smaller than that in the regular statistical model.

(1) When $|w_0| \to \infty$, $\Lambda(w_0)$ converges to $N/2$ in every case.
(2) If $N = 2$ or $N = 3$, then $\Lambda(w_0)$ is larger than $N/2$ when $|w_0|$ becomes large. Whereas, if $N \geq 4$, then the learning coefficients are always smaller than $N/2$. In other words, if the dimension of the parameter space is larger than 3, then singularities always make the generalization error smaller than in the regular statistical model. Also we can show that the learning coefficient $\Lambda(w_0)$ has an asymptotic expansion for $|w_0| \to \infty$,

$$2\Lambda(w_0) = N - \frac{(N-1)(N-3)}{|w_0|^2} + o\Big(\frac{1}{|w_0|^2}\Big).$$

Remark 7.10 (Shrinkage estimation) In regular statistical models, the shrinkage estimation and empirical Bayes method have the same property described in this section [85, 29]. Singularities of a model affect the learning process as the source of shrinkage estimation; moreover, the most appropriate singularities are selected by the *a posteriori* distribution for a given set of random samples.

7.8 Non-analytic learning machines

In Chapter 6, we assume that the log likelihood density function is a function-valued analytic function of the parameter. The reason why such an assumption is adopted is that our main purpose is to discover the universal theorems without regard to pathological examples.

Example 7.2 (Non-analytic model) The following is a non-analytic model

$$p(y|x, a) = \frac{1}{\sqrt{2\pi}} \exp\Big(-\frac{1}{2}\Big(y - \exp\Big(-\frac{x}{a^2}\Big)\Big)^2\Big),$$

for $a \neq 0$ and $p(y|x, 0)$ is defined by the standard normal distribution of y. If the true distribution is $q(y|x) = p(y|x, 0)$ and if $q(x)$ is the standard normal distribution, then the Kullback–Leibler distance is

$$K(a) = \frac{1}{2} \int \exp\Big(-\frac{2x}{a^2}\Big) q(x) dx$$
$$= \frac{1}{2} \exp\Big(-\frac{1}{a^4}\Big).$$

Hence $K(a) = 0$ if and only if $a = 0$ and $K(a)$ is a function of class C^∞ by the definition that the n times differential $K^{(n)}(a) = 0$ $(n = 1, 2, \ldots)$. There are no constants $c_1, c_2 > 0$ and no polynomials $K_1(a)$, $K_2(a)$ which satisfy

$$c_1 K_1(a) \leq K(a) \leq c_2 K_2(a)$$

in the neighborhood of the origin. Such a statistical model does not satisfy fundamental condition (I) or (II). In regular statistical theory, such pathological examples can be rejected by the condition that the Fisher information matrix should be positive definite. However, in singular learning theory, we cannot adopt such a condition.

Let us study the normal mixture.

$$p(x|w) = \sum_{h=1}^{H} \frac{a_h}{(2\pi)^{N/2}} \exp\Big(-\frac{|x-b_h|^2}{2}\Big), \tag{7.16}$$

where $x, b_h \in \mathbb{R}^N$, $a_h \in \mathbb{R}^1$, and $w = \{a_h, b_h\}$. The set of parameters is set as

$$W = \{(a_h, b_h); a_1 + \cdots + a_H = 1, a_h \geq 0, |b_h| \leq B\},$$

where $B > 0$ is a constant. Let $q(x)$ be a true distribution which is represented by $q(x) = p(x|w_0)$ for some parameter $w \in W$. The log density ratio function is not an $L^s(q)$-valued analytic function in general.

Example 7.3 Let $N = 1$ and

$$p_0(x|b) = \frac{1}{\sqrt{2\pi}} \exp(-\tfrac{1}{2}(x-b)^2).$$

For a mixture model

$$p(x|a) = (1-a)\ p_0(x|0) + a\ p_0(x|2)$$

and a true distribution $q(x) = p(x|0)$, the log density ratio function is equal to

$$\begin{aligned} f(x, a) &= \log(q(x)/p(x|a)) \\ &= -\log(1 + a\ (e^{2x-2} - 1)) \\ &= \sum_{j=1}^{\infty} \frac{a^j}{j}(1 - e^{2x-2})^j, \end{aligned}$$

where the last equation is a formal expansion. The convergence radius of this function as an $L^1(q)$-valued function is zero, because [109]

$$\lim_{j\to\infty} \sup\Big(\int \Big|\frac{\partial^j}{\partial a^j} f(x,0)\Big| q(x)dx\Big)^{1/j} = \infty.$$

This example shows that the normal mixture does not satisfy fundamental condition (I). However, since the four Main Theorems in Chapter 6 are universal formulas, we can improve them for a given statistical model if the model is not a pathological one. Let us introduce a function

$$S(t) = \begin{cases} \dfrac{-\log t + t - 1}{(t-1)^2} & (t \neq 1) \\ \frac{1}{2} & (t = 1). \end{cases}$$

Then $S(t)$ is an analytic function. A function $M(x)$ is introduced,

$$M(x) \equiv \sup_{w \in W} S\Big(\frac{p(x|w)}{q(x)}\Big).$$

We can show by the definition eq.(7.16) that there are $A_0, A_1, B_0, B_1 > 0$ such that

$$e^{-A_0|x|-A_1} \leq \frac{p(x|w)}{q(x)} \leq e^{B_0|x|+B_1} \quad (\forall (x, w) \in \mathbb{R}^N \times W).$$

Hence there exist $C_0, C_1 > 0$ such that

$$M(x) \leq C_0|x| + C_1.$$

The Kullback–Leibler distance is bounded by

$$\begin{aligned} K(w) &= \int q(x) \log \frac{q(x)}{p(x|w)} dx \\ &= \int q(x)\Big(-\log \frac{p(x|w)}{q(x)} + \frac{p(x|w)}{q(x)} - 1\Big)\, dx \\ &= \int q(x)\Big(\frac{p(x|w)}{q(x)} - 1\Big)^2 S\Big(\frac{p(x|w)}{q(x)}\Big) dx \\ &\leq \int \Big(\frac{p(x|w)}{q(x)} - 1\Big)^2 M(x) q(x) dx \equiv H(w). \end{aligned}$$

Here it is easy to show that $(p(x|w)/q(x) - 1)$ is an $L^s(Mq)$-valued analytic function for an arbitrary $s \geq 2$. Therefore we can apply the resolution theorem to $H(w)$, which means that there exists a resolution map $w = g(u)$ such that

$$K(g(u)) \leq H(g(u)) = u^{2k}.$$

There exists an $L^s(Mq)$-valued analytic function $a(x, u)$ such that

$$\frac{p(x|w)}{q(x)} - 1 = a(x, u) u^k,$$

where $a(x, u)$ is not equal to zero as an $L^s(Mq)$-valued function. On the other hand, for arbitrary compact set $C \in \mathbb{R}^N$,

$$K(g(u)) = \int q(x)a(x,u)^2 u^{2k} S\Big(\frac{p(x|w)}{q(x)}\Big)dx$$

$$\geq u^{2k} \int_C q(x)a(x,u)^2 S\Big(\frac{p(x|w)}{q(x)}\Big)dx.$$

By choosing the compact set C sufficiently large, there exists a constant $c_1 > 0$ such that

$$K(g(u)) \geq c_1 u^{2k}.$$

There exists a function $c_1(u) > 0$ of class C^∞,

$$K(g(u)) = c_1(u)u^{2k}.$$

In the same way as in Chapter 6, there exists $c_2(u, x)$ which is a function of class C^∞,

$$f(x, g(u)) = c_2(x, u)u^k.$$

The empirical process

$$\xi_n(u) = \frac{1}{\sqrt{n}} \sum_{j=1}^{n} \{c_2(X_j, u) - c_1(u)u^k\}$$

is a function of class C^∞ for u which converges in law to a Gaussian process [92] as a random variable of a Banach space defined by a sup-norm on a compact set. Therefore, we can obtain the same result for normal mixtures.

Remark 7.11 In this book, the main formulas are proved based on fundamental conditions (I) and (II). These conditions allow that the Fisher information matrix is not positive definite; however, this is a sufficient condition for the main formulas. The necessary and sufficient conditions for the main formulas are still unknown.

8
Singular statistics

In this chapter, we study statistical model evaluation and statistical hypothesis tests in singular learning machines. Firstly, we show that there is no universally optimal learning in general and that model evaluation and hypothesis tests are necessary in statistics. Secondly, we analyze two information criteria: stochastic complexity and generalization error in singular learning machines. Thirdly, we show a method to produce a statistical hypothesis test if the null hypothesis is a singularity of the alternative hypothesis. Then the methods by which the Bayes *a posteriori* distribution is generated are introduced. We discuss the Markov chain Monte Carlo and variational approximation. In the last part of this chapter, we compare regular and singular learning theories. Regular learning theory is based on the quadratic approximation of the log likelihood ratio function and the central limit theorem on the parameter space, whereas singular learning theory is based on the resolution of singularities and the central limit theorem on the functional space. Mathematically speaking, this book generalizes regular learning theory to singular statistical models.

8.1 Universally optimal learning

There are a lot of statistical estimation methods. One might expect that there is a universally optimal method, which always gives a smaller generalization error than any other method. However, in general, such a method does not exist.

Assumption. Assume that $\Phi(w)$ is the probability density function on $\mathbb{R}^d$, and that a parameter w is chosen with respect to $\Phi(w)$. After a parameter w

is chosen and fixed, random samples $d_n = \{x_1, x_2, \ldots, x_n\}$ are independently taken from a conditional distribution $P(x|w)$.

In the real world, we can measure only the set of random samples, and $\Phi(w)$ and $P(x|w)$ are unknown. Let $r(x|d_n)$ be an arbitrary conditional probability density function of x for a given set of samples d_n. The generalization error of $r(x|d_n)$ is defined by the Kullback–Leibler distance from the fixed distribution $P(x|w)$ to $r(x|d_n)$,

$$G(w, d_n) = \int dx\, P(x|w) \log \frac{P(x|w)}{r(x|d_n)}.$$

The mean generalization error $\mathcal{G}(r)$ is defined by the expectation over all w and d_n,

$$\begin{aligned}\mathcal{G}(r) &\equiv \int dw\ \Phi(w)\ E_{d_n}[G(w, d_n)]\\ &= \int dw\ \Phi(w) \prod_{i=1}^{n} \Big(\int P(x_i|w)dx_i\Big)\ G(w, d_n).\end{aligned}$$

Then $\mathcal{G}(r)$ is a functional of a given conditional probability $r(x|d_n)$. Let us study the minimization problem of $\mathcal{G}(r)$.

We define the predictive distribution $R(x|d_n)$ based on the true *a priori* probability density function $\Phi(w)$ and the true parametric model $P(x|w)$,

$$R(x|d_n) \equiv \frac{\displaystyle\int dw\ \Phi(w) P(x|w) \prod_{i=1}^{n} P(x_i|w)}{\displaystyle\int dw\ \Phi(w) \prod_{i=1}^{n} P(x_i|w)}. \tag{8.1}$$

Theorem 8.1 *The functional* $\mathcal{G}(r)$ *is minimized if and only if* $r(x|d_n) = R(x|d_n)$.

Proof of Theorem 8.1 The functional $\mathcal{G}(r)$ can be rewritten as

$$\begin{aligned}\mathcal{G}(r) &= \int dw \Phi(w) E_{d_n}\Big[\int dx\ R(x|d_n) \log \frac{R(x|d_n)}{r(x|d_n)}\Big]\\ &\quad + \int dw \Phi(w) \Big[\int dx\ P(x|w) \log P(x|w)\Big]\\ &\quad - \int dw \Phi(w) E_{d_n}\Big[\int dx\ P(x|w) \log R(x|d_n)\Big].\end{aligned}$$

The first term of the right-hand side is equal to the Kullback–Leibler distance from $R(x|d_n)$ to $r(x|d_n)$, and neither the second term nor third term depends on $r(x|d_n)$. Therefore, by the property of the Kullback–Leibler distance, the functional $\mathcal{G}(r)$ is minimized if and only if

$$r(x|d_n) = R(x|d_n).$$

Under the assumption as above, there is no better statistical estimation than $R(x|d_n)$. □

Let us introduce the mean entropy of $p(x|w)$ by using $E_w[\,] = \int \Phi(w)dw$,

$$S = -E_w\Big[\int P(x|w) \log P(x|w)dx\Big]$$

and the stochastic complexity by

$$F_n(d_n) = -\log \int dw \Phi(w) \prod_{i=1}^{n} P(x_i|w). \tag{8.2}$$

Then the minimum of $\mathcal{G}(r)$ is equal to

$$\min_r \mathcal{G}(r) = \int dw\ \Phi(w) E_{d_{n+1}}[F_{n+1}(d_{n+1}) - F_n(d_n) - S].$$

Remark 8.1 From Theorem 8.1, we obtain the following facts.
(1) If one knows the true *a priori* distribution $\Phi(w)$ and the true statistical model $P(x|w)$, there is no better statistical estimation than Bayes estimation using these.
(2) Even if one knows the true statistical model $P(x|w)$, the optimal predictive distribution depends on the *a priori* probability distribution, hence there is no universally optimal statistical estimation method.
(3) In practical applications, we design an *a priori* distribution $\varphi(w)$ and a model $p(x|w)$. If they are written as $\varphi(w|\theta_1)$ or $p_{\theta_2}(x|w)$ using parameters θ_1 and θ_2, such parameters are called hyperparameters. We have to optimize hyperparameters or the probability distribution of hyperparameters. If several possible models $p_1(x|w), p_2(x|w), \ldots, p_k(x|w)$ are compared and optimized, such a procedure is called statistical model selection.
(4) For hyperparameter optimization or statistical model selection, we need a statistical evaluation method of the pair (φ, p) for a set of random samples. The major concepts used in statistical model evaluation are stochastic complexity and generalization error.

8.2 Generalized Bayes information criterion

For a given set of training samples $\{X_i\}$, the stochastic complexity of $\varphi(w)$ and $p(x|w)$ with $p = 1$,

$$F_n(\varphi, p) = -\log \int \prod_{i=1}^{n} p(X_i|w)\varphi(w)dw,$$

can be understood as a likelihood of the pair (φ, p). Therefore, if $F_n(\varphi, p)$ is smaller, then the pair (φ, p) seems to be more appropriate for the given set of training samples. If the pair (φ, p) is too simple to approximate the true distribution $q(x)$, then

$$\begin{aligned} F_n(\varphi, p) &\approx -\sum_{i=1}^{n} \log p(X_i|w^*) \\ &\approx nK(q||p^*) - nS_n, \end{aligned}$$

where w^* is the parameter that minimizes $K(q||p^*)$, $p^*(x) = p(x|w^*)$, and S_n is the empirical entropy,

$$S_n = -\frac{1}{n}\sum_{i=1}^{n} \log q(X_i).$$

Since S_n does not depend on the pair (φ, p), the main term of $F_n(\varphi, p)$ is determined by the functional approximation error $K(q||p^*)$ in this case. On the other hand, if $K(q||p^*) << 1/n$, in other words, if the model can approximate the true distribution compared to the variance, then

$$F_n(\varphi, p) = nS_n + \lambda \log n - (m-1)\log\log n + R_1,$$

where R_1 is a random variable of constant order. In general, the more redundant the model is, the larger λ is. Therefore, the stochastic complexity is minimized when (φ, p) is appropriate.

If the set of parameters is compact, and the true distribution is contained in the model, then

$$nS_n = -\sum_{i=1}^{n} \log p(X_i|\hat{w}) + R_2,$$

where $\hat{w}$ is the maximum likelihood or *a posteriori* estimator, and R_2 is a random variable of constant order. Hence

$$F_n(\varphi, p) = -\sum_{i=1}^{n} \log p(X_i|\hat{w}) + \lambda \log n - (m-1)\log\log n + R_1 + R_2. \tag{8.3}$$

Therefore, the value $F_n(\varphi, p)$ can be approximated by using the maximum likelihood or *a posteriori* estimator. This is the generalized version of BIC.

Remark 8.2 (1) If a learning machine is a regular statistical model,

$$F_n(\varphi, p) \approx -\sum_{i=1}^{n} \log p(X_i|\hat{w}) + \frac{d}{2} \log n$$

is employed in model selection, which is called BIC or MDL. However, in singular learning machines, neither BIC nor MDL approximates the value of $F_n(\varphi, p)$.

(2) The stochastic complexity is the criterion for Bayes estimation. Even if the model is selected with respect to the stochastic complexity, it might not be appropriate for the maximum likelihood or *a posteriori* estimation. In other words, even if one chooses the model by minimization of eq.(8.3), the maximum likelihood or *a posteriori* estimation is not appropriate for singular learning machines.

(3) If the set of parameters is not compact, the above approximation eq.(8.3) fails in general. In normal mixtures or layered neural networks, the maximum likelihood estimator often diverges, hence eq.(8.3) cannot be calculated. To evaluate the stochastic complexity, other numerical methods such as Markov chain Monte Carlo are recommended.

8.3 Widely applicable information criteria

Based on Main Theorem 6.3, we establish new information criteria which can be used in both regular and singular learning machines. The criteria can be applied to both model selection and hyperparameter optimization.

Let X and $\{X_i\}$ be testing and training samples respectively which are independently subject to the unknown probability distribution $q(x)$. For a given set of an *a priori* distribution $\varphi(w)$ and a statistical model $p(x|w)$, we define Bayes generalization loss, Bayes training loss, Gibbs generalization loss, and Gibbs training loss respectively by

$$BL_{\mathrm{g}} = -E_X[\log E_w[p(X|w)]],$$

$$BL_{\mathrm{t}} = -\frac{1}{n}\sum_{i=1}^{n} \log E_w[p(X_i|w)],$$

$$GL_{\mathrm{g}} = -E_w E_X[\log p(X|w)],$$

$$GL_{\mathrm{t}} = -E_w\Big[\frac{1}{n}\sum_{i=1}^{n} \log p(X_i|w)\Big],$$

where $E_w[\ ]$ shows the expectation value over the Bayes *a posteriori* distribution. These losses are random variables. Both training losses BL_t and GL_t can be numerically calculated using training samples D_n and a learning machine $p(x|w)$ without any knowledge of the true density function $q(x)$. In fact, if one has a sequence of parameters $\{w_t; t = 1, 2, 3, \ldots, T\}$ which approximates the *a posteriori* distribution, in other words,

$$E_w[f(w)] \cong \frac{1}{T}\sum_{t=1}^{T} f(w_t)$$

for an arbitrary function f, then

$$BL_t \cong -\frac{1}{n}\sum_{i=1}^{n}\log\Big(\frac{1}{T}\sum_{t=1}^{T} p(X_i|w_t)\Big),$$

$$GL_t \cong -\frac{1}{T}\sum_{t=1}^{T}\frac{1}{n}\sum_{i=1}^{n}\log p(X_i|w_t).$$

Therefore, if one has a sequence of parameters, it is easy to calculate both Bayes and Gibbs training losses numerically. Let S_n be the empirical entropy of the true distribution,

$$S_n = \frac{1}{n}\sum_{i=1}^{n} -\log q(X_i),$$

and $S = E[S_n]$ be its expectation. Then

$$BL_g = B_g + S,$$

$$BL_t = B_t + S_n,$$

$$GL_g = G_g + S,$$

$$GL_t = G_t + S_n.$$

From Main Theorem 6.3 we obtain the equations

$$E[BL_g] = E[BL_t] + 2\beta(\ E[GL_t] - E[BL_t]\ ) + o\Big(\frac{1}{n}\Big),$$

$$E[GL_g] = E[GL_t] + 2\beta(\ E[GL_t] - E[BL_t]\ ) + o\Big(\frac{1}{n}\Big).$$

Let us define widely applicable information criteria (WAIC) by

$$WAIC_1 = BL_t + 2\beta(GL_t - BL_t),$$

$$WAIC_2 = GL_t + 2\beta(GL_t - BL_t).$$

Then the expectations of the two criteria are equal to the Bayes and Gibbs generalization losses respectively,

$$E[BL_{\rm g}] = E[WAIC_1] + o\Big(\frac{1}{n}\Big), \qquad (8.4)$$

$$E[GL_{\rm g}] = E[WAIC_2] + o\Big(\frac{1}{n}\Big). \qquad (8.5)$$

Recall that the empirical variance V is defined in eq.(6.23),

$$V = \sum_{i=1}^{n} \Big\{ E_w[\ (\log p(X_i|w))^2\ ] - (\ E_w[\log p(X_i|w)]\ )^2 \Big\}. \qquad (8.6)$$

Then, by Theorem 6.10, eq. (6.31), and eq. (6.32), we have an alternative representation,

$$WAIC_1 = BL_{\rm t} + \frac{\beta}{n} V + o_p\Big(\frac{1}{n}\Big), \qquad (8.7)$$

$$WAIC_2 = GL_{\rm t} + \frac{\beta}{n} V + o_p\Big(\frac{1}{n}\Big). \qquad (8.8)$$

Remark 8.3 (1) The relations eqs.(8.4) and (8.5) are obtained on the assumption that the true distribution is contained in the parametric model. When the true distribution is not contained in the parametric model and the function approximation error (bias) is much larger than statistical estimation error (variance), the main terms of the generalization errors are equal to the training errors, $E[B_{\rm g}] \cong E[B_{\rm t}]$ and $E[G_{\rm g}] \cong E[G_{\rm t}]$. In [120], $WAIC_1$ and $WAIC_2$ are equal to the Bayes and Gibbs generalization errors respectively, even if the true distribution is outside of the parametric model.
(2) Also when the bias is in proportion to the variance, it is strongly expected that $WAIC_1$ and $WAIC_2$ correspond to the Bayes and Gibbs generalization errors respectively. In fact, using a specific model, we proved in Theorem 7.5 that equations of states hold in such a case. It should be emphasized that AIC does not correspond to the generalization error when the bias is in proportion to the variance, even in regular statistical models. The criteria $WAIC_1$ and $WAIC_2$ give the indices for universal model evaluation.
(3) Figure 8.1 shows the behavior of WAIC in model selection. If the true distribution is contained in the finite-size statistical model, then WAIC is smallest as an expectation value. In singular learning machines, the generalization error by Bayes is not so large as the maximum likelihood method even if the statistical model is redundant.

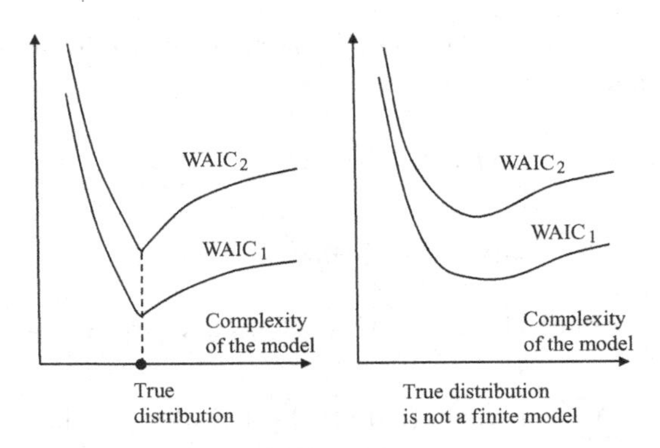

Fig. 8.1. Widely applicable information criteria

(4) If a model is regular and the true distribution is contained in the parametric model, the singular fluctuation is equal to

$$\beta(E[G_t^*] - E[B_t^*]) = \frac{d}{2}. \tag{8.9}$$

When $\beta \to \infty$ in regular models, both Bayes and Gibbs estimations result in the maximum likelihood method, hence WAIC mathematically contains *AIC*. In [120], WAIC also contains TIC.

(5) Assume that the true distribution is equal to a finite-size statistical model in the family of statistical models. A model selection criterion is said to have consistency if the probability that the true model is chosen goes to 1 as the number of training samples tends to infinity. It is well known that AIC in regular model selection does not have consistency. WAIC also does not have consistency. Let $WAIC_1(\varphi, p)$ be WAIC of $\varphi(w)$ and $p(x|w)$ and $\mathcal{F}$ be the family of all possible pairs (φ, p). Then

$$E[\min_{(\varphi,p)\in\mathcal{F}} WAIC_1(\varphi, p)] \neq \min_{(\varphi,p)\in\mathcal{F}} E[WAIC_1(\varphi, p)],$$

because $n \times WAIC_1$ converges not to a constant but to a random variable. On the contrary, the leading term of the stochastic complexity converges to a constant λ as the number of training samples tends to infinity. Hence the stochastic complexity has consistency but it does not correspond to generalization error. These are the general theoretical differences between the generalization error and the stochastic complexity.

(6) (Parameter can be understood as hyperparameter) Let $p(x|w)$ be a statistical model and $\varphi(w)$ be a fixed and localized distribution on a narrow neighborhood of the origin in the parameter space. For an arbitrary parameter w^*, the Bayes estimation using the *a priori* distribution $\varphi(w - w^*)$ can be optimized by WAIC

Table 8.1. *Experimental results*

H	Theory	B_g	$WAIC_1$	G_g	$WAIC_2$
1		1.677973	1.668674	1.688998	1.679448
2		0.826303	0.797272	0.853480	0.823484
3	0.013500	0.012696	0.014204	0.026243	0.027413
4	0.015000	0.014491	0.015340	0.030163	0.030554
5	0.016000	0.015600	0.016078	0.031975	0.032011
6	0.017000	0.016481	0.016687	0.033334	0.033048

by understanding that w^* is a hyperparameter. In other words, a parameter w^* can be optimized by minimization of the expected Bayes generalization error.

8.3.1 Experiments

We studied reduced rank regressions. The input and output vector is $x = (x_1, x_2) \in \mathbb{R}^{N_1} \times \mathbb{R}^{N_2}$ and the parameter is $w = (A, B)$ where A and B are $N_1 \times H$ and $H \times N_2$ matrices respectively. The learning machine is

$$p(x|w) = q(x_1)\frac{1}{(2\pi\sigma^2)^{N_2/2}} \exp\Big(-\frac{1}{2\sigma^2}|x_2 - BAx_1|^2\Big).$$

Since $q(x_1)$ has no parameter, it is not estimated. The true distribution is determined by matrices A_0 and B_0 such that rank $(B_0A_0) = H_0$. The algebraic variety of the true parameters $K(A, B) = 0$, where

$$K(A, B) \propto \|BA - B_0A_0\|^2,$$

has complicated singularities. We conducted experiments in a case when $N_1 = N_2 = 6$, $H_0 = 3$, $\beta = 1$, $n = 1000$, and $\sigma = 0.1$. The *a priori* distribution was $p(A, B) \propto \exp(-2.0 \cdot 10^{-5}(|A|^2 + |B|^2))$. Reduced rank regressions with hidden units $H = 1, 2, \ldots, 6$ were employed. The *a posteriori* distribution was numerically approximated by the Metropolis method, where the first 5000 steps were omitted and 2000 parameters were collected after every 200 steps. Expectation values B_g, $WAIC_1$, E_g, $WAIC_2$ were averaged by 20 trials, that is to say, 20 sets of training samples were independently taken from the true distribution. Theoretical values of $E[B_g]$ for $\beta = 1$ are given in Chapter 7. The results in Table 8.1 show the effectiveness of WAIC.

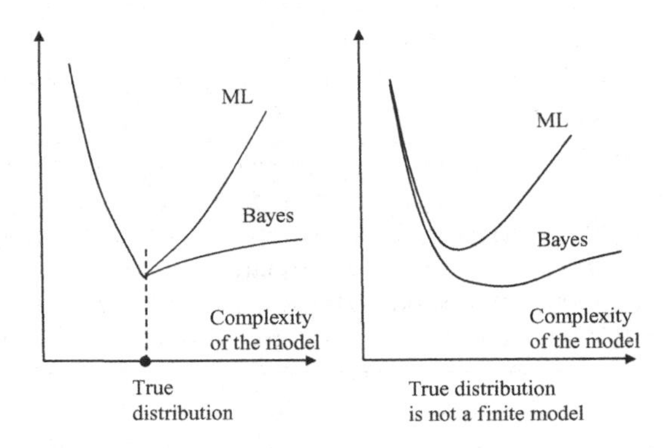

Fig. 8.2. ML and Bayes in moedel selection

8.4 Singular hypothesis test

8.4.1 Optimal hypothesis test

In this section we assume that there exists a probability distribution $\varphi(w)$ from which the parameter w is taken, and then $x_1, x_2, \ldots, x_n$ are independently taken from $p(x|w)$, where $x \in \mathbb{R}^N$ and $w \in \mathbb{R}^d$. Let $d_n = (x_1, x_2, \ldots, x_n)$ be a set of random samples.

In a hypothesis test, two probability distributions $\varphi_0(w)$ and $\varphi_1(w)$ on the set of parameters are prepared. We should choose one of the following hypotheses based on random samples.

(1) Null hypothesis (NH): "w is taken from $\varphi_0(w)dw$."

(2) Alternative hypothesis (AH): "w is taken from $\varphi_1(w)dw$."

Let $P(A|\varphi_0)$ and $P(A|\varphi_1)$ be the probabilities of an event A under the assumptions that $\varphi_0(w)$ and $\varphi_1(w)$ are the true distributions respectively.

There are many decision rules in choosing a hypothesis from d_n. A decision rule is represented by using a measurable function of d_n,

$$S(d_n) = S(x_1, x_2, \ldots, x_n).$$

The decision is made by the following rule,

$$S(d_n) \leq a \Longrightarrow \text{NH is chosen,}$$

$$S(d_n) > a \Longrightarrow \text{AH is chosen,}$$

where a is a parameter of the rule. Therefore the accuracy of the hypothesis test is determined by the function $S(d_n)$ and a.

For a given hypothesis test with (S, a), its level *Level*(S, a) is defined by the probability that AH is chosen under the assumption that the true distribution is $\varphi_0(w)$,

$$Level(S, a) = P(S(d_n) > a|\varphi_0).$$

The level represents the probability that the null hypothesis is rejected although it is true. In practical applications, a is determined so that the level is sufficiently small, for example, sometimes a is determined such that *Level*$(S, a) = 0.05$ or $L(S, a) = 0.01$. Conversely, a given level p, the event $\{d_n; S(d_n) > a\}$ such that $p = P(S(d_n) > a|\varphi_0)$ is said to be a rejection region. The power of a hypothesis test is defined by the probability that AH is chosen when the true distribution is $\varphi_1(w)$,

$$Power(S, a) = P(S(d_n) > a|\varphi_1).$$

The power represents the probability that the alternative hypothesis is chosen when it is true. Note that, if there is a monotone increasing function f such that $S_1(d_n) = f(S_2(d_n))$ $(\forall d_n)$, then $(S_1, f(a))$ and (S_2, a) give equivalent tests.

A test S_1 is said to be more powerful than a test S_2 if, for arbitrary a_1, a_2,

$$Level(S_1, a_1) = Level(S_2, a_2) \Longrightarrow Power(S_1, a_1) \geq Power(S_2, a_2)$$

holds. For a given set of hypotheses $\varphi_0(w)$ and $\varphi_1(w)$, there is an essentially unique test that is more powerful than any other test, which is called the most powerful test.

Theorem 8.2 (Neyman–Pearson) *The most powerful test is explicitly given by*

$$L(d_n) \equiv \frac{\int dw\, \varphi_1(w) \prod_{i=1}^{n} p(x_i|w)}{\int dw\, \varphi_0(w) \prod_{i=1}^{n} p(x_i|w)}. \tag{8.10}$$

Proof of Theorem 8.2 Assume that the level of the test $L(\ )$ is equal to that of an arbitrary test S, that is to say, there exist a, b such that

$$P(L(d_n) > b|\varphi_0) = P(S(d_n) > a|\varphi_0). \tag{8.11}$$

It is sufficient to prove that

$$P^* \equiv P(L(d_n) > b|\varphi_1) - P(S(d_n) > a|\varphi_1)$$

is nonnegative. Let A, B be two events defined by

$$A = \{d_n; S(d_n) > a\},$$
$$B = \{d_n; L(d_n) > b\}.$$

Then

$$\begin{aligned}
P^* &= \int \varphi_1(w)dw\Big[\int_B - \int_A\Big]\prod_{i=1}^n p(x_i|w)dx_i \\
&= \int \varphi_1(w)dw\Big[\int_{B\cap A^c} - \int_{A\cap B^c}\Big]\prod_{i=1}^n p(x_i|w)dx_i,
\end{aligned}$$

where A^c is the complementary set of A. Note that $B \cap A^c \subset B$ and $A \cap B^c \subset B^c$. By using the definition of $L(d^n)$ and eq.(8.11),

$$\begin{aligned}
P^* &\geq \int \varphi_0(w)dw\Big[\int_{B\cap A^c} - \int_{A\cap B^c}\Big]\prod_{i=1}^n p(x_i|w)dx_i \\
&= \int \varphi_0(w)dw\Big[\int_B - \int_A\Big]\prod_{i=1}^n p(x_i|w)dx_i \\
&= 0.
\end{aligned}$$

Therefore $L(d_n)$ is the most powerful test. □

Remark 8.4 If $\varphi_0(w) = \delta(w - w_0)$ and $\varphi_1(w) = \delta(w - w_1)$, then Neyman–Pearson theorem shows that the most powerful test results in the likelihood ratio test,

$$L(d_n) = \frac{\prod_{i=1}^n p(x_i|w_1)}{\prod_{i=1}^n p(x_i|w_0)}.$$

When the null hypothesis is defined by w_0 and the alternative hypothesis is $w \neq w_0$, then the following test is sometimes employed,

$$\hat{L}(d_n) = \frac{\prod_{i=1}^n p(x_i|\hat{w})}{\prod_{i=1}^n p(x_i|w_0),},$$

where $\hat{w}$ is the maximum likelihood or *a posteriori* estimator. However, this is not the most powerful test. In singular learning machines, this test has often very weak power in general.

8.4.2 Example of singular hypothesis test

To find the most powerful test, we need the probability distribution of

$$L(d_n) = \frac{\displaystyle\int dw\ \varphi_1(w) \prod_{i=1}^{n} p(x_i|w)}{\displaystyle\int dw\ \varphi_0(w) \prod_{i=1}^{n} p(x_i|w)},$$

where $d_n = (x_1, x_2, \ldots, x_n)$ are taken from w that is subject to $\varphi_0(w)$. If the null hypothesis test is given by one parameter, $\varphi_0(w) = \delta(w - w_0)$, then

$$L(d_n) = \int \exp(-nK_n(w))\ \varphi_1(w)\, dw,$$

where $K_n(w)$ is the log likelihood ratio function,

$$K_n(w) = \frac{1}{n} \sum_{i=1}^{n} \log \frac{p(x_i|w_0)}{p(x_i|w)}.$$

As we have already proved in Chapter 6, under the condition that the null hypothesis is true, $L(d_n)$ has an asymptotic expansion,

$$L(d_n) = \exp(-\lambda \log n + (m-1) \log\log n - F^R).$$

To construct a hypothesis test, we need the probability distribution of the random variable F^R.

Example 8.1 (Hypothesis test of changing point) Let $\{x_1, x_2, \ldots, x_n\}$ be a set of different, fixed points in $\mathbb{R}^1$ and $d_n = \{y_1, y_2, \ldots, y_n\}$ be a set of random variables which are subject to a conditional probability distribution. Let us construct a hypothesis test for a model,

$$p(y|x, a, b) = \frac{1}{\sqrt{2\pi\sigma^2}} \exp\Big(-\frac{1}{2\sigma^2}(y - aT(bx))^2\Big),$$

where $\sigma^2 > 0$ is a constant, and $T(x) = \tanh(x)$. The null and alternative hypotheses are respectively fixed as

$$\text{NH}: \varphi_0(a, b) = \delta(a)\delta(b),$$

$$\text{AH}: \varphi_1(a, b),$$

where $\varphi_1(a, b)$ is a fixed probability density function. The most powerful test is given by

$$L(d_n) = \frac{\int \exp\Big(-\frac{1}{2\sigma^2}\sum_{i=1}^{n}(y_i - aT(bx_i))^2\Big)\varphi_1(a, b)da\,db}{\exp\Big(-\frac{1}{2\sigma^2}\sum_{i=1}^{n} y_i^2\Big)}$$

$$= \int \exp\Big(-\frac{1}{2\sigma^2}\Big\{\sum_{i=1}^{n} a^2T(bx_i)^2 - 2\sum_{i=1}^{n} y_i aT(bx_i)\Big\}\Big)\varphi_1(a, b)da\,db.$$

Under the condition that the null hypothesis is true and $\varphi_1(0, 0) > 0$, the *a posteriori* distribution converges to $\delta(a)\delta(b)$ because the origin $(a, b) = (0, 0)$ is the singularity of $p(y|x, a, b) = p(y|x, 0, 0)$. By the Taylor expansion in the neighborhood of the origin, we have

$$a \tanh(bx) = abx + c_1ab^3x^3 + c_2ab^5x^5 + \cdots, \tag{8.12}$$

where c_1, c_2 are constants. Since $ab^3, ab^5, \ldots,$ are contained in the ideal $\langle ab\rangle$, they do not affect the asymptotic distribution. Therefore, if the null hypothesis is true, the most powerful test is asymptotically equal to

$$L^*(d_n) \cong \int e^{-nAa^2b^2+\sqrt{n}Bab}\varphi_1(a, b)da\,db, \tag{8.13}$$

where

$$A = \frac{(1/n)\sum_{i=1}^{n} x_i^2}{2\sigma^2}, \tag{8.14}$$

$$B = \frac{(1/\sqrt{n})\sum_{i=1}^{n} x_i y_i}{\sigma^2}. \tag{8.15}$$

Here A is a constant and B is a random variable which is subject to the normal distribution with mean zero and variance $2A$. Therefore, if the null hypothesis is true, the probability distribution of $L(d_n)$ is asymptotically equal to that of

$$L^*(g) \equiv \int e^{-nAa^2b^2+g\sqrt{2nA}ab}\varphi_1(a, b)da\,db, \tag{8.16}$$

where g is a random variable which is subject to the standard normal distribution. Let us study the case when the alternative hypothesis is given by

$$\varphi_1(a, b) = \begin{cases} 1/4 & (|a| \leq 1, |b| \leq 1, a \neq 0, b \neq 0) \\ 0 & \text{otherwise.} \end{cases} \tag{8.17}$$

Then the random variable $L^*(g)$ is equal to

$$L^*(g) = \int_{-1}^{1} dt \int_{[-1,1]^2} da\, db \frac{\delta(t-ab)}{4} \exp(-nAt^2 + g\sqrt{2nA}t)$$
$$= \int_{-\sqrt{n}}^{\sqrt{n}} dt \int_{-1}^{1} da \int_{-1}^{1} db \frac{1}{4\sqrt{n}}\, \delta\Big(\frac{t}{\sqrt{n}} - ab\Big)\, e^{-At^2+g\sqrt{2A}t}.$$

By using the Mellin transform as in Chapter 4, we can prove

$$\int_{-1}^{1} da \int_{-1}^{1} db\, \delta\Big(\frac{t}{\sqrt{n}} - ab\Big) = -2\log\frac{|t|}{\sqrt{n}}.$$

Therefore,

$$L^*(g) = \int_{-\sqrt{n}}^{\sqrt{n}} \frac{dt}{2\sqrt{n}}\Big(-\log\frac{|t|}{\sqrt{n}}\Big)e^{-At^2+g\sqrt{2A}t}$$
$$= \int_{0}^{\sqrt{n}} \frac{dt}{2\sqrt{n}}\Big(-\log\frac{|t|}{\sqrt{n}}\Big)e^{-At^2+g\sqrt{2A}t}$$
$$+ \int_{-\sqrt{n}}^{0} \frac{dt}{2\sqrt{n}}\Big(-\log\frac{|t|}{\sqrt{n}}\Big)e^{-At^2+g\sqrt{2A}t}$$
$$= \int_{0}^{\sqrt{n}} \frac{dt}{\sqrt{n}}\Big(-\log\frac{|t|}{\sqrt{n}}\Big)e^{-At^2}\Big(\frac{e^{g\sqrt{2A}t} + e^{-g\sqrt{2A}t}}{2}\Big)$$
$$= \int_{0}^{\sqrt{n}} \frac{dt}{\sqrt{n}}\Big(-\log\frac{|t|}{\sqrt{n}}\Big)e^{-At^2}\cosh(g\sqrt{2A}t).$$

By using this equation, we can construct a test for a given level. Let $P_0(\ )$ be the standard normal distribution. For a given level $\epsilon > 0$, the function $f(\epsilon)$ is determined so that

$$\epsilon = P_0(|g| \geq f(\epsilon))$$

is satisfied. For example, if $\epsilon = 0.05$, then $f(\epsilon) \approx 1.96$, or if $\epsilon = 0.01$, then $f(\epsilon) \approx 2.56$. Then the most powerful test is determined by

$$L(d_n) \leq a(\epsilon) \Longrightarrow \text{NH}$$
$$L(d_n) > a(\epsilon) \Longrightarrow \text{AH},$$

where $a(\epsilon)$ is a function determined by

$$a(\epsilon) \equiv \int_{0}^{\sqrt{n}} \frac{dt}{\sqrt{n}}\Big(-\log\frac{|t|}{\sqrt{n}}\Big)e^{-At^2}\cosh(f(\epsilon)\sqrt{2A}\,t).$$

Remark 8.5 In the above example, if the null hypothesis is true, then the approximation $L(d_n) \cong L^*(d_n)$ holds. Hence asymptotically

$$\begin{aligned} L(d_n) > a(\epsilon) &\Longleftrightarrow L^*(d_n) > a(\epsilon) \\ &\Longleftrightarrow B > f(\epsilon)\sqrt{2A} \end{aligned}$$

holds. In other words, the asymptotic test upon the null hypothesis can be done by $B > b(\epsilon)\sqrt{2A}$ for a given level ϵ, which has the same level as the most powerful test. This test does not depend on the *a priori* probability distribution. However, it is not so powerful as the most powerful test. In fact, if the alternative hypothesis is true, then the approximation $L(d_n) \approx L^*(d_n)$ does not hold. When we determine the reject region, we can use the approximation $L(d_n) \approx L^*(d_n)$ because the reject region is determined by the null hypothesis. However, when we test the hypotheses, we should use $L(d_n)$; we cannot use $L^*(d_n)$.

8.5 Realization of *a posteriori* distribution

In singular statistical estimation, Bayes estimation is recommended because the *a posteriori* distribution has more information about singularities than one point estimation such as the maximum likelihood or *a posteriori* estimation. However, in singular statistical models, the *a posteriori* distribution has very singular shape which cannot be easily approximated. A good algorithm is necessary to construct the *a posteriori* distribution.

8.5.1 Markov chain Monte Carlo

For a model $p(x|w)$ ($x \in \mathbb{R}^N$, $w \in \mathbb{R}^d$) and an *a priori* distribution $\varphi(w)$, a function $H(w)$ is defined by

$$H(w) = -\sum_{i=1}^{n} \log p(X_i|w) - \frac{1}{\beta} \log \varphi(w).$$

Then the *a posteriori* distribution is rewritten as

$$\begin{aligned} p(w) &= \frac{1}{Z_n} \varphi(w) \prod_{i=1}^{n} p(X_i|w)^{\beta} \\ &= \frac{1}{Z_n} \exp(-\beta H(w)). \end{aligned}$$

In the Metropolis algorithm, parameters $\{w_t \in \mathbb{R}^d; t = 1, 2, 3, \ldots\}$ are sampled by the following procedure.

Metropolis Algorithm

(1) The initial point w_1 is set. Let $t = 1$.
(2) For a given w_t, the new trial parameter w' is sampled by

$$w' = w_t + \mathcal{N},$$

where $\mathcal{N}$ is some fixed and symmetrical random variable, for example, a d-dimensional normal distribution.
(3) If $H(w') \leq H(w)$, then $w_{t+1} = w'$. Otherwise, $w_{t+1} = w'$ with probability $\exp(-\beta(H(w') - H(w)))$, or $w_{t+1} = w_t$ with probability $1 - \exp(-\beta(H(w') - H(w)))$.
(4) $t := t + 1$ and go to (2).

Then the detailed balance condition

$$p(w_b|w_a)p(w_a) = p(w_a|w_b)p(w_b) \quad (\forall w_a, w_b \in \mathbb{R}^d)$$

is satisfied for every step, and the *a posteriori* distribution is a fixed point of this probabilistic iteration, which means that the probability distribution

$$\frac{1}{T}\sum_{t=1}^{T} \delta(w - w_t)$$

converges to the *a posteriori* distribution $p(w)$ when $n \to \infty$. The mean of a function $f(w)$ by the *a posteriori* distribution is approximated by

$$E_w[f(w)] \approx \frac{1}{T}\sum_{t=1}^{T} f(w_i).$$

The evidence or stochastic complexity can be numerically calculated. The evidence can be given by

$$Z(\beta) = \int e^{-\beta \hat{H}(w)} \varphi(w) dw,$$

where

$$\hat{H}(w) = -\sum_{i=1}^{n} \log p(X_i|w).$$

Then, by using $Z(0) = 1$, the evidence is equal to

$$Z(1) = \prod_{j=0}^{J-1} \Big(\frac{Z(\beta_{k+1})}{Z(\beta_k)} \Big)$$

$$= \prod_{j=0}^{J-1} E_w^{(\beta_k)} \Big[e^{-(\beta_{k+1}-\beta_k)\hat{H}(w)} \Big], \tag{8.18}$$

where $E^{(\beta)}[\ ]$ shows the *a posteriori* distribution with the inverse temperature $\beta > 0$. The stochastic complexity is given by $F_n = -\log Z(1)$. To calculate $E^{(\beta)}[\ ]$, the Metropolis algorithm with

$$H(w) = \hat{H} - \frac{1}{\beta} \log \varphi(w)$$

is employed.

Remark 8.6 (1) Let $F(\beta) = -\log Z(\beta)$. Then

$$F(1) = \int_0^1 \frac{dF}{dt}(\beta) d\beta$$

$$= \int_0^1 E^{(\beta)}[\hat{H}(w)] d\beta.$$

Equation (8.18) is essentially equivalent to this calculation.
(2) In calculation of the evidence or the stochastic complexity, expectations with different $\beta_1, \ldots, \beta_{J-1}$ are necessary:

$$E^{(\beta_1)}[\ ],\ E^{(\beta_2)}[\ ], \ldots, E^{(\beta_{J-1})}[\ ].$$

In other words, a simultaneous probability distribution of $W = (w_1, w_2, \ldots, w_{J-1})$,

$$P(W) = \prod_{j=1}^{J} p^{(\beta_j)}(w_j),$$

is needed, where

$$p^{(j)}(w) = \frac{1}{Z(\beta_j)} \exp(-\beta_j \hat{H}(w)).$$

For the purpose of sampling from the distribution $P(W)$, there is a method called the tempered or exchange Monte Carlo method. This method consists of two kinds of Markov chain Monte Carlo procedures. One is the ordinary process of Monte Carlo in each sequence in the distribution of β_j. The other is the exchanging process of w_j in β_j-distribution and w_{j+1} in β_{j+1}-distribution. Two parameters w_j and w_{j+1} are exchanged with probability

$$\exp\{-(\beta_{j+1} - \beta_j)(\hat{H}(w_j) - \hat{H}(w_{j+1}))\}.$$

The exchanging process with this probability satisfies the detailed valance condition and makes $P(W)$ invariant, hence $P(W)$ is approximated as the limiting distribution by these two processes. In general, the equilibrium state for small β can be realized rather easily. The exchanging process is expected to make the Monte Carlo process for large β faster than the conventional method using the process with smaller β. By using singular learning theory, the exchange probability between β_1, β_2 ($\beta_2 > \beta_1$) is asymptotically equal to

$$P(\beta_1, \beta_2) = 1 - \frac{1}{\sqrt{\pi}} \frac{\beta_2 - \beta_1}{\beta_1} \frac{\Gamma(\lambda + 1/2)}{\Gamma(\lambda)},$$

where λ is the learning coefficient [63]. In the design of the sequence of the temperatures $\{\beta_j\}$, the exchange probabilities need to satisfy $P(\beta_1, \beta_2) = P(\beta_2, \beta_3) = \cdots$. For such a purpose, the geometrical progression is optimal for $\{\beta_j\}$.

Remark 8.7 The Markov chain Monte Carlo method is important in Bayesian estimation, and a lot of improved algorithms are being studied. The *a posteriori* distribution of a singular statistical model provides a good target distribution for the purpose of comparing several Monte Carlo methods, because the probability distributions used in practical applications are singular distributions or almost singular distributions. Singular learning theory is useful to construct a base on which Markov chain Monte Carlo methods are optimized, because the evidence is theoretically clarified. We can compare the numerical results by several Monte Carlo methods with theoretical results.

8.5.2 Variational Bayes approximation

To find the maximum likelihood estimator of a mixture model, the expectation and maximization algorithm (EM) algorithm is sometimes employed. However, in singular statistical models, the maximum likelihood estimator (MLE) often does not exist or diverges. Even if it exists, the generalization error of MLE is very large, and hypothesis testing using MLE is very weak. Therefore, in singular learning machines, MLE is not appropriate for statistical estimation and hypothesis testing. Recently the EM algorithm was improved from the point of Bayes estimation, which is called the variational Bayes approximation or mean field approximation. Let us study a normal mixture model of $x \in \mathbb{R}^M$,

$$p(x|w) = \sum_{k=1}^{K} \frac{a_k}{\sqrt{2\pi}^M} \exp\Big(-\frac{1}{2}|x - b_k|^2\Big), \tag{8.19}$$

where K is the number of mixtures, and $w = (a, b) = \{(a_k, b_k); k = 1, 2, \ldots, K\}$ is the set parameters which satisfies $0 \le a_k \le 1, a_1 + a_2 + \cdots + a_K = 1$, and $b_k \in \mathbb{R}^M$. In variational Bayes approximation, we employ the conjugate prior

$$\varphi(w) = \varphi_1(a)\varphi_2(b),$$

$$\varphi_1(a) = \frac{\Gamma(K\phi_0)}{\Gamma(\phi_0)^K} \delta\Big(\sum_{k=1}^{K} a_k - 1\Big) \prod_{k=1}^{K} a_k^{\phi_0 - 1},$$

$$\varphi_2(b) = \Big(\frac{\beta_0}{2\pi}\Big)^{KM/2} \prod_{k=1}^{K} \exp\Big(-\frac{\beta_0}{2}|b_k - b_0|^2\Big),$$

where $\phi_0 > 0$, $\beta_0 > 0$, and $b_0 \in \mathbb{R}^M$ are hyperparameters.

Let us introduce a random variable $Y = (Y^1, Y^2, \ldots, Y^K)$ which takes values in the set

$$C = \{(1, 0, 0, \ldots, 0), (0, 1, 0, \ldots, 0), \ldots, (0, 0, 0, \ldots, 1)\}.$$

In other words, only one of $\{Y_k\}$ is equal to 1, and the others are equal to zero. Such a random variable is said to be a competitive random variable. By introducing a probability distribution of (x, y),

$$p(x, y|w) = \prod_{k=1}^{K} \Big(\frac{a_k}{\sqrt{2\pi}^M} \exp\Big(-\frac{|x - b_k|^2}{2}\Big)\Big)^{y^k},$$

where $y = (y^1, y^2, \ldots, y^K) \in C$, it follows that $p(x|w)$ is equal to the marginal distribution,

$$p(x|w) = \sum_{y \in C} p(x, y|w),$$

where $\sum_y$ shows the sum of y over the set C. Therefore Y can be understood as a hidden variable in $p(x, y|w)$.

The sets of random samples are respectively denoted by

$$d_n = \{x_i \in \mathbb{R}^M; i = 1, 2, \ldots, n\},$$

$$h_n = \{y_i \in C; i = 1, 2, \ldots, n, \},$$

where $d_n \in \mathbb{R}^{Mn}$ is the set of data and $h_n \in C^n$ is the set of corresponding hidden variables. The simultaneous probability density function of (d_n, h_n, w) is given by

$$P(d_n, h_n, w) = \varphi(w) \prod_{i=1}^{n} p(x_i, y_i | w). \tag{8.20}$$

The conditional probability distribution of (h_n, w) for a given set of data is equal to

$$P(h_n, w | d_n) = \frac{1}{Z_n} P(d_n, h_n, w),$$

where Z_n is a constant,

$$\begin{aligned} Z_n &= \sum_{h_n \in C^n} \int dw \; P(d_n, h_n, w) \\ &= \int dw \; \varphi(w) \prod_{i=1}^{n} p(x_i | w) \end{aligned}$$

is the evidence of the pair $p(x|w)$ and $\varphi(w)$.

Let us introduce the Kullback–Leibler distance $\mathcal{K}$ which is defined for the probability distributions on the set of hidden variables and the set of the parameters, $C^n \times \mathbb{R}^d$. The Kullback–Leibler distance from an arbitrary probability density $q(h_n)r(w)$ to the target probability density $P(h_n, w|d_n)$ is equal to

$$\mathcal{K}(q, r) = \sum_{h_n \in C^n} \int dw \; q(h_n) r(w) \log \frac{q(h_n) r(w)}{P(h_n, w | d_n)}.$$

The set of all probability density functions on $C^n \times \mathbb{R}^d$ in which the hidden variable and the parameter are independent is denoted by

$$\mathcal{S} = \{q(h_n) r(w)\}.$$

The probability density function $P(h_n, w|d_n)$ is not contained in $\mathcal{S}$ in general. However, in the variational Bayes approximation, $P(h_n, w|d_n)$ is approximated by finding the optimal $(q, r) \in \mathcal{S}$ that minimizes $\mathcal{K}(q, r)$. The optimal (q, r) is said to be the variational Bayes approximation or mean field approximation. Minimization of $\mathcal{K}(q, r)$ is equivalent to minimization of

$$\mathcal{F}(q, r) = \sum_{h_n \in C^n} \int dw \; q(h_n) r(w) \log \frac{q(h_n) r(w)}{P(h_n, w, d_n)}.$$

If $P(h_n, w|d_n)$ is contained in $\mathcal{S}$, equivalently if $\mathcal{K}(q, r)$ can be made to be zero, then the minimal value of $\mathcal{F}(q, r)$ is equal to $F_n = -\log Z_n$, which is the stochastic complexity of the original pair $p(x|w)$ and $\varphi(w)$. In general, the stochastic complexity of the variational Bayes is defined by

$$\hat{F}_n \equiv \min_{(q,r)\in\mathcal{S}} \mathcal{F}(q,r).$$

It follows that $\hat{F}_n$ gives the upper bound of the Bayes stochastic complexity,

$$\hat{F}_n \geq F_n.$$

Remark 8.8 (1) (Gibbs variational principle) In general, for a given function $H(x)$, a functional of a probability distribution $p(x)$ defined by

$$\mathcal{F}(p) = \int q(x) \log q(x) dx + \beta \int q(x) H(x) dx$$

is minimized if and only if

$$p(x) = \frac{1}{Z} \exp(-\beta H(x)).$$

This is called the Gibbs variational principle. It is well known in statistical mechanics that the equilibrium state $p(x)$ is uniquely characterized by the minimization of the free energy.
(2) The difference $\hat{F}_n - F_n$ is equal to the Kullback–Leibler distance from the variational Bayes to the true Bayes. Hence the difference can be understood as an index of how precise the variational Bayes is.
(3) As is shown in Chapter 1, the Bayes generalization error B_g with $\beta = 1$ is equal to the increase of the mean stochastic complexity,

$$E[B_g] = E[F_{n+1}] - E[F_n]$$

for arbitrary natural number n. However, the variational Bayes generalization error $\hat{B}_g$ does not satisfy the same relation. More explicitly,

$$E[\hat{B}_g] \neq E[\hat{F}_{n+1}] - E[\hat{F}_n].$$

If the true parameter is originated from the *a priori* distribution, then Bayes estimation gives the smallest generalization error, hence $E[B_g] \leq E[\hat{B}_g]$ with $\beta = 1$. Otherwise, both cases $E[\hat{B}_g] \geq E[B_g]$ and $E[\hat{B}_g] \leq E[B_g]$ happen in general.
(4) In general, the optimal q that minimizes $K(q||p)$ is a more localized distribution than p. In other words, the variational Bayes or the mean field approximation gives the localized approximated distribution compared to the target. In singular learning machines, the *a posteriori* distribution is not localized. Hence the variational Bayes has different stochastic complexity and generalization error from the true Bayes estimation.

To derive the variational Bayes learning algorithm, we need the Dirichlet distribution.

Remark 8.9 (Dirichlet distribution) The Dirichlet distribution of $a = (a_1, a_2, \ldots, a_K) \in [0, 1]^K$ is defined by

$$\varphi(a_1, a_2, \ldots, a_K) = \frac{1}{Z}\delta\Big(1 - \sum_{k=1}^{K} a_k\Big)\prod_{k=1}^{K} a_k^{\phi_k - 1},$$

where $\{\phi_k > 0\}$ is a set of hyperparameters and Z is the normalizing constant,

$$Z = \frac{\prod_{k=1}^{K}\Gamma(\phi_k)}{\Gamma\big(\sum_{k=1}^{K}\phi_k\big)}.$$

Then two averages satisfy

$$\int a_j p(a_1, a_2, \ldots, a_K)\prod_{k=1}^{K} da_k = \frac{\phi_j}{\sum_{k=1}^{K}\phi_k},$$

$$\int (\log a_j)p(a_1, a_2, \ldots, a_K)\prod_{k=1}^{K} da_k = \psi(\phi_j) - \psi\Big(\sum_{k=1}^{K}\phi_k\Big),$$

where $\psi(x) = (\log\Gamma(x))'$ is a digamma function.

Let us derive an algorithm by which the variational Bayes approximation can be numerically found. Let the Lagrangian be

$$\mathcal{L}(q, r, \alpha_1, \alpha_2) = \mathcal{F}(q, r) + \alpha_1\Big(\sum_{h_n} q(h_n) - 1\Big) + \alpha_2\Big(\int r(w)dw - 1\Big).$$

The optimal $q(h_n)$ and $r(w)$ should satisfy the variational equations, $\mathcal{L}(q + \delta q, r) = \mathcal{L}(q, r + \delta r) = 0$ and $\partial\mathcal{L}/\partial\alpha_1 = \partial\mathcal{L}/\partial\alpha_2 = 0$. Then the optimal $q(d_n)$ and $r(w)$ should satisfy the relations

$$q(h_n) = \frac{1}{C_1}\exp\Big(E_r[\log P(d_n, h_n, w)]\Big), \tag{8.21}$$

$$r(w) = \frac{1}{C_2}\exp\Big(E_q[\log P(d_n, h_n, w)]\Big), \tag{8.22}$$

where $E_r[\ ]$ and $E_q[\ ]$ are expectations over $r(w)$ and $q(h_n)$ respectively, and $C_1, C_2 > 0$ are normalizing constants. By using eq.(8.20), for $P = P(d_n, h_n, w)$,

$$\begin{aligned}
\log P &= \sum_{k=1}^{K} (\log a_k) \Big\{ \sum_{i=1}^{n} y_i^k + \phi_0 - 1 \Big\} \\
&\quad - \frac{1}{2} \sum_{k=1}^{K} \Big\{ \sum_{i=1}^{n} y_i^k + \beta_0 \Big\} \\
&\quad + \sum_{k=1}^{K} b_k \cdot \Big\{ \sum_{i=1}^{n} y_i^k x_i + \beta_0 b_0 \Big\} + A \qquad (8.23) \\
&= \sum_{i=1}^{n} \sum_{k=1}^{K} \Big\{ \log a_k - \frac{1}{2} |b_k - x_i|^2 + \log \frac{1}{\sqrt{2\pi}} \Big\} + B, \qquad (8.24)
\end{aligned}$$

where A is a constant for parameter w and B is a constant for y_i^k. Let $\overline{y}_i^k = E_q[y_i^k]$ for a given $q(d_n)$. Then by eq.(8.22) and eq.(8.23)

$$r(a, b) \propto \prod_{k=1}^{K} (a_k)^{S_k - 1} \exp\Big(-\frac{T_k}{2} |b_k - U_k|^2 \Big),$$

where S_k, T_k and U_k are determined by $\overline{y}_i^k$,

$$\begin{aligned}
S_k &= \sum_{i=1}^{n} \overline{y}_i^k + \phi_0, \\
T_k &= \sum_{i=1}^{n} \overline{y}_i^k + \beta_0, \\
U_k &= \frac{1}{T_k} \Big\{ \sum_{i=1}^{n} \overline{y}_i^k x_i + \beta_0 b_0 \Big\}.
\end{aligned}$$

Then by eq.(8.21) and eq.(8.24), $q(h_n)$ can be obtained using $r(a, b)$:

$$q(h_n) \propto \prod_{i=1}^{n} \prod_{k=1}^{K} \exp(y_i^k L_i^k),$$

where

$$L_i^k = \psi(S_k) - \frac{1}{2}\left\{\frac{M}{T_k} + |x_i - U_k|^2\right\},$$

$$\overline{y}_i^k = \frac{\exp\left(L_i^k\right)}{\sum_{j=1}^K \exp\left(L_i^j\right)}.$$

Hence we obtain the recursive formula,

$$(S_k, T_k, U_k) \Leftarrow \overline{y}_i^k,$$

$$\overline{y}_i^k \Leftarrow (S_k, T_k, U_k).$$

From some appropriate initial point, if this algorithm converges to the global minimum of $\mathcal{F}(q, r)$, then

$$\begin{aligned}
\hat{F}_n = & -\frac{nM}{2}\log(2\pi) - \sum_{i=1}^{n}\log\Big(\sum_{k=1}^{K} e^{L_i^k}\Big) \\
& + \log\frac{\Gamma(n + K\phi_0)\Gamma(\phi_0)^K}{\Gamma(K\phi_0)\Gamma(S_1)\cdots\Gamma(S_K)} \\
& + \sum_{k=1}^{K}\psi(S_k)(S_k - \phi_0) - \frac{MK}{2}(1 + \log\beta_0) \\
& + \sum_{k=1}^{K}\left\{\frac{M}{2}\log T_k + \frac{\beta_0}{2}\Big(\frac{M}{T_k} + |U_k - b_0|^2\Big)\right\}.
\end{aligned}$$

There is a theoretical bound of this value.

Theorem 8.3 *Assume that the true distribution is a normal mixture in eq.(8.19) with* K_0 *components. Then the variational stochastic complexity* $\hat{F}_n$ *satisfies the inequality*

$$S_n + \lambda_1 \log n + nK_n(\hat{w}) + c_1 < \hat{F}_n < S_n + \lambda_2 \log n + c_2,$$

where S_n *is the empirical entropy of the true distribution,* $K_n(\hat{w})$ *is the log likelihood ratio of the optimal parameter of* $\hat{w}$ *with respect to the variational Bayes,* c_1, c_2 *are constants, and* λ_1, λ_2 *are respectively given by*

$$\lambda_1 = \begin{cases} (K-1)\phi_0 + M/2 & (\phi_0 \le (M+1)/2) \\ (MK + K - 1)/2 & (\phi_0 > (M+1)/2) \end{cases}$$

$$\lambda_2 = \begin{cases} (K - K_0)\phi_0 + (MK_0 + K_0 - 1)/2 & (\phi_0 \le (M+1)/2) \\ (MK + K - 1)/2 & (\phi_0 > (M+1)/2). \end{cases}$$

Remark 8.10 (1) The proof of this theorem is given in [96]. This theorem shows that the variational Bayes has phase transition with respect to the hyper parameter $\phi_0 = (M+1)/2$.
(2) If the Dirichlet distribution is employed as the *a priori* distribution, then the learning coefficient λ of Bayes estimation is given by eq.(7.4) with $K = H$, $K_0 = H_0$, $M = N$. The difference between Bayes stochastic complexity $F_n \cong S_n + \lambda \log n$ and the variational one $\hat{F}_n$ is not so large.
(3) There are theoretical results of variational Bayes for general mixture models [97], hidden Markov models [42], and reduced rank regressions [65]. The generalization error by the variational Bayes estimation is still an open problem. In [65], the behavior of the variational generalization error is different from that of stochastic complexity.

8.6 From regular to singular

In this book, we have shown that singular statistical theory is established by functional methods in modern mathematics. The reasons why functional methods are necessary are as follows.

(1) The problem of singularities is resolved in the functional space.
(2) Asymptotic normality of the limiting process made by training samples is shown only in the functional space.
(3) On the functional space, statistical learning theory is completely described by its algebraic structure.

Let us compare regular statistical theory with singular learning theory. The well-known conventional statistical theory of regular models consists of the following parts.

(1) The set of the true parameters is one point.
(2) The Fisher information matrix at the true parameter is positive definite.
(3) The log likelihood ratio function can be approximated by a quadratic form.
(4) The Bayes *a posteriori* distribution converges to the normal distribution.
(5) The distribution of the maximum likelihood estimator also converges to the normal distribution.
(6) Both the learning coefficient and the singular fluctuation are $d/2$, where d is the dimension of the parameter space.

On the other hand, singular learning theory consists of the following properties.

Table 8.2. *Correspondence between regular and singular*

	Regular	Singular
Algebra	Linear algebra	Ring and ideal
Geometry	Differential geometry	Algebraic geometry
Analysis	Real-valued function	Function-valued func.
Probability theory	Central limit theorem	Empirical process
Parameter set	Manifold	Analytic set
Model and parameter	Identifiable	Nonidentifiable
True parameter	One point	Real analytic set
Fisher inform. matrix	Positive definite	Semi-positive def.
Cramer–Rao inequality	Yes	No meaning
Asymptotic normality	Yes	No
ML estimator	Asymptotic efficient	Not efficient
Bayes *a posteriori*	Normal distribution	Singular dist.
Standard form	Quadratic form	Normal crossing
Basic transform	Isomorphic	Birational
Learning coefficient	$d/2$	λ
Singular fluctuation	$d/2$	ν
Stochastic complexity	$(d/2)\log n$	$\lambda \log n$
Information criterion	AIC	WAIC
Phase transition	No	Yes
Examples	Normal, binomial Linear regression Linear prediction	Mixtures Neural networks Hidden Markov

(1) The set of true parameters consists of a real analytic set with singularities.
(2) The Kullback–Leibler distance is made to be normal crossing by resolution of singularities.
(3) The log likelihood ratio function can be approximated by the standard form as in Main Theorem 6.1.
(4) The Bayes *a posteriori* distribution converges to the singular distribution.
(5) The distribution of the maximum likelihood estimator converges to that of the maximum value of the Gaussian process.
(6) Neither the learning coefficient nor the singular fluctuation is equal to $d/2$, in general.

Table 8.2 shows the correspondence between regular and singular learning theory.

Regular statistical theory was built via the probability distribution on parameter space.

$$\text{Samples} \rightarrow \text{Parameter space} \rightarrow \text{Prediction}.$$

Singular learning theory is established via the probability distribution on functional space.

$$\text{Samples} \rightarrow \text{Functional space} \rightarrow \text{Prediction}.$$

In regular statistical theory, "statistic" is defined as a function from samples to parameter space. However, in singular learning theory it should be defined as a function from samples to functional space. On the functional space, asymptotic normality holds in both regular and singular cases.

We have proved that, even in singular learning theory, there are universal formulas, which are mathematically beautiful and statistically useful. We expect that these are the case for the future study of statistical learning theory.

Remark 8.11 In mathematical physics, quantum field theory and statistical mechanics are characterized by the probability distribution of $\exp(-\beta H(x))$ where $H(x)$ is a Hamiltonian function. It is well known in [12] that physical problems are determined by the algebraic structure of $H(x)$. Statistical learning theory can be understood as mathematical physics where the Hamiltonian is a random process defined by the log likelihood ratio function. This book clarified that the algebraic geometrical structure of the log likelihood ratio function determines the learning process.

Bibliography

[1] H. Akaike. A new look at the statistical model identification. *IEEE Transactions on Automatic Control*, Vol. 19, pp. 716–723, 1974.

[2] H. Akaike. Likelihood and Bayes procedure. In *Bayesian Statistics*, ed. J. M. Bernald. University Press, Valencia, Spain, 1980, 143–166.

[3] S. Amari. *Differential Geometrical Methods in Statistics*. Springer Lecture Notes in Statistics. Berlin: Springer-Verlag, 1985.

[4] S. Amari. A universal theorem on learning curves. *Neural Networks*, Vol. 6, No. 2, pp. 161–166, 1993.

[5] S. Amari and H. Nagaoka. *Methods of Information Geometry*. Oxford: AMS and Oxford University Press, 2000.

[6] S. Amari and N. Murata. Statistical theory of learning curves under entropic loss. *Neural Computation*, Vol. 5, pp. 140–153, 1993.

[7] S. Amari. Natural gradient works efficiently in learning. *Neural Computation*, Vol. 10, pp. 251–276, 1998.

[8] S. Amari and H. Nakahara. Difficulty of singularity in population coding. *Neural Computation*, Vol. 17, No. 4, pp. 839–858, 2005.

[9] S. Amari, H. Park, and T. Ozeki. Singularities affect dynamics of learning in neuromanifolds. *Neural Computation*, Vol. 18, No. 5, pp. 1007–1065, 2006.

[10] M. Aoyagi and S. Watanabe. Stochastic complexities of reduced rank regression in Bayesian estimation. *Neural Networks*, Vol. 18, No. 7, pp. 924–933, 2005.

[11] M. Aoyagi and S. Watanabe. Resolution of singularities and generalization error with Bayesian estimation for layered neural network. *IEICE Transactions*. Vol. J88-D-II, No. 10, pp. 2112–2124, 2005.

[12] H. Araki. *Mathematical Theory of Quantum Fields*. International Series of Monographs on Physics. Oxford: Oxford University Press, 1999.

[13] H. Araki. Relative entropy of states of von Neumann algebras. *Publications of the Research Institute for Mathematical Sciences*, Vol. 11, No. 3, pp. 809–833, 1975.

[14] M. F. Atiyah. Resolution of singularities and division of distributions. *Communications of Pure and Applied Mathematics*, Vol. 13, pp. 145–150, 1970.

[15] A. R. Barron. Approximation and estimation bounds for artificial neural networks. *Machine Learning*, Vol. 14, No. 1, pp. 115–133, 1994.

[16] I. N. Bernstein. The analytic continuation of generalized functions with respect to a parameter. *Functional Analysis and Applications*, Vol. 6, pp. 26–40, 1972.
[17] J. E. Björk. *Rings of Differential Operators.* Amsterdam: North-Holland, 1970.
[18] J. Bocknak, M. Coste, and M. -F. Roy. *Real Algebraic Geometry.* Berlin: Springer Verlag, 1998.
[19] G. Bodnár and J. Schicho. A computer program for the resolution of singularities. In *Resolution of Singularities*, Progress in Mathematics, Vol. 181, ed. H. Hauser. Basel: Birkhäuser, 1997, pp. 231–238.
[20] H. Chernoff. On the distribution of the likelihood ratio. *Annals of Mathematical Statistics*, Vol. 25, pp. 573–578, 1954.
[21] D. A. Cox, J. B. Little, and D. O'sea. *Ideals, Varieties, and Algorithms*, 3rd edn. New York: Springer, 2007.
[22] H. Cramer. *Mathematical Methods of Statistics.* Princeton, NJ: Princeton University Press, 1949.
[23] D. Dacunha-Castelle and E. Gassiat. Testing in locally conic models, and application to mixture models. *Probability and Statistics*, Vol. 1, pp. 285–317, 1997.
[24] D. A. Darling and P. Erdös. A limit theorem for the maximum of normalized sums of independent random variables. *Duke Mathematics Journal*, Vol. 23, pp. 143–155, 1956.
[25] B. Davies. *Integral Transforms and their Applications.* New York: Springer, 1978.
[26] P. Diaconis and B. Sturmfels. Algebraic algorithms for sampling from conditional distributions. *The Annals of Statistics*, Vol. 26, No. 1, pp. 363–397, 1998.
[27] W. Donoghue. *Distributions and Fourier Transforms.* New York: Academic Press, 1969.
[28] M. Drton, B. Sturmfels, and S. Sullivant. Algebraic factor analysis: tetrads, pentads and beyond. *Probability Theory and Related Fields*, Vol. 138, No. 3–4, pp. 1432–2064, 2007.
[29] B. Efron and C. Moriis. Stein's estimation rule and its competitors – an Empirical Bayes approach. *Journal of American Statistical Association*, Vol. 68, pp. 117–130, 1973.
[30] K. Fujiwara and S. Watanabe. Hypothesis testing in singular learning machines and its application to time sequence analysis. *IEICE Transactions*, Vol. J91-D, No. 4, pp. 889–896, 2008.
[31] K. Fukumizu. Likelihood ratio of unidentifiable models and multilayer neural networks. *The Annals of Statistics*, Vol. 31, No. 3, pp. 833–851, 2003.
[32] I. M. Gelfand and G. E. Shilov. *Generalized Functions.* San Diego, CA: Academic Press, 1964.
[33] I. I. Gihman and A. V. Shorohod. *The Theory of Stochastic Processes*, Vols. 1, 2, 3. Berlin: Springer-Verlag, 1974.
[34] I. J. Good. *The Estimation of Probabilities*, Cambridge, MA: MIT Press, 1965.
[35] K. Hagiwara. On the problem in model selection of neural network regression in overrealizable scenario. *Neural Computation*, Vol. 14, pp. 1979–2002, 2002.
[36] J. A. Hartigan. A failure of likelihood asymptotics for normal mixtures. In *Proceedings of the Berkeley Conference in Honor of J. Neyman and J. Kiefer*, Vol. 2, ed. L. LeCam and R. A. Olshen. Belmoant, CA: Wadsworth, 1985, pp. 807–810.

[37] H. Hauser. The Hironaka theorem on resolution of singularities (Or A proof we always wanted to understand). *Bulletin of the American Mathematical Society*. Vol. 40, No. 3, pp. 323–403, 2003.

[38] D. Haussler and M. Opper. Mutual information, metric entropy and cumulative relative entropy risk. *The Annals of Statistics*. Vol. 25, No. 6, pp. 2451–2492, 1997.

[39] T. Hayasaka, M. Kitahara, and S. Usui. On the asymptotic distribution of the least-squares estimators in unidentifiable models. *Neural Computation*, Vol. 16, No. 1, pp. 99–114, 2004.

[40] H. Hironaka. Resolution of singularities of an algebraic variety over a field of characteristic zero. *Annals of Mathematics*, Vol. 79, pp. 109–326, 1964.

[41] L. Hörmander. *An Introduction to Complex Analysis in Several Variables*. Princeton, NJ: Van Nostrand, 1966.

[42] T. Hosino, K. Watanabe, and S. Watanabe. Stochastic complexities of variational Bayesian hidden Markov models. *Proceedings of 2005 IEEE International Joint Conference on Neural Networks*, Vol. 2, pp. 1114–1119, 2005.

[43] H. Hotelling. Tubes and spheres in n-spaces, and a class of statistical problems. *American Journal of Mathematics*, Vol. 61, pp. 440–460, 1939.

[44] M. Inoue, H. Park, and M. Okada. On-line learning theory of soft committee machines with correlated hidden units – Steepest gradient descent and natural gradient descent. *Journal of Physical Society of Japan*, Vol. 72, No. 4, pp. 805–810, 2003.

[45] S. Janson. *Gaussian Hilbert Space*. Cambridge University Press, 1997.

[46] M. Kashiwara. B-functions and holonomic systems. *Inventiones Mathematicae*, Vol. 38, pp. 33–53, 1976.

[47] M. Knowles and D. Siegmund. On Hotelling's approach to testing for a nonlinear parameter in regression. *International Statistical Review*, Vol. 57, pp. 205–220. 1989.

[48] J. Kollár. The structure of algebraic thresholds – an introduction to Mori's program. *Bulletin of the American Mathematical Society*, Vol. 17, pp. 211–273, 1987.

[49] J. Kollár. *Lectures on Resolution of Singularities*. Princeton, NJ: Princeton University Press, 2007.

[50] J. Kollór, S. Mori, C. H. Clemens, and A. Corti. *Birational Geometry of Algebraic Varieties*. Cambridge Tract in Mathematics. Cambridge University Press, 1998.

[51] F. Komaki. On asymptotic properties of predictive distributions. *Biometrika*, Vol. 83, No. 2, pp. 299–313, 1996.

[52] S. Kuriki and A. Takemura. Tail probabilities of the maxima of multilinear forms and their applications. *The Annals of Statistics*, Vol. 29, No. 2, pp. 328–371, 2001.

[53] M. R. Leadbetter, G. Lindgren, and H. Rootzén. *Extremes and Related Properties of Random Sequences and Processes*. Berlin: Springer-Verlag, 1983.

[54] E. Levin, N. Tishby, and S. A. Solla. A statistical approaches to learning and generalization in layered neural networks. *Proceedings of IEEE*, Vol. 78, No. 10, pp. 1568–1674, 1990.

[55] X. Liu and Y. Shao. Asymptotics for likelihood ratio tests under loss of identifiability. *The Annals of Statistics*, Vol. 31, No. 3, pp. 807–832, 2003.

[56] D. J. Mackay. Bayesian interpolation. *Neural Computation*, Vol. 4, No. 2, pp. 415–447, 1992.
[57] G. McLachlan and D. Peel. *Finite Mixture Models.* New York: John Wiley, 2000.
[58] H. N. Mhaskar. Neural networks for optimal approximation of smooth and analytic functions. *Neural Computation*, Vol. 8, pp. 164–177. 1996.
[59] E. Miller and B. Sturmfels. *Combinatorial Commutative Algebra.* Graduate Texts in Mathematics, vol. 227, New York: Springer-Verlag, 2005.
[60] D. Mumford. *The Red Book of Varieties and Schemes*, 2nd edn. Berlin: Springer-Verlag, 1999.
[61] N. Murata, S. Yoshizawa, and S. Amari. Network information criterion – determining the number of hidden units for an artificial neural network model. *IEEE Transactions on Neural Networks*, Vol. 5, No. 6, pp. 797–807, 1994.
[62] M. Mustata. Singularities of pairs via jet schemes. *Journal of the American Mathematical Society*, Vol. 15, pp. 599–615, 2002.
[63] K. Nagata and S. Watanabe. Asymptotic behavior of exchange ratio in exchange Monte Carlo method. *Neural Networks*, Vol. 21, No. 7, pp. 980–988, 2008.
[64] S. Nakajima and S. Watanabe. Generalization performance of subspace Bayes approach in linear neural networks. *IEICE Transactions, Information and Systems*, Vol. E89-D, No. 3, pp. 1128–1138, 2006.
[65] S. Nakajima and S. Watanabe. Variational Bayes solution of linear neural networks and its generalization performance. *Neural Computation*, vol. 19, no. 4, pp. 1112–1153, 2007.
[66] K. Nishiue and S. Watanabe. Effects of priors in model selection problem of learning machines with singularities. *Electronics and Communications in Japan* (Part II: Electronics), Vol. 88, No. 2. pp. 47–58, 2005.
[67] T. Oaku. Algorithms for b-functions, restrictions, and algebraic local cohomology groups of D-modules. *Advances in Applied Mathematics*, Vol. 19, pp. 61–105, 1997.
[68] T. Oaku. Algorithms for the b-function and D-modules associated with a polynomial. *Journal of Pure Applied Algebra*, Vol. 117–118, pp. 495–518, 1997.
[69] K. Oka. *Sur les functions analytiques de plusieurs variables.* Tokyo: Iwanami shoten, 1962.
[70] M. Opper and D. Haussler. Bounds for predictive errors in the statistical mechanics of supervised learning. *Physical Review Letters*, Vol. 75, No. 20, pp. 3772–3775, 1995.
[71] L. Pachter and B. Sturmfels. *Algebraic Statistics for Computational Biology*, Cambridge University Press, 2005.
[72] P. C. B. Pillips. Partially identified econometric models. *Econometric Theory*, Vol. 5, pp. 181–240, 1989.
[73] K. R. Parthasarathy. *Probability Measures on Metric Spaces.* New York: Academic Press, 1967.
[74] G. Pistone, E. Riccomagno, and H. Wynn. *Algebraic Statistics: Computational Commutative Algebra in Statistics.* Boca Raton, FA: Chapman and Hall/CRC, 2001.
[75] J. Rissanen. Stochastic complexity and modeling. *Annals of Statistics*, Vol. 14, pp. 1080–1100, 1986.
[76] D. Ruelle. *Thermodynamic Formalism.* Reading, MA: Addison Wesley, 1978.

[77] D. Rusakov and D. Geiger. Asymptotic model selection for naive Bayesian network. *Journal of Machine Learning Research.* Vol. 6, pp. 1–35, 2005.
[78] M. Saito. On real log canonical thresholds, *arXiv:0707. 2308v1*, 2007.
[79] M. Saito, B. Sturmfels, and N. Takayama. Gröbner deformations of hypergeometric differential equations. *Algorithms and Computation in Mathematics*, Vol. 6. Berlin: Springer, 2000.
[80] M. Sato and T. Shintani. On zeta functions associated with prehomogeneous vector space. *Annals of Mathematics*, Vol. 100, pp. 131–170, 1974.
[81] G. Schwarz. Estimating the dimension of a model. *Annals of Statistics*, Vol. 6, No. 2, pp. 461–464. 1978.
[82] I. R. Shafarevich. *Basic Algebraic Geometry.* Berlin: Springer-Verlag, 1974.
[83] R. Shibata. An optimal model selection of regression variables. *Biometrika*, Vol. 68, pp. 45–54, 1981.
[84] K. E. Smith, L. Kahanpäa, P. Kekäläinen, and W. Traves. *An Invitation to Algebraic Geometry.* New York: Springer, 2000.
[85] C. Stein. Inadmisibility of the usual estimator for the mean of a multivariate normal distribution. In *Proceedings of the 3rd Berkeley Symposium on Mathematical Statistics and Probabilities.* Berkeley, CA: University of California Press, 1956, pp. 197–206.
[86] B. Sturmfels. *Gröbner Bases and Convex Polytopes.* University Lecture Series. American Mathematical Society, 1995.
[87] M. Sugiyama and K.-R. Müller. The subspace information criterion for infinite dimensional hypothesis spaces. *Journal of Machine Learning Research*, Vol. 3, pp. 323–359, 2003.
[88] M. Sugiyama and H. Ogawa. Optimal design of regularization term and regularization parameter by subspace information criterion. *Neural Networks*, Vol. 15, No. 3, pp. 349–361, 2002.
[89] N. Takayama. An algorithm for constructing cohomological series solutions of holonomic systems, *Journal of Japan society for symbolic and algebraic computation*, Vol. 10, No. 4, pp. 2–11, 2003.
[90] A. Takemura and T. Kuriki. On the equivalence of the tube and Euler characteristic methods for the distribution of the maximum of the gaussian fields over piecewise smooth domains. *Annals of Applied Probability*, Vol. 12, No. 2, pp. 768–796, 2002.
[91] K. Tsuda, S. Akaho, M. Kawanabe, and K. -R. Müller. Asymptotic properties of the Fisher kernel. *Neural Computation*, Vol. 16, No. 1, pp. 115–137, 2004.
[92] A. W. van der Vaart and J. A. Wellner. *Weak Convergence and Empirical Processes.* New York: Springer,1996.
[93] V. N. Vapnik. *Statistical Learning Theory.* New York: John Wiley, 1998.
[94] S. Veres. Asymptotic distributions of likelihood ratios for overparameterized ARMA processes. *Journal of Time Series Analysis*, Vol. 8, No. 3, pp. 345–357, 1987.
[95] R. Walter. *Principles of Mathematical Analysis.* International Series in Pure and Applied Mathematics. New York: McGraw-Hill, 1976.
[96] K. Watanabe and S. Watanabe. Stochastic Complexities of Gaussian Mixtures in Variational Bayesian Approximation. *Journal of Machine Learning Research*, Vol. 7, No. 4, pp. 625–644, 2006.

[97] K. Watanabe and S. Watanabe. Stochastic complexities of general mixture models in variational Bayesian learning. *Neural Networks*, Vol. 20, No. 2, pp. 210–217, 2007.

[98] S. Watanabe. Generalized Bayesian framework for neural networks with singular Fisher information matrices. In *Proceedings of the International Symposium on Nonlinear Theory and Its Applications*, 1995, pp. 207–210.

[99] S. Watanabe. Algebraic analysis for singular statistical estimation. *Algorithmic Learning Theory*, Lecture Notes on Computer Sciences, Vol. 1720. Springer, 1999, pp. 39–50.

[100] S. Watanabe. Algebraic analysis for nonidentifiable learning machines. *Neural Computation*, Vol. 13, No. 4, pp. 899–933, 2001.

[101] S. Watanabe. Algebraic geometrical methods for hierarchical learning machines. *Neural Networks*, Vol. 14, No. 8, pp. 1409–1060, 2001.

[102] S. Watanabe. Learning efficiency of redundant neural networks in Bayesian estimation. *IEEE Transactions on Neural Networks*, Vol. 12, No. 6, 1475–1486, 2001.

[103] S. Watanabe. Algebraic information geometry for learning machines with singularities. *Advances in Neural Information Processing Systems*, Vol. 13, pp. 329–336, 2001.

[104] S. Watanabe. Algebraic geometry of singular learning machines and symmetry of generalization and training errors. *Neurocomputing*, Vol. 67, pp. 198-213, 2005.

[105] S. Watanabe. *Algebraic Geometry and Learning Theory.* Tokyo: Morikita Publishing, 2006.

[106] S. Watanabe. Generalization and training errors in Bayes and Gibbs estimations in singular learning machines. *IEICE Technical Report*, December, 2007.

[107] S. Watanabe. Equations of states in singular statistical estimation. *arXiv:0712.0653*, 2007.

[108] S. Watanabe. A forumula of equations of states in singular learning machines. In *Proceedings of IEEE World Congress in Computational Intelligence*, 2008.

[109] S. Watanabe. On a relation between a limit theorem in learning theory and singular fluctuation. IEICE Technical Report, No. NC2008–111, pp. 45–50, 2009.

[110] S. Watanabe and S. Amari. Learning coefficients of layered models when the true distribution mismatches the singularities. *Neural Computation*, Vol. 15, No. 5, pp. 1013–1033, 2003.

[111] S. Watanabe, K. Yamazaki, and M. Aoyagi. Kullback information of normal mixture is not an analytic function. *IEICE Technical Report*, NC2004-50, 2004, pp. 41–46.

[112] H. Wei, J. Zhang, F. Cousseau, T. Ozeki, and S. Amari. Dynamics of learning near singularities in layered networks. *Neural Computation*, Vol. 20, No. 3, pp. 813–843, 2008.

[113] H. Weyl. On the volume of tubes. *American Journal of Mathematics*, Vol. 61, pp. 461–472, 1939.

[114] K. Yamanishi. A decision-theoretic extension of stochastic complexity and its applications to learning. *IEEE Transactions on Information Theory.* Vol. 44, No. 4, pp. 1424–1439, 1998.

[115] K. Yamazaki and S. Watanabe. A probabilistic algorithm to calculate the learning curves of hierarchical learning machines with singularities. *IEICE Transactions*, Vol. J85-D-II, No. 3, pp. 363–372, 2002.

[116] K. Yamazaki and S. Watanabe. Stochastic complexity of Bayesian networks. *Proceedings of International Conference on Uncertainty in Artificial Intelligence*, 2003.

[117] K. Yamazaki and S. Watanabe. Singularities in mixture models and upper bounds of stochastic complexity. *Neural Networks*, Vol. 16, No. 7, pp. 1029–1038, 2003.

[118] K. Yamazaki and S. Watanabe. Singularities in Complete bipartite graph-type Boltzmann machines and upper bounds of stochastic complexities. *IEEE Transactions on Neural Networks*, Vol. 16, No. 2, pp. 312–324, 2005.

[119] K. Yamazaki and S. Watanabe. Algebraic geometry and stochastic complexity of hidden Markov models. *Neurocomputing*, Vol. 69, pp. 62–84, 2005.

[120] K. Yamazaki, M. i Kawanabe, S. Watanabe, M. Sugiyama, and K. -R. Müller. Asymptotic Bayesian generalization error when training and test distributions are different. In *Proceedings of International Conference on Machine Learning*, Corvalis, Oregon, 2007, pp. 1079–1086.

Index